Noise
Control
Management

Noise Control Management

Howard K. Pelton

Pelton, Marsh, Kinsella, Inc.

Dallas, TX

VNR VAN NOSTRAND REINHOLD
New York

Library of Congress Catalog Card Number 92-18950
ISBN 0-442-00763-9

Van Nostrand Reinhold
115 Fifth Avenue
New York, New York 10003

Chapman and Hall
2-6 Boundary Row
London, SE1 8HN, England

Thomas Nelson Australia
102 Dodds Street
South Melbourne 3205
Victoria, Australia

Nelson Canada
1120 Birchmount Road
Scarborough, Ontario MIK 5G4, Canada

16 15 14 13 12 11 10 9 8 7 6 5 4 3 2 1

Library of Congress Cataloging-in-Publication Data

Pelton, Howard K.
    Noise control management / by Howard K. Pelton.
        p.   cm.
    Includes bibliographical references and index.
    ISBN 0-442-00763-9
    1. Noise control.   2. Industrial noise.   3. Noise control—Case
studies.   I. Title.
TD892.P45   1992
620.2′3—dc20

                                                    92-18950
                                                    CIP

for

SUZANNE

Contents

Preface

Experience has shown that personnel assigned the responsibility for noise control may not have the time, training, or inclination required to solve noise control problems. These types of problems are usually outside the area of expertise of plant managers, safety personnel, project engineers, industrial hygienists, plant engineers, industrial nurses, hearing conservationists, and others.

The field of noise control has been well established for at least 50 years and goes back almost 100 years.

There are many excellent books on the subject of noise control, a number of which are used as texts and references. *Noise Control Management*, however, treats the subject from the point of view of the individual who must set up the program and solve the problem within a specific time period and budget. It is divided into three parts.

Part I, "The Noise Control Management Approach," describes the basic approach of the noise control management problem, the maintenance of noise control devices after installation, the specification of equipment, and general guidelines for noise control. A guide for hearing conservation is also included. This should be considered part of a noise management program. Part II, "Terminology, Criteria, Measurements, and Instrumentation," is provided for those desiring further detail about sound and the various noise criteria for the plant environment. Instrumentation and measurement procedures are also covered. This part provides a general overview and is not intended as a rigorous treatment of the subject, which can be found in the references cited. Part III, "Noise Control Case Histories," provides examples, based on the author's experience, of typical noise control problems found in industry and the general approach used to reduce noise. The appendix offers a general glossary of terms, the OSHA Noise Regulation, acoustical standards, a generic hearing conservation program, a noise control specification for a typical project, and a noise and vibration control material buyer's guide.

By most conventions noise is defined as unwanted sound. In a plant there are many noise sources. The task is to visualize and define the problem. This book takes the approach of providing a foundation in the form of basic information about noise, guidelines for noise and vibration control, noise-measuring instruments, and procedures leading to the development of a noise control management program. Most noise problems can be solved and the noise levels can be controlled for a long time through proper maintenance. The noise management approach provides the road map for this process.

In the industrial environment a sound level above 85 dBA is classified as noise, or unwanted sound, since it may damage hearing or interfere with speech. The questions of how, when, and why the sound is unwanted should be answered for a complete definition. For example, did the sound become unwanted due to bad design or improper maintenance?

When a sound becomes noise it should be reduced to an acceptable level. This is done with control methods at the source, along the path the noise travels, or at the receiver of the noise.

Reduction of noise at the source is usually the best method; however, this may require a great deal of research by the company building the device. Unless there are incentives to reduce noise at the source the more standard approaches of path control may be used. They include enclosures, barriers, silencers, and vibration isolation among others.

Receiver control protects the individuals or group hearing the noise. This can be accomplished by hearing protection or quiet control rooms as examples. Thus noise control deals with defining noise and developing a method of control. This process must be managed.

What is management? Several authorities define management as the process that plans, organizes, and controls the operation of all the resources, such as personnel, machines, funds, methods, and markets, to provide direction, coordination, and leadership to human efforts that will achieve the objectives of the business. For noise control a similar definition and process can be used.

Objectives are established for a noise abatement plan. The plan is organized to control and direct the operation of the resources available to meet the objectives. Most noise abatement plans will cut across departmental lines; therefore, cooperation, coordination, and leadership are needed to meet the plan objectives within the budget.

The major objective of this book is to provide the tools that will result in a "quieter" plant environment through proper management of the noise levels over a long period of time. Management support is fundamental to any project, and noise control is not so different in that respect. The approaches provided in this book will be of assistance in selling management on the need for noise control and how the problems can be solved in the most cost-effective manner.

The last portion of the book is devoted to noise control case histories, which will allow the reader to visualize the concepts. In a plant environment, with many noise sources, the prospect of defining the problem can be overwhelming. Therefore, use of a systematic approach based on a familiar methodology can build confidence. If the concepts can be visualized, then the noise management approach outlined in this book will suggest the overall approach. Thus, increasingly more complex problems can be solved in a systematic manner.

Our environment should be free from distractions and noise. This is certainly not easy to find these days. In addition, the ability to use and enjoy our natural ability to hear into later years should not be hampered by the bombardment of high noise levels either in the workplace or in everyday life. While the major thrust of this book is directed toward the industrial setting, there are many applications that will apply to architecturally related problems in a variety of commercial buildings.

Acknowledgments

I would like to acknowledge and thank several people who have been a great influence in my career. First of all is Mr. Al Pease, now retired from the Maxim Silencer Division of Beaird Industries in Shreveport, Louisiana. He was a patient teacher who really helped me acquire a good foundation when I entered this field. He also provided practical insights into the field of industrial silencing.

Next, I would like to acknowledge Dr. J. J. Thigpen, now retired dean of engineering at Louisiana Tech University, Ruston, Louisiana. While working on my master's degree, he was head of the school of mechanical engineering. He encouraged and allowed special studies in the field of industrial noise control and research in silencer design for my master's thesis.

Herman Blum (deceased), founder of Blum Consulting Engineers, deserves grateful acknowledgment for the freedom he allowed me to establish a solid consulting practice within his engineering organization over a ten-year period. In addition, many thanks are due to my current partners Gary Kinsella and Dave Marsh for encouraging me to develop the noise and vibration control practice that is very meaningful and fulfilling to me. I would also like to thank Lesley Klem for help provided in preparing the initial and final parts of the manuscript, and Nancy Edwards, technical illustrator, for help in preparing a number of CAD graphics and graphs for the book. Thanks also to a good friend and colleague, Jack Randorff, Ph.D., for his thoughtful review of the manuscript and helpful suggestions.

Finally, thanks must go to Suzanne. She has put up with all the long hours, travel schedules, and discussions about my work over the years. She has been very encouraging in developing this book, and has allowed me the time to do so.

Part I

The Noise Control Management Approach

Chapter 1

Noise Control Management: A Systems Approach

INTRODUCTION

The success of any noise control program hinges on support from the company's management. To enlist that support, the noise control program must be implemented in a systematic and cost-effective manner. The following sections describe the development of such an industrial noise program from its start—from the design process through its implementation in the plant, to a final measurement of its success several months later.

BACKGROUND

Industrial noise has been a problem, recognized or not, for many years. Experts in the fields of hearing conservation, audiology, and otology have proven that prolonged exposure without proper protection will cause a premature hearing loss. However, recent advances in acoustical engineering make protection available for workers in nearly every industry.

Has industry used these developments in noise control technology? Despite federal regulations requiring worker protection (29 CFR 1910.95) and easy access to noise control programs, far too many companies are still reluctant to implement these programs. The primary reason, according to extensive research, is cost. In the early days of noise control, too many programs were designed with too little thought concerning cost-effectiveness.

Additionally, negative or noncommittal attitudes toward the programs are frequently encountered among management ranks. Many companies feel that their industry's "unique character" excludes them from compliance with federal regulations. Others simply wait until they are forced to comply—a "don't fix it until it's broken" approach. In most cases, lack of accurate and complete information regarding regulations, compliance requirements, and noise control solutions is the foundation for the negative attitudes.

No doubt this is a tough audience to win over. But by addressing this audience in terms that they understand, respect, and appreciate, one can elicit support, and in many cases enthusiasm, for a noise control management program.

For example, instead of one person or department making the proposal for changes, it is more effective to include several key people: supervisors, personnel director, safety director, and other engineers, for example. Draw them into the process by asking for their suggestions, using their suggestions, and including them in meetings and communications surrounding the project.

Once a noise control program is implemented, everyone wants to know if it works. A crucial component of every program is a system for measuring results. Noise reduction and exposure to noise are two primary items to consider, but other factors are equally important in obtaining an accurate picture of the program's success:

1. Is a noise control section included in the company's procurement specifications? (Please refer to Chapter 3 for more on this subject.)
2. Is noise control discussed in safety meetings and on the job?
3. Is noise control hardware being maintained with the same vigilance as other machinery?
4. Are employees giving meaningful suggestions for improving the noise control program?
5. Does the noise control program have support from management?

Obviously, input from management may add other considerations to this list.

It is also important for managers to realize the latitude they have in involving themselves with the noise control program. "Off-the-shelf" solutions are adequate for some companies, while others choose to involve themselves deeply in researching, developing, and implementing noise control measures. The following section describes one approach.

APPROACH TO NOISE CONTROL MANAGEMENT

Drawing a complete picture—taking all factors into account—is essential when designing a noise control program. The general steps for a noise management program [DOL, 1971] are:

1. Identify noise hazard areas.
2. Establish attainable goals.
3. Conduct feasibility studies.
4. Select methods and materials, including the design and installation of prototypes, if necessary.
5. Evaluate noise control methods and modify them.
6. Implement any final changes and modifications.
7. Evaluate the system for compliance with applicable regulations.

Identifying noise hazard areas and establishing attainable goals are two of the most important phases. These data help determine the company's needs. Some of the constraints to be considered include:

1. Production—the manner in which a product is produced may restrict which noise control techniques can be utilized.
2. Maintenance—the noise control technique chosen must not prohibit or hamper maintenance around a given machine.
3. Physical restraints—in an existing installation, there are usually pipes, ducts, and other items that can prohibit full utilization of all noise control methods available.
4. Cost—this factor must be balanced against the other three considerations to help develop the most cost-effective approach possible.

Feasibility studies encompass two areas, technical and economic requirements. The technical requirements are:

1. The method used must be proven successful and readily available.
2. The performance levels must be definable.
3. The noise control solutions and materials must be compatible with the particular operation whose noise they are controlling.
4. The materials must be of high quality.
5. The performance levels must be maintainable.
6. Alternatives must be developed to respond to existing constraints.

Two basic factors determine economic feasibility: direct and hidden costs. Direct costs include materials, their installation, and engineering consultation. Many times these are the only items upon which the cost of a noise management system is based. To obtain a total picture, all potential hidden costs must also be included, when applicable:

1. Increased cost of production, product losses
2. Additional setup time and labor
3. Maintenance costs, both direct and hidden
4. Cost of capital for noise control treatment
5. Equipment relocation costs, including utilities relocation
6. Plant expansion and modernization
7. Increased number of personnel
8. Inflationary factors for prolonged abatement programs
9. Inside engineering costs
10. Additional safety costs
11. Downtime costs

Downtime cost is a good example of hidden costs. Downtime costs include money spent when machinery, normally operated continuously, is shut down to allow the installation of noise control materials. If it costs $250,000 in revenue to install $50,000 in noise management solution, the economic feasibility of that system is significantly reduced.

Another economic factor to consider is the possible increase in production costs that can occur as a result of the noise control program. This increase can affect the company's ability to price its products competitively.

Conversely, by improving the efficiency of a manufacturing operation, noise control measures often can increase a company's competitive position. For example, if a noise management system includes the installation of an automatic feeding system in a machine previously hand operated, not only is the machine operator's noise exposure reduced, but productivity likely will increase as well. The increased efficiency could, in fact, produce an impressive return on investment period and result in long-term profitability increases.

Once the feasibility of an overall program is determined the individual elements of a program can be evaluated and priorities assigned. Specifically, this provides an objective method to rank order priorities of engineering controls for the areas within a plant, and sources within each area. This procedure establishes a logical sequence for implementation of the program.

ESTABLISHMENT OF PRIORITIES

The methodology described below has been successfully used by the author for at least the last 15 years. The priority rating equation was published some years ago and republished as cited below. This approach seems to provide a valid guide for the development of long-range noise abatement programs. It has been accepted by management, engineering, and outside authorities to track the progress of the program. It is flexible and can change as situations require. This approach also assists in forecasting the levels of funding required.

The highest noise source is not necessarily the number one priority. Factors other than noise level, such as source noise level, number of employees affected, time exposure to noise sources, cost, and feasibility factors are included. An empirical equation [Dear, 1987] for priority number assignments is as follows:

$$PN = \frac{(n)L_p(t_i)}{C(F_e)^2}$$

where:

PN = priority number (the higher the number the higher the relative priority for the given noise source candidate for utilization of engineering controls)
n = number of employees
t_i = time exposed to source
L_p = sound pressure level
C = cost/$10,000 (in some large plants C can be raised to a power K ($K = 1.3$) to obtain a better distribution)
F_e = feasibility factor:
    1 = currently available "off the shelf"
    2 = materials available with some engineering design

3 = feasible with major engineering required and prototype demonstration

4 = technical feasibility not proven—greater than 6 months and basic R&D required to demonstrate feasibility

Note: This equation along with all its components can be easily programmed using an electronic spread sheet to allow a variety of "what if" scenarios to be tried. An example in Table 1.1 illustrates the use of this approach for developing priorities.

For each area or source the proper selection of the feasibility factor (F) is very important and includes practical considerations in the total effort. This is really based on the judgment and experience of the individual developing the noise abatement plan. The plan, of course, should take into account the various constraints discussed earlier: production, maintenance, and physical.

The cost factor should include the expected life of the noise control equipment. As dis-

TABLE 1.1   Typical Noise Abatement Plan Setting Priorities

Plant Location	L_p (dBA)	Number Employees Normally Exposed (n)	Time Exposed to Source (t)	Feasibility Factor[a] (F_e)	Cost Factor (C)	Priority Factor $\left(P = \dfrac{(n * t * L_p)}{C(F_e * F_e)}\right)$
Process area A						
Predryer	97	2	8	1	13	119
Water pump and chiller	101	2	8	1	1.2	1347
"Roots" blower level 1	92	2	8	1	12	123
Heaters level 1	95	8	8	3	6	113
"Roots" blower level 2	99	2	8	2	10	40
Furnace	90	12	8	2	15	144
Final dryer	90	8	8	2	3	480
Ventilation blower 1	94	2	8	2	7.5	50
Centrifugal compressor 1	92	2	8	2	7.5	49
Air exhaust fans	84	2	8	1	7.5	179
Steam nozzles	95	6	8	4	10	29
Heaters level 2	91	2	8	3	11	15
Process area B						
Fans A and B	98	8	8	2	15	105
Fans C and D	97	8	8	2	15	103
Fans E and F	96	8	8	2	15	102
Vibratory separator	92	8	8	1	3	1963
Blower A	90	8	8	2	7.5	192
Blower B	89	8	8	2	7.5	190
Water pumps near blower B	90	8	8	2	1.2	1200
Predryer	91	8	8	3	13	50
Heaters	88	8	8	3	11	57
Utility plant						
Boiler feed pump 1	92	2	8	1	3	491
Air compressor 1	105	2	8	2	8	53
Air compressor 2	104	2	8	2	8	52
Vacuum pumps	98	2	8	3	15	12
Air dryer vents	97	2	8	2	1	388
Boiler F.D. fan	92	2	8	2	5	74
Boiler feed pump 2	89	2	8	2	3	119

[a] 1 = Available "off the shelf"; 2 = materials available with some engineering; 3 = materials available with custom design required; 4 = feasible with R&D; 5 = technical feasibility not proven—basic R&D required.

cussed in Chapter 2 the noise control equipment should be designed for a life expectancy of at least ten years.

Another factor that should be included is the time value of money. The cost to capitalize or finance the equipment is part of the real cost, as with any piece of capital equipment. For example, a series of large enclosures that cost $25,000 each is not an expense item. Small equipment may be expensed through a maintenance or safety budget.

Priority ranking is usually done on a manufacturing-area basis where noise studies have identified the hazardous areas. The areas are usually divided by manufacturing areas of responsibility. Within each area subdivisions can be made as appropriate.

Once the priority areas have been established they should be reviewed by the area supervisors and evaluated from a commonsense standpoint. Some items may be rearranged, combined, or eliminated based on the individual manufacturing area's short- and long-term plans. This information is important in the process of developing a final plan.

For example, in the author's experience in developing a plan for a process plant, certain process furnaces and dryers were going to be replaced in two years. This was in the middle of a four-year plan. Thus, there was an opportunity to develop and specify a noise control system with the new equipment at a reduced cost.

The empirical equation outlined above is a valid starting point. Over the years it has been found very useful in developing a noise abatement program and simplifying a very complex task. The program can be as flexible as required.

An example of this priority method is illustrated in Fig. 1.1, which describes a manufacturing facility with a number of noise sources. These sources were analyzed using this equation to arrive at a priority ranking for each noise source and then develop an abatement plan.

Top priority was air noise reduction, which in this case could have an impact on the greatest number of people for the least cost. Reducing the noise called for the installation of off-the-shelf equipment.

The second priority was the ventilation fan. With a feasibility factor of 1, this solution could affect a large number of people through a simple and inexpensive procedure—reducing the fan's speed and changing the blade pitch.

Lower on the priority list were the large punch presses, with a feasibility factor of 3; this rating was due to the solution's high cost, more complex nature, and relatively minor impact on employees.

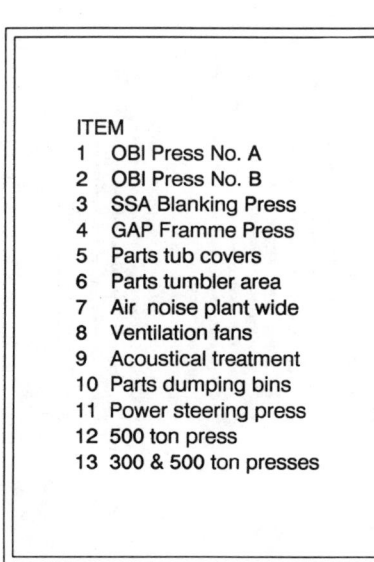

ITEM
1 OBI Press No. A
2 OBI Press No. B
3 SSA Blanking Press
4 GAP Framme Press
5 Parts tub covers
6 Parts tumbler area
7 Air noise plant wide
8 Ventilation fans
9 Acoustical treatment
10 Parts dumping bins
11 Power steering press
12 500 ton press
13 300 & 500 ton presses

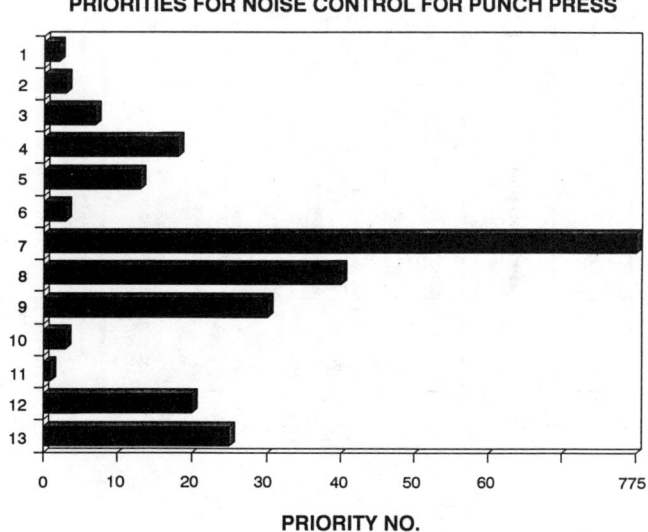

PRIORITIES FOR NOISE CONTROL FOR PUNCH PRESS

PRIORITY NO.

FIGURE 1.1   Noise control priorities for a metal-working plant.

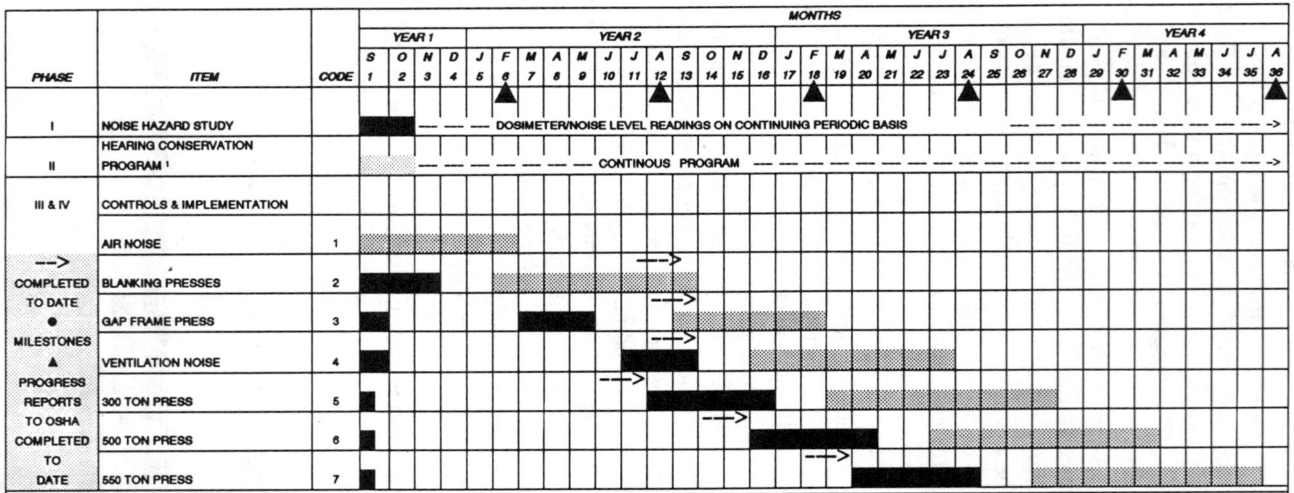

MONTHS

PHASE	ITEM	CODE	YEAR 1												YEAR 2												YEAR 3												YEAR 4						
			S	O	N	D	J	F	M	A	M	J	J	A	S	O	N	D	J	F	M	A	M	J	J	A	S	O	N	D	J	F	M	A	M	J	J	A	S	O	N	D	J	F	M	A	M	J	J	A
			1	2	3	4	5	6	7	8	9	10	11	12	13	14	15	16	17	18	19	20	21	22	23	24	25	26	27	28	29	30	31	32	33	34	35	36
I	NOISE HAZARD STUDY																																						
II	HEARING CONSERVATION PROGRAM I																																						
III & IV	CONTROLS & IMPLEMENTATION																																						
	AIR NOISE	1																																						
	BLANKING PRESSES	2																																						
	GAP FRAME PRESS	3																																						
	VENTILATION NOISE	4																																						
	300 TON PRESS	5																																						
	500 TON PRESS	6																																						
	550 TON PRESS	7																																						

DOSIMETER/NOISE LEVEL READINGS ON CONTINUING PERIODIC BASIS

CONTINUOUS PROGRAM

Legend:
- -→ COMPLETED TO DATE
- ● MILESTONES
- ▲ PROGRESS REPORTS TO OSHA
- COMPLETED TO DATE

I. NOISE HAZARD STUDY

A. Measure the sound pressure levels of equipment operation at each employee work position for the purpose of evaluating employee exposure.

B. Evaluate each employee exposure pattern of exposure to the noise and design of the machine and environment.

C. Develop a priority list for noise controls based on the needs of the largest number of employees for protection.

D. Set design goals for the engineering or administrative noise control priorities to meet current occupational noise exposure standards.

III. ENGINEERING NOISE CONTROLS (PRIORITY ORDER)

SELECT FIRST PRIORITY

A. Engineering evaluation of the potential noise sources that can contribute to the level at the exposure position based on the design of the machine and environment.

B. Develop the potential corrective measure applicable to each noise source. If noise control solutions are found not to be feasible, go to administrative procedures and hearing protection.

C. Make detailed measurements of the existing sources to confirm or deny their importance.

D. Preliminary engineering of multiple noise controls specifically applicable to the confirmed sources that meet the production and maintenance constraints.

E. Plant personnel engineering evaluation to select those controls which meet the production and maintenance requirements of the plant.

F. Provide drawings and specifications of selected noise control devices.

SELECT NEXT PRIORITY AND REPEAT (A)

II. HEARING CONSERVATION PROGRAM

A. Evaluation of audiometric program.

B. Conduct baseline audiometric tests on present employees.

C. Conduct annual employee education training program.

D. Evaluation and modification of present hearing protection program.

E. Establish annual audiometric monitoring program and correlate hearing losses to noise hazard data.

F. Re-evaluation of design goals based on the results of correlation study (in "E" above).

IV. IMPLEMENTATION OF NOISE CONTROL PROGRAM

A. Preparation of bid package.

B. Out for bid.

C. Selection of suppliers.

D. Procurement of materials and load time for equipment.

E. Installation of noise control materials.

F. Evaluation and acceptance of modification design.

G. Modification, if required.

H. Final testing and acceptance.

FIGURE 1.2 Noise abatement plan for a metal-working plant.

Next, a sequence can be established for implementing the noise control program (as shown in Fig. 1.2). While the priority scheme quantifies the order in which problem areas are to be addressed, other factors may intervene to determine the actual implementation schedule. The order in which the solutions appear was ultimately determined, in part, to allow greater flexibility in the implementation process.

Figure 1.2 contains the detailed steps of each phase recommended for the projects. Each company's approach to these projects may be different, but the overall goal is to reduce area noise levels and employee noise exposure.

PROGRAM ELEMENTS

The overall program is divided into four phases, each with a logically sequenced approach with specific steps to follow. Below the description of each item are the steps recommended and shown in Fig. 1.2.

Noise Hazard Study

This phase should be conducted by company personnel using noise dosimeters, identifying individuals and/or job types exposed to high or hazardous noise levels. These people/job types would be included in a hearing conservation program. This noise hazard study, as well as a noise survey, should be conducted on at least an annual basis. The following elements are recommended:

1. Measure the sound pressure levels of equipment operation at each employee work position for the purpose of evaluating employee exposure.
2. Evaluate each employee pattern of exposure to the noise and determine if there is a hearing hazard.
3. Develop a priority list for noise controls based on the needs of the largest number of employees for protection.
4. Set design goals for the engineering or administrative noise control priorities to meet current occupational noise exposure standards.

Hearing Conservation Program

This phase can be coordinated by the human resources department. It consists of conducting audiograms, which measure employee hearing acuity; choosing hearing protection devices and training employees to use them properly; and conducting general training related to hearing conservation. The following elements are recommended:

1. Evaluate audiometric program.
2. Conduct baseline audiometric tests on present employees.
3. Conduct annual employee education and training program.
4. Evaluate and modify present hearing protection program.
5. Establish annual audiometric monitoring program and correlate hearing losses to noise hazard data.
6. Reevaluate design goals based on results of correlation study (in 5 above).

Engineering Noise Controls

In this phase, more detailed noise measurement data are produced; spectral analyses, which identify the noise sources, are performed; and the methods, materials, and costs for priority setting are determined. The sequence as outlined can be repeated as necessary when new information is developed. The priorities that are established, whether by the method outlined or by other methods, are then placed on the Gantt (bar) chart with the estimated time required for detailed engineering and implementation. The following elements are recommended:

1. Engineering evaluation of the potential noise sources that can contribute to the level at the exposure position based on the design of the machine and environment.
2. Development of the potential corrective measure applicable to each noise source. If noise control solutions are found not to be feasible, go to administrative procedures and hearing protection.

3. Detailed measurement of the existing sources to confirm or deny their importance.

4. Preliminary engineering of multiple noise controls specifically applicable to the confirmed sources that meet the production and maintenance constraints.

5. Plant personnel engineering evaluation to select those controls that meet the production and maintenance requirements of the plant.

6. Provide of drawings and specifications of selected noise control devices.

When this sequence is completed the next priority is selected and the steps are repeated.

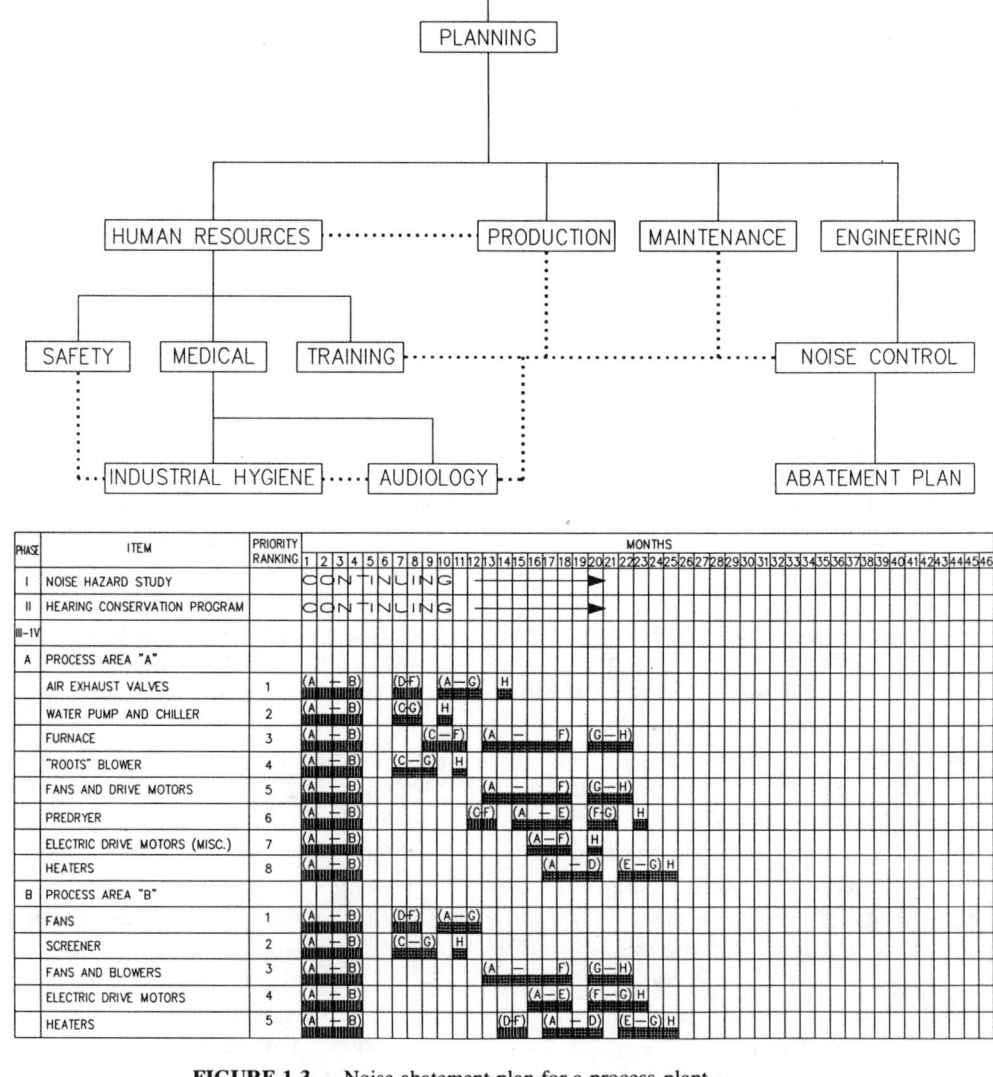

FIGURE 1.3   Noise abatement plan for a process plant.

Detailed Engineering and Installation

After the problem has been defined, the detailed engineering phase begins. This phase may include performing additional noise and vibration data analyses to define the sources in greater detail and to create a detailed design and final analysis, or it may simply entail ordering the appropriate noise control materials. The step-by-step sequence should be followed on each project to allow for the logical flow of events. This approach follows the same project engineering steps, except with different materials and terminology. The following elements are recommended:

1. Preparation of bid package
2. Out for bid
3. Selection of suppliers
4. Procurement of materials and lead time for equipment
5. Installation of noise control materials
6. Evaluation and acceptance of modification design
7. Modification, if required
8. Final testing and acceptance

Another example of a noise control program is illustrated in Fig. 1.3. This includes an organizational chart and a noise abatement plan that is divided into four parts as described above. Each part can be a separate project following the sequence described above. This provides the road map to guide the overall program. From this it is easy to develop, using an electronic spread sheet, a method to follow costs, material procurement delivery, installation, and results. The organizational chart illustrates the general responsibilities for the noise abatement plan.

SUMMARY

Upon completion of the design, and development of drawings and specifications, the various projects can be released for bid or negotiated with a contractor. In some cases, plant maintenance personnel can install the noise control materials. After installation follow-up, noise measurements, including noise hazard studies, should be conducted. If modifications are required, previous steps in this phase can be repeated until the goal, or at least a satisfactory result, is achieved.

Since the noise control plan may run many months, or even several years, changes in equipment and processes may occur and can affect the anticipated results. This problem requires that the plant project engineering personnel remain in close communication with the production personnel to determine what, if any, changes will impact the noise control program.

In 1979, OSHA estimated that the annual cost to industry for complying with the Hearing Conservation Amendment would be $53 per employee included in a hearing conservation program. Using an average annual inflation rate of 5 percent, one can estimate a current cost per employee. The primary benefits that justify this expense are the protection of employee hearing and prevention of occupationally related cases of hearing impairment. As a result, the potential for workers' compensation claims due to hearing loss can be significantly reduced through an effective hearing conservation program.

The cost estimates provided by OSHA take into consideration the tangible expenses for hearing protection devices, annual hearing tests, instrumentation, and the conducting of periodic noise surveys. However, there are also several intangible, hidden costs that are not taken into consideration in the estimate, yet are certainly associated with maintaining an ongoing hearing conservation program. Most companies will incur a sizable expense when conducting annual audiometric examinations. Besides the actual unit cost per hearing test and its subsequent evaluation, there is a hidden productivity cost relating to the workers being away from their jobs during the examinations.

Equally important are the administrative time and cost for medical, safety, and industrial hygiene personnel and resources that are required to maintain an effective hearing conservation program. There are also intangible costs associated with health- and safety-related accidents and injuries that might occur because of poor communication or speech interference problems when employees work in areas with high noise levels.

There is no guarantee that a hearing conservation program will protect all employees. Problems associated with the use and care of hearing protection devices—especially their consistent use—can still result in a small percentage of workers who will incur noise-induced hearing loss. The best long-term solution, therefore, and usually the most cost-effective means of protecting employees from hazardous noise is the implementation of feasible engineering noise control [OSHA, 1974] measures that can minimize or even eliminate the risk of injury.

An order of magnitude cost for noise control can be developed. Usually a majority of the employees in the present conservation program will be eliminated if the noise control program is adopted. Further study of the cost impacts for noise control, hearing conservation cost over a 10- to 15-year period—initial life of noise control material—and possible compensability costs for hearing loss claims should continue to be evaluated.

References

Dear, T. A. 1987. Controlling Noise at Its Source Can Help Protect Workers' Hearing. *Occupational Health and Safety*, pp. 60–64.

DOL/OSHA Bulletin 334. 1971. Guidelines to the Department of Labor's Occupational Noise Standards. Bureau of National Affairs Reference Files, Nov. 1980, 41:3001–41:3006.

OSHA. 1974. Occupational Noise Exposure Regulation. U.S. Federal Register (39FR37773), 10/24/74 rc:Part 29CFR1910.95.

Chapter 2

Maintaining Noise Levels After Installation Through Proper Maintenance

INTRODUCTION

By installing new equipment, or modifying existing equipment, to meet specifications such as those presented in this book, the safety-conscious company has taken the first step toward providing the most acoustically safe environment possible for its employees. But the commitment to safety should not end there. Along with a hearing conservation program (in appropriate cases), every company should implement a comprehensive equipment acoustical maintenance program. This should also include or be a part of a predictive maintenance program, which is discussed in more detail later in this chapter.

Hand in hand with general mechanical maintenance—which improves the performance and life span of any piece of equipment—an equipment acoustical maintenance program will ensure that the equipment remains within the limits specified by the company. It will also help ensure the reliability and longevity of the equipment. This type of program will keep the noise level in the range designed. It is possible through proper acoustical maintenance to keep the noise controlled over a 10- to 15-year period.

As defined in this chapter, the acoustical maintenance program requires that each machine operate within 2 dBA of the minimum sound level of which it is capable. (Refer to Chapter 6 and Appendix A for dBA and A-weighted sound level.) It also requires that each machine remain in good general maintenance condition to allow minimum operator attention, reducing the time required for the operator to be in the direct sound field of the machine.

The equipment acoustical maintenance program should include the following elements:

1. An initial baseline sound level survey should be conducted for each machine while it operates under normal conditions and while it is in good maintenance condition. This should consist of recording the A-weighted sound level at a fixed location for each machine. The minimum air pressure or other items necessary for proper operation of each machine should also be recorded.

2. Periodically—bimonthly, if possible—a maintenance noise survey of each machine should be performed:
   a. The operating sound level should be compared with the baseline sound level.
   b. The machine should be inspected visually to identify components that may be generating excessive noise.
   c. If noise-producing elements are identified, the machine should be tagged and the appropriate repairs made.

d. Maintenance personnel should maintain a "noise awareness." They should become familiar with the mechanisms of noise generation of each machine and with the visual inspection procedures. In the course of their normal responsibility, they should identify and repair improperly operating machine elements that may cause excessive noise, even though the machine may be operating properly.

The following sections explain each of these program elements in greater detail and provide sample forms for each of the tests and record-keeping functions described.

BASELINE NOISE SURVEY PROCEDURES

An initial sound level survey should be conducted with each machine in good maintenance condition and while each machine is operating under normal conditions. Any Type 1 or Type 2 sound level meter may be used. Sound level readings should be performed on the A-weighting scale of the meter. The microphone should be located at the operator's position or employee hearing zone. This location should be maintained as the reference measurement point for all maintenance surveys. The sound level for each machine should be documented and should serve as the baseline. All future measurements should be compared to these data. When machinery is changed or noise modifications made, new baseline acoustical data should be established.

The minimum air pressures and other items necessary for proper machine operation should also be established, and this information should be recorded.

MAINTENANCE NOISE SURVEY

If a predictive maintenance program (described later in the chapter) is being used, it is important to coordinate the acoustical maintenance with the program. This will not be a duplication of effort, but each will enhance the other. After all, equipment that operates in a well-maintained condition is quieter and may have a longer production life.

Periodically (bimonthly, if possible) a maintenance person, technician, or engineer should perform a noise survey to compare operating noise levels with the baseline data. The measurements should be performed at the same location, using the same instrument as the baseline measurements. If the sound level exceeds the baseline data by 2 dBA or more, maintenance may be required. The reason for a higher level should be determined.

In addition to the sound level measurements, the operating air pressures and other items should be checked during the periodic survey. A visual inspection of the machine should also be performed (see "Visual Inspection").

The National Machine Tool Builders Association (NMTBA) has a well-defined set of data collection charts that have been reproduced in part from its manual in various figures in this chapter. These have been supplemented by other materials that are also useful.

A log book for machine and assessment data is recommended throughout the program. It is important that all maintenance problems related to noise be documented in the log book. In the event that a maintenance-related problem occurs during an OSHA inspection, this documentation will serve to verify that temporary high noise levels are inherent in the operations and cannot feasibly be avoided, but that proper attention is being given to this problem. The noise measurement data and location forms from the National Machine Tool Builders Association [NMTB, 1986] can be used as a guide to acquire information that should be recorded. These are found in Figs. 2.1, 2.2, and 2.3.

While general in nature, the following data should be reported as a minimum:

1. Load conditions.
2. Test points corresponding to those labeled on the equipment outline.
3. The ambient noise level at each test point.
4. The minimum and maximum noise levels occurring during the equipment cycle. This will show the character of the noise, steady or nonsteady.

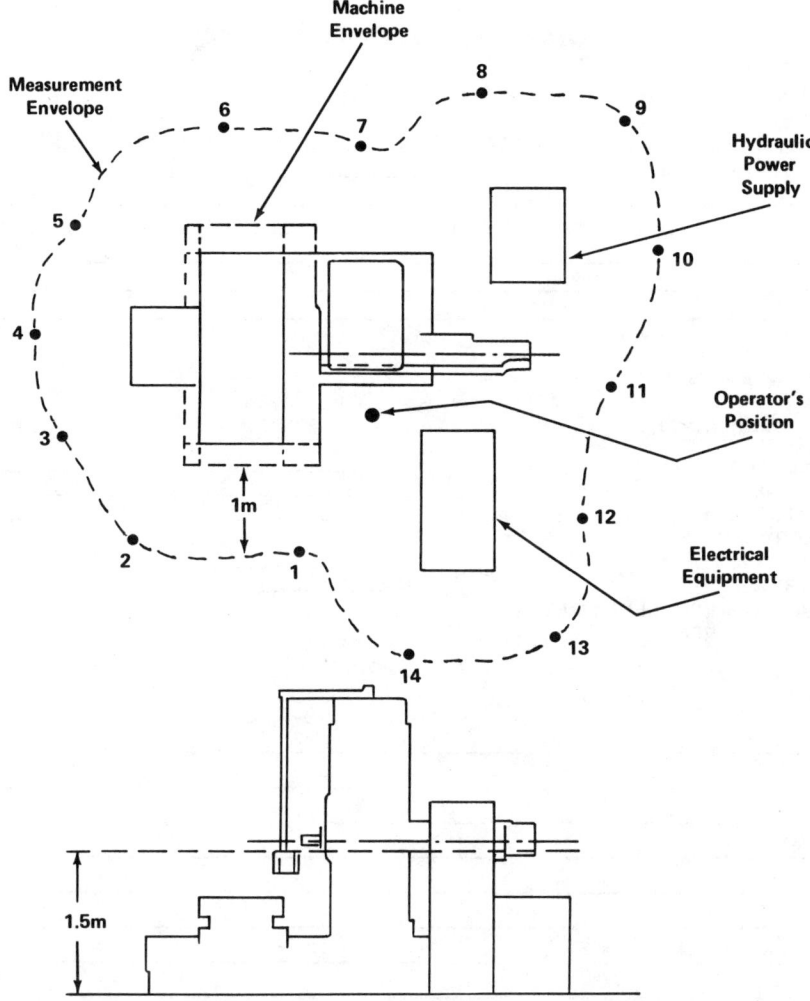

FIGURE 2.1   Noise measurement location sheet.   (Reprinted by permission from National Machine Tool Builders Association.)

5. Corrected minimum and maximum dBA shall be recorded when correction factors are applied.
6. Amplitude and frequency of occurrence of impulsive noise.
7. A time-weighted average of nonsteady noise, which may be measured directly (noise dosimeter or equivalent sound level meter) or may be calculated from data obtained during the machine cycle in a manner similar to that shown in OSHA CFR 1910.95(c) Appendix B (Non-Mandatory) Temporal Sampling Procedures for Use with a Sound Level Meter.

VISUAL INSPECTION

In conjunction with the periodic maintenance noise surveys, a visual inspection of each machine should be performed to identify components generating excessive noise. In addition, all maintenance personnel should become familiar with the inspection procedures and should correct noise-producing problems as a normal part of their daily activities. The equip-

Machine Identification No. _____

Company Name _____

NOISE MEASUREMENT DATA

A. MACHINE SPECIFICATIONS

BUILDER _____ BUILDER'S NO. _____ BUYER'S P.O. NO. _____

EQUIPMENT SPECIFICATION: TYPE _____ MODEL _____

SERIAL NO. _____ SIZE _____ CAPACITY_____

SPEED_____ HORSEPOWER _____ AUXILIARIES _____

_____

_____

_____

B. INSTRUMENTATION

INSTRUMENT	MODEL	SERIAL NO.
SOUND LEVEL METER	_____	_____
MICROPHONE	_____	_____
CALIBRATOR	_____	_____
TAPE RECORDER	_____	_____
IMPACT METER	_____	_____
OSCILLOSCOPE	_____	_____
INTEGRATING SOUND LEVEL METER	_____	_____
GRAPHIC LEVEL RECORDER	_____	_____

C. COMMENTS

_____

_____

_____

_____

_____

_____

_____

_____

_____

_____

By:

NAME _____

POSITION _____

COMPANY_____

DATE _____

Machine Identification No. _____

D. SKETCH OF MACHINE AND MEASUREMENT ENVELOPES

TEST LOCATION _____

TEST SPACE DESCRIPTION _____

_____

_____

FIGURE 2.2 Noise measurement data sheet. (Reprinted by permission from National Machine Tool Builders Association.)

Machine Identification No. ______________

E. TEST DATA – TYPE: ______________________________

OBSERVER: ______________________________ DATE: ______________

LOAD CONDITIONS (DESCRIBE) ______________________________

TEST POINT	AMBIENT dB(A) (SLOW)	MINIMUM dB(A) (SLOW)	CORRECTED MINIMUM dB(A)	MAXIMUM dB(A) (SLOW)	CORRECTED MAXIMUM dB(A)	IMPULSE PEAK dB	FREQUENCY – IMPULSES PER HOUR	dB(A) STEADY STATE / TIME-WEIGHTED AVERAGE*

REMARKS: ______________________________

*BASIS FOR TIME-WEIGHTING; ________ dB(A) per doubling of time.

FIGURE 2.3 Noise measurement octave band noise data sheet 1. (Reprinted by permission from National Machine Tool Builders Association.)

ment operators who are most familiar with the machines should call the maintenance personnel if repairs are required between inspections.

A sample machine periodic checklist is shown in Table 2.1 to assist in recording the data. This checklist was developed by a plant with a large number of small punch presses to produce the company's product and was part of their noise abatement program.

The following information should be included in each visual inspection report.

1. A check should be made with each operator to identify any machine problems that may be causing excessive time to be spent at the machine.
2. All loose parts should be secured.
3. All machine controls should be checked for proper setting.
4. Air leaks should be identified.
5. All moving components should be checked for alignment.
6. Rotating parts should be checked for shaft alignment and imbalance.

7. The adjustment of all air cylinders should be checked to ensure that excessive impact forces do not occur.
8. Air mufflers should be checked to ensure that they are in place and not damaged or clogged. This is usually the first item to be thrown out when clogged and not replaced. Thus, the noise goes up.
9. The inspector (or operator) should listen for unusual noises that may indicate component wear or other problems.

When a noise-producing problem is identified during the inspections, the problem should be corrected immediately if it involves only a minor malfunction or adjustment. If the problem requires more extensive attention, tags may be placed at those locations. A sample noise maintenance identification tag is shown in Table 2.2. This procedure ensures the correction of simple and often overlooked problems and

TABLE 2.1   Machine Periodic Checklist

_____ Dept.

Date	dBA Operating Conditions	Inspector's Observation	OK	Visual Dynamics

TABLE 2.2   Noise Maintenance Identification Tag Punch Press Dept.

Alignment _____

Damping _____

Air _____

Lube _____

Dynamics _____

Acoustics _____

Date _____ Equip. name _____

By _____

Do not remove until repaired

_____

When repaired—remove tag and forward to maintenance office

Date repaired _____

Repaired by _____

keeps production and maintenance personnel continuously aware of noise. The checklist and/or tag can be used for any type of equipment. The acoustic maintenance can be tied into the predictive maintenance program (see the last section of this chapter) for computer data base tracking.

MEASUREMENT PARAMETERS FOR NEW EQUIPMENT

These techniques provide a method of acquiring accurate baseline noise level data on each piece of equipment. The measurements must be made in a suitable test space under the control of the equipment manufacturer or purchaser's representative (see noise test procedures discussed in Chapter 9). Suitability of the test space should be left to the manufacturer's discretion, because errors due to high ambient sound levels and acoustic reflections, while prejudicial to the noise level test results, may not warrant the effort needed to eliminate them.

To obtain an accurate measure of the sound level produced by the equipment, the ambient sound level at all microphone locations should be at least 10 dBA lower than the level occurring when the equipment is operating. When the ambient sound is at this lower level, general correction factors may be applied to the sound level measurements. A test procedure is provided in Appendix E. This procedure includes the appropriate corrections for background noise.

If the application of correction factors is appropriate, the ambient noise level should remain steady within ± 1 dBA for the duration of the test. Should the ambient level vary, the minimum applicable correction factor should be applied.

A plan of the equipment being tested and measurement locations similar to the one shown in Fig. 2.1 may be included on the noise measurement data form. It should illustrate the test area, including all major reflecting surfaces such as walls, cabinets and control panels, within 5 ft of the "measurement envelope." A brief description of the major reflecting surface materials should be included. The measurement envelope should extend 3 ft from the pro-

jected floor plan of the equipment. Test points should be located on the envelope at a height of 5 ft above floor level when the machine is installed. They should also be located at the operator's position. Operator positions are defined as the operator's hearing zone during normal equipment operation.

Additionally, noise level measurements should be made at a sufficient number of locations on the measurement envelope to determine the directional characteristics of the equipment noise. In particular, data must be taken at the measurement location at which the highest sound level exists. Additional noise level data may be taken at operators' positions and at other locations on the boundary of the measurement envelope to satisfy particular requirements. In all cases, the observer shall remain a minimum of 2 ft away from the microphone.

Measurements should be made with the equipment operating in one or more of the ways outlined below as agreed upon by the manufacturer and the owner:

1. At the required production rate and performing specified operations when the equipment is purchased for a specified purpose
2. In the unloaded manner of operation that generates maximum noise levels, when the machine is purchased without specified tooling
3. At typical simulated loads equipment performance when the equipment will be used for a broad range of purposes

Field calibrations of the measuring system should be performed at the beginning and at the end of each measurement session. If the difference between the two calibrations shows more than a 1-dB deviation, the data acquired may be considered suspect and should be rechecked. It is recommended that the accuracy of the calibrator be certified annually or in accordance with the manufacturer's specifications.

The microphone should be oriented to have the incidence, either normal or grazing, that provides the flattest frequency response for the microphone being used. No microphone frequency response corrections should be used.

When noise levels are tape-recorded for record or measurement purposes, a comparison between the original and recorded noise levels should be made to determine that distortion did not occur.

The minimum and maximum A-weighted noise levels, dBA, should be measured with the meter on slow response at each microphone location. Noise that is essentially constant in level, having a range of variation of 3 dBA or less with the meter set on slow response, is considered *steady noise*. Noise that varies in level over a range greater than 3 dBA, with the meter set on slow response, is considered *nonsteady noise*.

Where noise levels change in discrete increments, such as during equipment duty cycle, the levels, durations, and associated events should be recorded for each distinct interval in order that a time-weighted average can be calculated. The time-weighted average may also be measured directly with a noise dosimeter.

Impulse noise can be defined as a burst of noise of less than 0.5 sec (500 msec) and a rise time of not more than 35 msec to peak sound pressure level. This can be either as a single event or repetitive events with greater than 1-sec intervals between them. The OSHA regulations state that the peak sound pressure level shall not exceed 140 dB with the sound level meter set on fast response [OSHA, 1983]. The current standard is silent on the number of occurrences allowed. The proposed number of occurrences is 100 per day, increasing by a factor of 10 for each 10-dB decrease in peak sound pressure level [Lord, 1987]. A standard on impulse noise published by the American National Standards Institute (ANSI), S12.7-1986, Methods for Measurement of Impulse Noise, provides more detail on this subject. In the absence of a meter to measure impulse noise a standard sound level set on the C scale fast response will provide an indication of the problem. If the level exceeds 125 dBC_f then the 140-dB peak sound pressure level has probably been exceeded.

INSTRUMENTATION

This section provides guidelines on specific instruments recommended for use in the equipment acoustical maintenance program.

For all measurements, a condenser or electret microphone should be used to provide the accuracy, stability, and frequency response. Where a cable length of three meters or more is required to couple a ceramic-type microphone to the instrumentation, a preamplifier, located at the microphone, should be used.

Sound levels should be measured with a type 1 or type 2 sound level meter meeting the requirements of ANSI S1.4-1971, except that it is not required to have a B-weighting filter. Suitable meters for measuring time-weighted averages (noise dosimeters or equivalent sound level meter) should have similar accuracy.

Peak sound pressure level measurements may be taken with any peak reading sound level meter system having a rise time of 100 microseconds or less. Peak measurements may also be taken with an oscilloscope. The oscilloscope should be of the memory type, or photographs should be taken of the oscilloscope trace.

PREDICTIVE MAINTENANCE

As dominant sources are modified in specific locations and the overall noise levels are reduce, auxiliary equipment will start to influence the overall noise environment more. Regular maintenance of this equipment will provide the first level of control. In addition, maintenance of the modifications recommended will be necessary to ensure proper performance for years of operation.

Maintenance for the control of noise is not a new concept, but it has only in the last ten years made sufficient progress to instrument and monitor the "machinery health" with nontechnical individuals. Predictive maintenance is where maintenance actions are determined and planned from an assessment of machine condition. This differs from preventive maintenance, where machine maintenance is regularly scheduled from calendar days or hours run, and breakdown maintenance, where it is fixed when it breaks.

TABLE 2.3   Factors Affecting the Selection of a Predictive Maintenance Program

Process criticality
Capital cost
Operating efficiency
Safety
Historical reliability
Monitoring feasibility

A maintenance program is anticipated to maintain present and future noise control activities. The most cost-effective program for the company will likely be a combination of the three types of maintenance described above. The type of maintenance program that is appropriate will depend on many factors. Some of these are illustrated in Table 2.3.

The most important question will be whether or not this program can save the company money in gained production efficiency and uptime. The answer is probably yes, and the most cost-efficient program for the optimum gain in production can be determined through a historical study of breakdown records, cost of lost operation, and feasibility of effective monitoring.

During the time when priority rating schemes and noise control management efforts are being considered, a detailed review of production history and maintenance procedures should be undertaken to determine how predictive maintenance will impact process productivity and reduce noise. Intuitively, one knows they go hand in hand. An employee who only works on one machine knows if operation is "normal" by listening, but in a complex environment with many sounds occurring simultaneously, that diagnostic "ear" may need to be an electronic monitoring measurement. Such activity doesn't have to be complicated, and it may likely be relatively easy as well as profitable.

References

American National Machine Tool Builders Association. 1970. *NMTB Noise Measurement Techniques*. Revised 1986. McLean, VA.

Lord, Harold W., Gately, William S., and Evenson, Harold A. 1987. *Noise Control for Engineers*. Malabar, FL: Robert E. Krieger.

OSHA. 1983. Occupational Noise Exposure Regulation. U.S. Federal Register (39FR37773), RC:Part 29CFR1910.95.

Chapter 3

Noise and Vibration Specifications

INTRODUCTION

The preparation necessary to ensure that the design of a new plant or the retrofit of equipment in an existing plant meets criteria for allowable noise limits must begin in the early stages of planning. As a part of the noise abatement plan described in Chapter 1 design specifications must be developed for establishing criteria of noise and vibration control equipment and its installation.

The procedure outlined in Chapter 1 is primarily for remedial noise control projects. However, noise control must be considered for new facilities to be constructed. The most effective and economical approach for new construction is to designate certain noise and vibration control features as an integral part of the equipment to be purchased. Such an approach is most efficiently handled by proper use of performance and design specifications.

For new facilities construction performance specifications require that the proposed equipment will satisfy the selected criteria in the noise abatement procedures and guidelines provided at the end of this chapter. Examples of project-specific design specifications with specific noise and vibration control features are found in Appendix E.

Both performance and design specifications are discussed in this chapter. Performance specifications are provided as a general guideline. They can be reformatted to meet individual requirements. The design specifications are provided in Construction Specification Institute (CSI) format.

Most companies purchase equipment on a regular basis for new facility construction or for a specific noise abatement program. Thus the equipment should be purchased using a noise level requirement. The phrase ''The equipment must meet the OSHA Noise Standard'' is simply not enough to specify the equipment or system. The material in this chapter will provide a framework for developing a realistic specification(s) for a variety of projects.

THE PROCESS

The owner or CEO will usually have or will develop as part of the policy and procedures a plan, overall idea, or formalized guideline for a project. This may or may not include a section on noise abatement. If a new facility or renovation of a facility is contemplated these will be given to an architect or engineer/architect, or a design-build construction firm, for development into detailed plans and specifications. The guideline should describe in detail the various criteria on which the project is based. In order to focus on the noise issues the guidelines should provide not only required employee noise exposure criteria and maximum equipment noise levels, but also environmental noise criteria for the plant boundaries. The designers can then develop a set of detailed specifications for the project from which bids can be obtained for the construction phase.

The following general steps are recommended for a project:

1. The owner identifies and defines the problem.
2. A noise abatement procedure is developed.
3. Project specifications are developed for the engineering contractor or architect.
4. The project is engineered and drawings and specifications are prepared for bidding.
5. The project is bid.
6. Equipment is purchased and installed.
7. Results are checked against specifications and corrections made if required.

SPECIFICATIONS

A specification should be a clear, concise statement and description of what is required for a particular project. It should outline what the contractor is to provide, the quality of the materials and workmanship, and how the project will be judged to be complete and correct.

The specification should not contain "legalese" or have broad or ambiguous statements. The one who prepares the specification must bear in mind that the owner and contractor will be held to the document and its provisions, good or bad. In addition there is another document that will be a part of the total package. This is the contract drawings. While it is beyond the scope of this discussion to cover contract drawings, it should be pointed out that they are an integral part of the total contract documents. Thus the specifications and drawings should be referenced to each other.

A final point should be made about the contract documents; that is, they are not intended to show all the details of the work. They are intended to describe and illustrate the character and intent of the work under contract. Further

they may be supplemented from time to time, either by the engineer or by the contractor with the engineer's approval. The changes become a part of the contract documents.

As an example of the differences in what the drawing and specifications will generally include, consider an acoustical enclosure to house a piece of noisy equipment. The drawings will show the following:

1. Overall dimensions
2. How the enclosure will look
3. Size of any openings and access doors
4. Thickness of wall and how the materials fit together
5. Any other dimensions and information that will clearly illustrate the intent

The specification, on the other hand, will describe the following:

1. A defined scope of work
2. Quality and quantity of the materials of construction
3. Detailed description of the materials and their performance
4. Overall performance requirements of the enclosure
5. The execution of the installation
6. How the results will be obtained and reported

Thus the drawings will give a visual picture of the enclosure and the specification will describe, in words, what is on the drawing in sufficient detail that it can be fabricated and installed. Numbers, words, and illustrations are used to demonstrate the end result [Ranous, 1964].

PURCHASE SPECIFICATIONS

Many companies have equipment purchase specifications relative to noise that state: "The equipment must comply with OSHA requirements." This is meaningless since the OSHA standard for occupational noise exposure CFR1910.95 set forth the limits for occupational noise exposure of employees and not the noise generated by individual pieces of equipment [Harris, 1979]. To be sure, the purchaser must take into account the noise levels of the individual piece of equipment so as not to overexpose the employees. The noise levels determined by the company should be part of a purchase specification to be installed. This is nothing more than a description of all the particulars that are of significance [Ranous, 1964], spelling out the upper limits of the noise levels, how any measurements are to be made, and the applicable test standards [Harris, 1979].

Any purchase specification should at least have the following elements:

1. An objective—a statement outlining the maximum noise levels
2. A definition of the equipment covered in the specification
3. A testing procedure developed from a national standard
   a. Instrumentation
   b. Testing area
   c. Operational requirements
   d. Measurement procedures
   e. Calculation procedures for ambient noise
   f. Information to be reported
4. A statement that the purchaser will assist the vendor in meeting the noise levels if the vendor does not know how
5. A requirement to quote the additional cost for providing a quiet product, if available
6. A citation of a specific code or standard for testing the material being purchased

For example, the Society of Automotive Engineers (SAE) has numerous standards and procedures for testing of various equipment. An example of this is the SAE recommended practice J1166, "Operator Station Sound Level Measurement Procedure for Powered Mobile Earthmoving Machinery—Work Cycle Test." Another example is the Compressed Air and Gas Institute (CAGI) for testing procedures of a variety of pneumatic tools and equipment such as rock drills and hand tools. Using a recommended practice or procedure such as that illustrated will provide the purchaser with all the information to adequately specify the purchase of equipment.

DESIRABLE DESIGN FEATURES FOR NOISE REDUCTION

In the selection of equipment for a new plant or for replacements or additions to an existing facility, satisfactory noise limits often can be obtained by properly attending to specific design features of the equipment. For example, for totally enclosed fan-cooled (TEFC) electric motors, unidirectional cooling fans or intake silencers should be considered standard equipment. A summary tabulation of such design features is presented in Table 3.1.

NOISE ABATEMENT PERFORMANCE SPECIFICATIONS AND GUIDELINES

A typical performance noise specification or guideline can be used for a variety of facilities. These guidelines describe both buyer and contractor (or vendor) responsibilities to assist in setting reasonable equipment noise performance limits, ensure that these limits are satisfied, and obtain the required information by

TABLE 3.1   Noise Control Design Features of Electric Motors

Equipment	Source of Noise	Noise Control Options
TEFC motor	TEFC cooling fan	Acoustical fan shroud, unidirectional fan, and/or fan air intake silencer
Weather-protected (WPII) motor	Cooling air openings, mechanical and electrical noise	Lined duct or silencers, enclosures

which informed decisions can be made. For this discussion we will focus on a gas-processing plant and a manufacturing plant. For these two types of facilities the only difference, as it relates to the noise guidelines, is the type of equipment. Therefore, the difference in the guideline is the list of typical equipment. Furthermore, the required noise levels can be modified from that shown to suit the individual needs.

At the end of this chapter is an example of noise abatement performance specifications and guidelines. Before that, however, it will be well to consider some specifics of the guideline.

OWNER'S RESPONSIBILITY

Generally, the owner (buyer) should specify absolute noise level requirements for all operating equipment. Requests for a specific degree of noise reduction should be employed preferably only when the order is for noise-attenuating devices; in this case, the required noise reduction should be specified in dBA and in octave bands in the range 63 Hz to 8000 Hz, if appropriate.

When performance levels are specified, care should be taken to provide relevant design information, such as gas flow, gas velocity, gas pressure, maximum permissible temperature rise, and equipment operating speeds. For example, in specifying mufflers and silencers, the buyer should require a guarantee that regenerated noise produced by the flow within the muffler itself will not limit the performance of any offered unit.

The project engineer is strongly urged to enlist the aid of the contracts and/or purchasing department to ensure that the purchase contract incorporates means for requiring the supplier to comply with the specified noise limits. Company procedures designed to enact guarantees should be consulted and followed.

VENDOR'S AND/OR CONTRACTOR'S RESPONSIBILITY

The contractor or equipment vendor should submit all the data requested and in the form required. The procedures used for measurement of equipment noise and/or calculation of reported data should be provided by the vendor. Testing standards should be identified, along with any independent contractor performing the tests.

If the vendor provides data measured by industry standards, but due to limitations during testing omits items normally associated with the equipment, they should be measured and reported as separate items. An example of this is a drive motor required on the test stand that is different from the motor supplied with the equipment.

The vendor should not supply sound levels measured from equipment having mufflers or other noise-reducing devices attached, unless these items are supplied by the vendor as part of the equipment or these data are requested by the buyer. The vendor should supply sound level measurements made with the equipment operating under normal conditions or in the working range that generates the highest noise.

If an equipment item in standard form will not meet the noise specification, then the vendor should explain how it intends to meet the specification. This should be accomplished with drawings and engineering analysis, as appropriate, and should include revised estimated noise levels. The vendor should also report the difference in bid price between the standard and quieted version and any effect on the promised delivery time.

If a vendor cannot offer specific measured noise levels for a reduced item but proposes instead to offer estimated data, the following information should be included so the buyer can determine the acceptability of such data:

1. The technical basis on which the estimates of the noise level have been made (i.e., similar equipment, field installation in other plants, competitor's claims for essentially identical equipment, theoretical calculations, and empirical extrapolation)
2. The claimed accuracy of the noise level estimates
3. How the vendor intends to make good

any deviation from the promised specification after the equipment is installed and operating.

ASSESSING THE VENDOR'S DESIGN

Procuring an item of equipment should be a complete program involving cooperation between management and the engineering and purchasing departments. The specified requirements for meeting a noise criterion should receive due consideration along with other requirements. A detailed evaluation must be made of all proposed vendor's equipment. This should contain performance data, including noise data, as well as cost considerations. All other items being equal, the quietest equipment should be selected.

As pointed out earlier, simply specifying equipment noise levels is of little value unless the onus of compliance is placed upon the vendor. Thus, a clear understanding between buyer and vendor on the method of acoustically qualifying the purchased equipment should be achieved early in the procurement program. Included in this understanding should be a clear statement of field test procedures.

Some items of equipment can be suitably qualified in the vendor's shop, whereas others, such as furnaces, large turbines, and compressors, can be tested only after they have been installed in the plant. Previous data from other installations should also be requested for evaluation. Regardless of the chosen location or manner of qualification, a buyer representative should be present to ensure the validity of measurements and data obtained. Measurement techniques described in Chapter 9 should be consulted for evaluating measurement procedures or following a procedure developed for this equipment.

Should equipment fail to meet the specified requirements, the buyer should enforce the terms of the contract and specification. To facilitate enforcement, penalties for noncompliance should be defined in the contract. Such a procedure can eliminate or reduce additional and costly in-plant noise control efforts with respect to the purchased equipment and will establish a policy for future purchases. The vendor should be required to provide noise control measures that will enable its equipment to comply with the noise specification.

DESIGN SPECIFICATIONS

The Construction Specification Institute (CSI) [CSI, 1985] is a master format system of numbers and titles for organizing construction information into a regular standard order and sequence. This system provides for all parties involved a standard method of compiling information about the project to improve communication between those parties. It is particularly effective in assisting the owner in communicating with architects, engineers, and contractors. Since its beginning in 1963 CSI has evolved into a widely accepted industry standard in the United States and Canada. This system was developed and endorsed by such organizations as the American Institute of Architects, the National Society of Professional Engineers, and the Associated General Contractors of America. The author's experience has been with this type of specification in the architectural and engineering field. Other forms of specifications might be used depending upon an individual's or firm's experience.

The CSI format is divided as follows (CSI, 1985):

Division: There are 16 unchanged divisions that are set within a rigid format. Three examples are: site work; concrete; and mechanical. Within the mechanical section sound, vibration and seismic control for mechanical equipment are covered. There is also a special section to deal with special construction of sound, vibration, and seismic control as it relates to manufactured components for the reduction of undesirable sound and vibration within a structure.

Section: A portion of project specifications covering one portion of the total work or requirements.

Part: An organizational device to divide a specification into three distinct groups of related information as follows:

Part 1, General—Defines specific administrative and procedural requirements unique to the section.

Part 2, Products—Describes, in detail, the quality of items that are required for incorporation into the project under this section.

Part 3, Execution—Describes, in detail, preparatory actions and how the products are to be incorporated into the project.

Article: A major subject consisting of related paragraphs within a part of a specification system.

Paragraph: One or more sentences dealing with a particular item or point, separated from the preceding text by beginning a new line. Groups of related paragraphs constitute an article.

An example of this type of specification for noise control is included in the next section, which is a case history to illustrate how the total process works using additions to an existing facility. Specifications include floating floors, noise control for engine noise, and vibration isolation of equipment.

DESIGN SPECIFICATION CASE HISTORY

The Problem

A company desires to place three emergency electrical generators on the ninth floor of its building. This is a mechanical equipment space. Office space will be located above, below, and adjacent to the mechanical equipment space. Office spaces located on another floor will be relocated to the space below the mechanical space. In addition, the emergency generators will be used to provide power during peak electrical demand periods of the day. Thus, the noise generated will occur during the day when adjacent offices are occupied. The generators will be driven by diesel engines. The mechanical room also has two 1000-hp gas turbines that are used only for emergency backup power in the building.

Noise Control Evaluation

The owner and noise control engineer mutually agree on the noise criteria. This is the noise level of the currently occupied space. Employees will be relocated below the mechanical equipment space. Thus noise measurements are required to determine the noise levels in the occupied space. The space will be evaluated to determine any changes that might occur in the new space.

Since three large diesel engine driven generators are to be located in this equipment space it is necessary to determine the amount of noise reduction of the existing 10-in. floor slab. Based on past experience it is known that additional noise control will be required, but how much will not be known until the floor slab noise reduction can be determined.

One approach would be to use known test data for this type of floor slab and estimate the noise reduction. Another approach is to test the floor slab to determine noise reduction. This will provide the most accurate results. Since the two gas turbines are located in the space, the owner has offered to operate them, affording the opportunity to have a high noise source for the test. Measurements are made in the equipment space and in the space below to determine the noise reduction. This is simply a matter of subtracting the two measurements. When the space is finished out, the ceiling and furnishings will provide some more noise reduction. Therefore, the test provides a very good conservative measure of the noise reduction.

Figure 3.1 illustrates the space and the relationship of the equipment and the occupied spaces. Tests are also made to determine the noise reduction for the adjacent spaces. Measurements made in the occupied space that serves as the criterion for the adjacent spaces are shown in Fig. 3.2. By comparing the noise levels in the occupied space and the new space, one can immediately determine the amount of additional noise reduction.

From this simple analysis it is determined that an acoustical enclosure will be required to contain the engine noise. Since most of the equipment room floor will be protected by the enclosure, some thought must be given to pro-

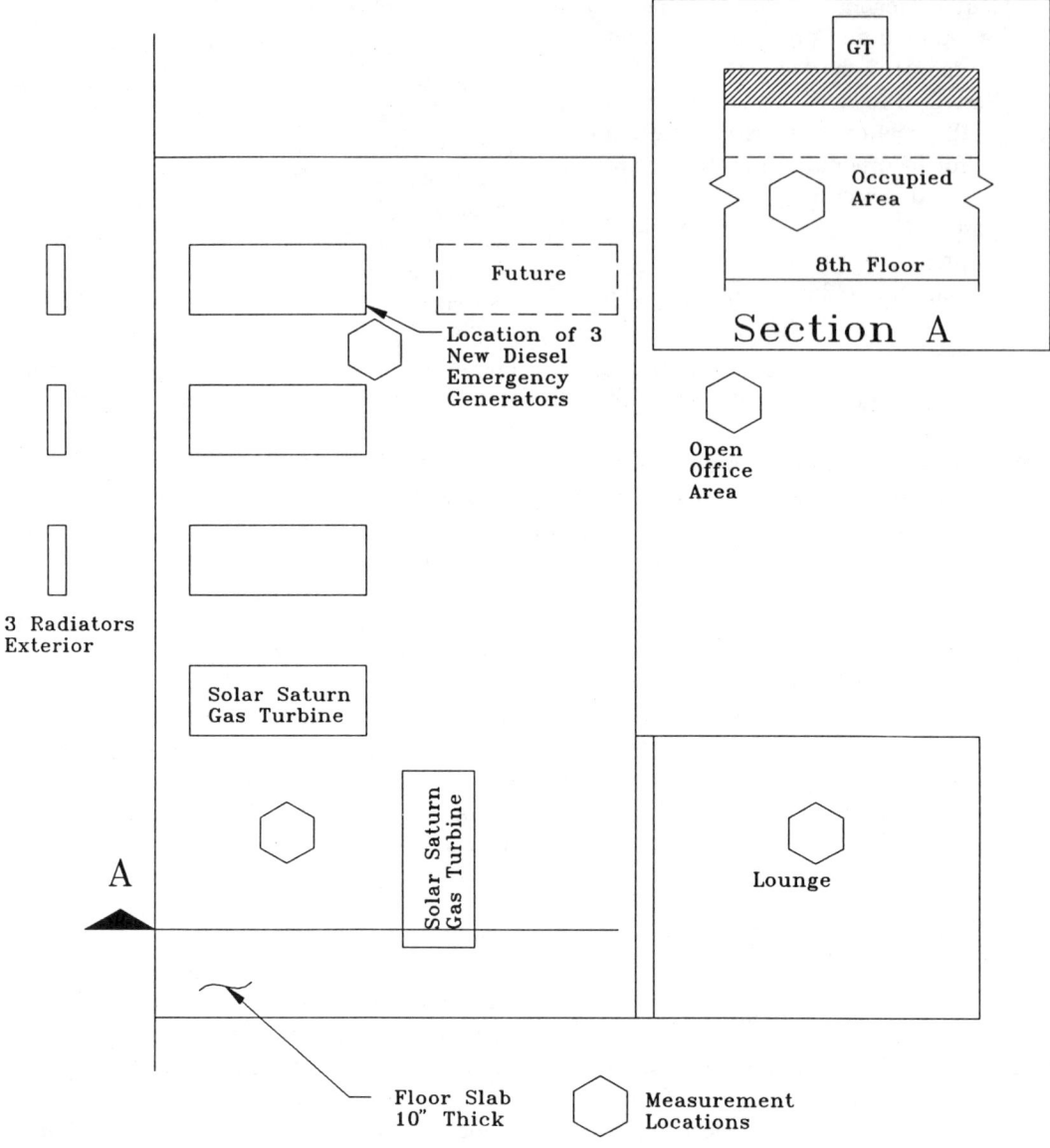

FIGURE 3.1   Emergency generators and adjacent occupied space.

viding more noise reduction to the space within the enclosure. If the floor slab is doubled in thickness this will only reduce the noise level 6 dB, and much more is required. It is possible to take advantage of the two separated slabs with an air space using the principle of a floating slab. Figure 3.3 illustrates this type of design, which has been used successfully for many years.

In addition to the enclosure there are several other items that must be considered to complete the design before the specification can be developed. These are:

1. Ventilation of the enclosure requiring silencers for the engine noise and cooling fans
2. Engine exhaust silencing
3. Vibration isolation of the engine, piping, and radiators

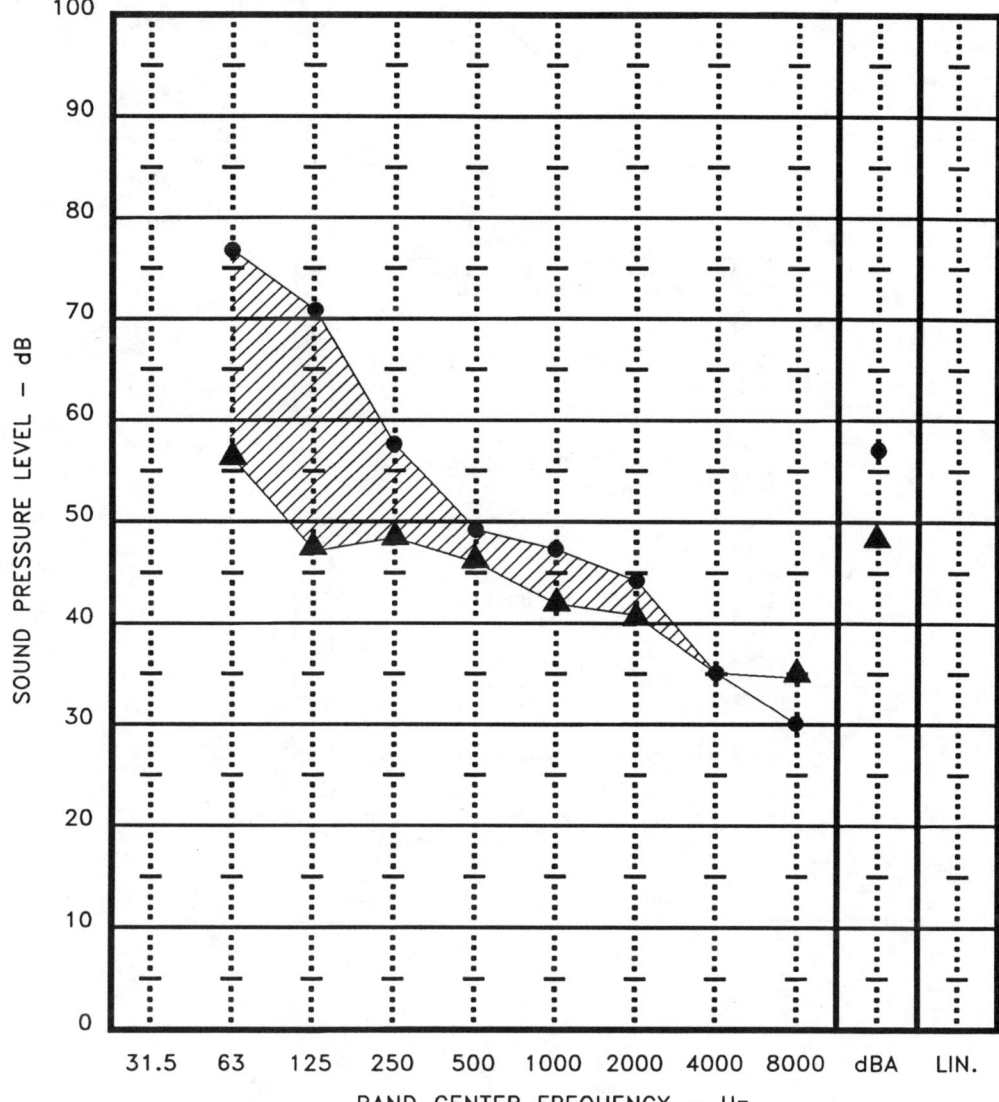

FIGURE 3.2   Noise level measurements.

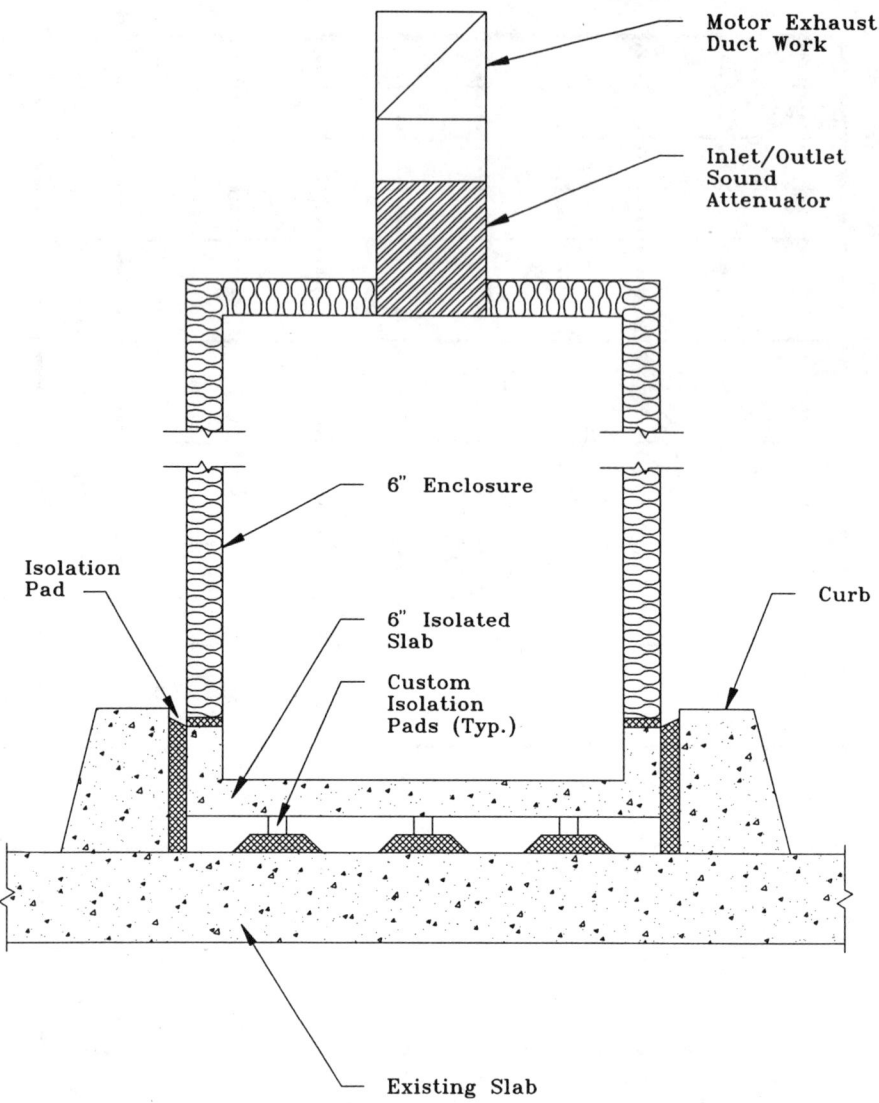

FIGURE 3.3   Noise control concepts.

Once these items have been considered, the specifications can be written with the equipment described and the system noise reduction identified.

EXAMPLES OF NOISE ABATEMENT GUIDELINES, PROCEDURES, AND SPECIFICATIONS

This section contains a general noise abatement procedures and guidelines format that an owner can use to describe a project to an architect, engineer/architect, or design-build contractor. Appendix E includes sample design specifications, sample equipment sound measurement procedures, and vibration control specifications used for the emergency generator installation discussed in the previous section. This will demonstrate how these procedures, guidelines, and specifications might be used.

Note again that the sample design specification describes the specifications required for the addition of a newly constructed area of a plant. The sample equipment sound measurement procedure refers to the addition of a new piece of equipment to an existing facility. The vibration control specification is for a related piece of equipment.

Each company will have needs that require changes to these examples to suit its specific purchasing requirements, but the examples will serve adequately as a starting point for meeting all of those requirements. The American Petroleum Institute has Standard 615, *Sound Control of Mechanical Equipment for Refinery Services*, which outlines a methodology of obtaining the required noise levels for new equipment [API, 1980]. Other trade organizations such as the Society of Automotive Engineers and the American Society of Mechanical Engineers, have similar standards. The type of plant or noise source may dictate the approach to be used in developing a noise abatement guideline.

Noise Abatement Performance Specification and Guidelines

A. Scope
  1. It is the intent of this specification to provide the procedures to be followed by contractors, architects, and/or engineers in providing a plant where the noise will comply with the owner's requirements set out below.
  2. The purpose of this specification is:
     a. To assure that the level of noise in any plant ordinarily occupied by operating and maintenance personnel will be below the level for hearing impairment
     b. To limit noise in control rooms and other work areas to levels that promote efficient working conditions; i.e., low speech interference levels
     c. To make sure that noise from the owner's facilities will not intrude into the surrounding community
  3. It is the responsibility of the contractor to provide plant equipment with sound pressure levels that comply with the requirements of this specification. The contractor shall obtain from the equipment seller a completed typical vendor equipment noise data sheet (Fig. 3.4) for all plant equipment that is thought to produce noise of a nature as described above.
  4. This guideline sets forth the permissible noise level limits for the equipment and outlines the procedures for testing, reporting, and guaranteeing the noise levels for the equipment supplied.

B. References
The latest edition of the following standards shall apply unless otherwise stated:
ANSI S1.1—Acoustical Terminology
ANSI S1.2—Method for the Physical Measurement of Sound
ANSI S1.4—Specifications for General-Purpose Sound Level Meters
ANSI S1.11—Octave, Half-Octave, and Third-Octave Filter Sets
ANSI S1.13—Methods for the Measurement of Sound Pressure Levels
OSHA   29CFR1910.95—Occupational Noise Exposure

C. Noise levels (equipment)
  1. The maximum noise levels for each piece of equipment shall not exceed the levels shown in Table 3.2.
  2. If the measured reported levels in any one octave band are higher than those in the two adjacent bands by more than 5 dB, the allowable noise level for that band shall be 5 dB less than the permissible level.
  3. The allowable noise levels shall apply to conditions as installed with the equipment operating at the design load. If noise control treatment(s) or special methods of installation are proposed to reduce noise to the allowable levels, then the allowable levels may refer to these special conditions provided the noise control treatment(s) are fully documented and incorporated into the final installation.

D. Reporting
  1. The equipment supplier shall state in his

1.   Noise Data

MAXIMUM Lp DESIGN LEVELS TABLE 7.1	OCTAVE CENTER BAND FREQUENCY (Hz)	MAXIMUM Lp GUARANTEED BY VENDOR AT LOCATION DEFINED IN SPECIFICATION, dB		
		STOCK ITEM	SPECIAL DESIGN (See 4. Below)	ACOUSTICAL TREATMENT (See 5. Below)
	dBA			
	31.5			
	63			
	125			
	250			
	500			
	1000			
	2000			
	4000			
	8000			

2.   Basis of Noise Data ——————————————————————————

3.   Test Conditions:

	Area - Fr²(M²)	Material
Wall		
Ceiling		
Floor		

4.   Special Design:——————————————————————————
——————————————————————————
—————————————————— Cost:————————

5.   Acoustical Treatment:——————————————————————
—————————————————— Cost:————————

6.   Exceptions Taken:———————————————————————
—————————————————— Cost:————————

7.   Name of Vendor:————————————————————————
Represent by:——————————— Title:——————————
Signature:——————————— Date:——————————

FIGURE 3.4   Typical vendor equipment noise data sheet.

quotation the sound pressure levels of each item, by octave bands, in the space provided on noise data sheets. This noise performance will be considered in the evaluation of the quotations and in the issuing of a purchase order.

2. The sound pressure levels on the noise data sheet shall be based on one of the following:
   a. Actual measured data taken in the supplier's shop carried out in accordance with the standards referenced
   b. Noise test data measured on a simi-

lar unit running under similar conditions in an existing plant
   c. Noise test data obtained on a similar unit in the supplier's shop measured in accordance with the standards referenced
   d. An unconditional guarantee that the equipment when operating under design conditions will not produce noise exceeding the allowable levels specified

3. All data shall be certified by persons knowledgeable and experienced in noise

TABLE 3.2   Typical Maximum Permissible Noise Levels (dB re 20 micro Pa) for Process and Power Plants

Frequency (Hz)	Continuous Source[a]		Intermittent Source[b]	
	At Grade	Property Line	At Grade	Property Line
63	97	67	102	71
125	90	60	95	64
250	85	54	90	58
500	80	49	85	54
1000	78	46	83	51
2000	76	44	81	49
4000	76	43	81	48
8000	77	42	82	47
dBA	85	53	90	58

[a]Continuous source: A noise source that extends over a long period of time without interruption. This applies to any location that operating personnel routinely visit during the workday.
[b]Intermittent source: A noise that routinely occurs on a predictable basis (e.g., venting during start-up or shutdown).

Source: Author's private files and technical reports.

measurements, and/or the noise test shall be witnessed by the owner's representative.

4. If requested, and determined feasible by mutual agreement, the sound power level and the directivity of the equipment should be reported as well.

E. Deviations

1. If, in the opinion of the equipment supplier, the specified noise levels cannot be met without extensive reworking of standard equipment, requiring an increased price, the supplier may so indicate by quoting the equipment with and without noise reduction equipment, and reporting the noise levels it is prepared to guarantee for the equipment to be supplied. The supplier shall quote the additional cost to meet the allowable noise levels. The owner shall at his or her option assist the supplier with noise control treatment or design to ensure that the allowable levels can be met.

2. All deviations from the specification shall be clearly described in the bid. The absence of such a list shall be construed to indicate complete compliance with the specification and standards referenced.

3. The contractor shall make recommendations regarding the use of special equipment or alternate solutions when a vendor's proposal for noise reduction appears impractical or involves excessive costs.

F. Guarantee

Except when deviations are specifically noted in the quotation, the seller shall guarantee to meet the noise level requirements of this specification and the noise data sheet. Any remedial work performed by either the owner or the seller as a result of the latter's failure to meet the guaranteed noise levels shall be at the expense of the seller.

G. Reporting of results

The contractor shall prepare a noise contour map or grid coordinate map with noise levels in dBA predicting the resulting noise levels from combined noise sources with plant operating under design conditions for all areas within the battery limits or other designated boundary lines. Such contours shall be drawn at 5-dBA intervals starting at 80 dBA. A preliminary contour map shall be prepared using specification noise levels to show potential high-noise areas, and a final map shall be made using vendor-quoted levels. For any area where the level is excessive, the contractor shall propose possible methods to reduce noise and advise the owner of incremental costs.

H. Noise sources

1. Equipment

Complete noise data sheets shall be provided for all equipment items producing noise. This must include but not be limited to the following:

a. Typical process and utility plant noise sources

i. All motor-driven items (pumps, compressors, blowers, air coolers, etc.)

ii. All electric motors, engines, and turbines

iii. Burners (fired heaters, boilers, incinerators)

iv. Control valves
v. Flares
b. Typical manufacturing plant noise sources
i. Manufacturing machinery and equipment
ii. Machine tools
iii. Material-handling systems and devices
iv. Pumps
v. Hydraulic systems
vi. Gear drives
vii. Washers and other cleaning equipment
viii. Foundry equipment
ix. All over equipment that generates noise
2. Piping systems
Piping systems are to be designed to meet the noise requirement provided in this specification. In addition, atmospheric vents or suctions shall be investigated by the contractor for noise levels produced and be provided with silencers as necessary to comply with sound pressure levels required in this specification.

I. Property line noise levels
The noise at the property line shall not exceed the values given in this specification.

J. Noise measurements
The owner at his or her option shall measure the noise levels in and around the plant to determine compliance with allowable noise levels. Measurements shall be made in accordance with a sound level meter conforming to ANSI S1.4 standard. For octave band analysis the linear setting shall be used. For overall measurements both C- and A-weighted data shall be reported.

References

API Standard 615. 1980. *Sound Control of Mechanical Equipment for Refinery Services*. American Petroleum Institute.

Construction Specification Institute. 1985. [Master Format—An Overview]. Alexandria, VA.

Harris, Cyril R., Ph.D., ed. 1979. *Handbook of Noise Control*. New York: McGraw-Hill.

Ranous, Charles, R. 1964. *Communication for Engineers*. Boston: Allyn & Bacon.

Chapter 4

Noise Control Guidelines and Materials

INTRODUCTION

Walking into a noisy manufacturing or process plant can be very confusing. It is difficult to communicate, it is annoying, and usually you just want to finish your business or pass on through as quickly as possible. If the plant manager gives you the responsibility of developing a noise abatement plan for the plant, then you must go into the plant with a different viewpoint, or there will still be confusion and annoyance.

In order to have some historical perspective on the field of noise control it will be of interest to know that engine silencers and vibration isolation were being designed, built, and installed by the Maxim Silencer Company and Korfund Dynamics before 1915. In 1937 the Celotex Corporation published *A Manual for Architects*, part of which dealt with industrial and architectural noise control issues. The manual illustrated and gave data on how to isolate machinery vibration with Celotex materials and how to use Celotex materials as part of a noise control enclosure [Celotex, 1937]. These same techniques have been used over the years and are still used today to achieve the good noise control results.

Sound level meters were being built in the mid-1930s for measurement purposes. The field of room acoustics and the use of absorption materials was well understood and established in the 1930s as well. After World War II the field of noise control became more popular and there were publications that established this as a field of engineering. For example, one technical magazine, *Noise Control*, published by the Acoustical Society of America, had articles, advertisements, and information about this field. When it ceased publication, *Sound and Vibration* magazine took on the same role in the mid-1960s up to the present. There are many resources available and this field is not new. It has been well established for at least 50 years and goes back almost 100 years.

The purpose of this chapter is to bring order out of chaos. As you walk into the plant the next time it will be with a different attitude, but you must be able to *listen* to what you hear. That is to say, connect the ear to the brain and determine what you are hearing. Take the sound level meter and watch the display. What activity is occurring and what does the meter read at that time?

From your perspective this is as if a large jigsaw puzzle has been dumped out of the box onto the table, and your mission is to put it together so it looks like the picture on the box; only there is no picture. Where do you start? With the border pieces that have straight edges.

This might be a simple noise survey to define the outside edges—the levels at columns and operator positions. The survey will help you begin to visualize what the picture might look like. You will begin to listen, really listen, to what is going on around you in terms of noise. Where is that impact coming from, or what causes that whirring noise, or where is that rumble?

All of this is just the beginning of visualization and developing the concepts. It is now time to undertake the task of defining the problem, determining how to solve it for a reasonable cost, buying and installing the equipment, then keeping it working for an extended period of time. The first step is to define the problem.

GENERAL GUIDELINES FOR NOISE CONTROL

Defining the Problem and Determining Feasibility

A noise problem can be divided into three parts. These are:

Noise source
Noise path
Noise receiver

If the source can be identified and modified this is usually much less disruptive than if the noise path must be modified with an enclosure, for example. The noise receiver, the employee, can use protective equipment or be placed in a control room to control noise exposure. The end result may be similar, but the ability to produce a product, maintain the equipment, and control the cost will be different. How all of this is handled in the end will determine the feasibility for noise control. The remainder of this section will help in reaching the first goal of defining the problem and determining feasibility.

Identifying the Noise Sources

In his book *Sound, Noise and Vibration Control*, Lyle F. Yerges presents a tabulation that provides, in a very succinct form, not only a subjective description of noise sources, but what the peak components are in the "inherent" or natural characteristics of the noise spectrum as well as the "incidental" or less significant characteristic parts of the noise spectrum. This has been reproduced by permission and with some minor changes in Table 4.1 so that the flavor of the original table is preserved [Yerges, 1978]. Being able to describe the noise source in words is very helpful in communicating it to others, since that is how most people will describe the noise source to you; but more is required.

A sound is generated when the frequency of the disturbance is in the audible range from 20 to 20,000 Hz. An overall measurement in dBA will not usually help determine the source of noise. Thus a frequency spectrum must be obtained. At a minimum, an octave band analysis is required to determine where the noise energy is located in the audible spectrum. While we can "hear" this without the actual octave band frequency analysis it can only be described as low-, medium-, or high-frequency noise. More information about the source can be obtained by measuring with a narrower bandwidth. For example, a one-third octave band will divide each octave band into three parts, providing much more information; or if a one-twelfth octave band is used, even more information is obtained.

Examples of full and one-third spectra for the same source are illustrated in Fig. 4.1. When the spectra of the various sources, such as pieces of equipment, are compared, a picture may begin to emerge about the overall spectra that have been measured.

In order to have these comparisons it may be necessary to turn off various pieces of equipment, measure again, and compare the spectra. In Fig. 4.2 such a comparison is illustrated. In a process plant, for example, the equipment cannot be turned off at the particular time the measurement is required. But it might be possible to obtain data when the plant is coming down for a turnaround or on start-up. Process and utility plants are brought on line in stages over a period of time. This provides time for the noise control engineer to obtain useful data that will help in identifying and quantifying noise sources.

TABLE 4.1   Mechanical Equipment Noise Characteristics of Sound or Vibration

Source and Type of Motion	Inherent	Incidental
Rotating: Fans Blowers Turbines Centrifugal Pumps Centrifugal Compress Centrifugal Chillers	A tone of frequency (broadband) $f = \dfrac{rpm}{60} \times$ number of blades on wheel or impeller (and higher harmonics)	Aerodynamic "roar" Dynamic imbalance with vibration frequency: $f = \dfrac{rpm}{60}$ and higher harmonics
Cooling Towers	Fan Noise and Water Splash	Same
Motors/Generators	"Whine" of frequency: $f = \dfrac{rpm}{60}$ or some multiple	Same and cooling fan noise
Gears	"Whine" of frequency: $f = \dfrac{rpm}{60} \times$ number of teeth; or some multiple	Vibration with frequency similar to inherent noise. High-speed impact and sliding noise; broadband "grinding" and "screeching."
Bearings	"Squeal" of frequency: $f = \dfrac{rpm}{60} \times$ some multiple	Same
Grinders	"Grinding" noise. Essentially broadband, often with a relatively strong pure tone of frequency: $f = \dfrac{rpm}{60} \times$ some multiple	Same
Saws Planers Routers Shapers	"Scream" of frequency: $f = \dfrac{rpm}{60} \times$ some multiple teeth, blades, or cutters	Same and "ring" of saw
Reciprocating Internal combustion engines	A "roar" of frequency firing rate: $f = \dfrac{rpm}{60} \times$ multiple number of cylinders Exhaust noise	Cooling fan and pump noise valve "clatter," air noise Dynamic imbalance with vibration frequency: $f = \dfrac{rpm}{60}$ and higher harmonics
Compressors Pumps	Same Intake and exhaust noise	Same Pressure pulses in gas and fluid lines; frequency related to inherent sound frequency
Vibrating Transformers Ballasts Rectifiers Light filaments	Relatively pure "tones" of frequency: $f = 2 \times$ cps of AC current or some harmonic sound	Sympathetic vibration housing, casings and attachments, some multiple of inherent

TABLE 4.1   (*Continued*)

Source and Type of Motion	Inherent	Incidental

Vibrating (*Cont.*)
   High speed vibrators
   Vibrating conveyors
   Vibrators
   High-speed shakers
	"Vibration" or "buzz" or "rattling"
	Essentially broadband
		Same

Bells Buzzers Vibrating horns	Relatively pure tone; frequency related to method of generation	Same

Impact
   Presses
   Hammers
   Shears
   Riveters
   Punches
   Tumblers
   Shake-out devices
   Office machines
   Printing and duplicating
      equipment
   Accounting machines
   Print-out equipment
	"Hammering, rattling, pounding, thumping, banging," etc.
	Essentially broadband
		Vibration, broadband in frequency

Flow noise
   Airflow in ducts
   Fluid flow in pipes
	"Rushing" or "flow" sound; relatively broadband	Same
		Cavitating pumps or supply/return imbalance create squeals and pulsations

Valves and metering devices	"Rushing, whooshing, swishing" type sound	Same Sound often travels long distances down
Throttling devices Dampers Flash tanks Orifices Nozzles	Often a strong, almost pure "tone" or "scream" or "screech" Often very high frequency	Ducts or pipes via walls or via fluid or gas Strong pulsation or hammering when valves or throttling devices close or open

Source: Yerges, Lyle F. 1978. *Sound, Noise, and Vibration Control*. New York: Van Nostrand Reinhold.   Reprinted by permission.

As an example of how this can work to the noise control engineer's advantage, the following firsthand example is provided from the author's experience.

During a second visit to a large process plant, after the initial survey had been made and data reviewed, just prior to starting the second survey the local power utility had a major transformer substation go out. This cut off power to the plant and safety valves and flares started going off to shut down the plant for an emergency condition. After about two hours some power was restored and the plant could be started back on line. In the meantime, the plant's emergency power system was operating to keep essential equipment running.

After it was determined that it was safe to be in the plant, additional noise data were obtained for comparison to the original data. This plant used high-pressure steam in the process and it was a dominant source in many locations. In addition, as with most process plants, there were numerous pumps with drive motors located in the main aisle. Because the main noise source—steam—was off during this period it was possible to accurately measure the levels of the pumps and their drive motors all along the main aisles. This led to confirmation

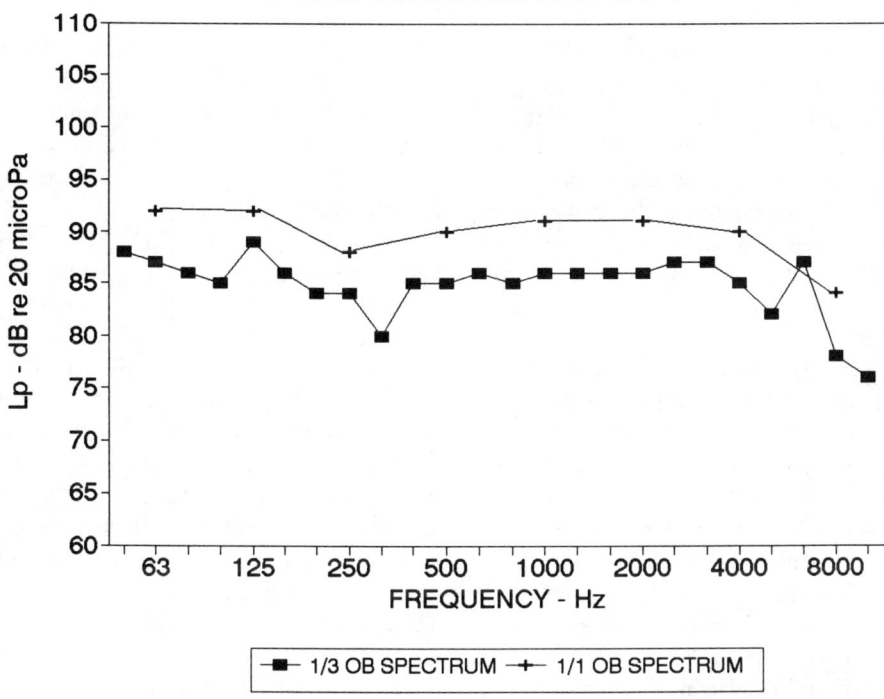

FIGURE 4.1   Comparison of a full octave and a one-third octave band spectrum of a coal crusher.

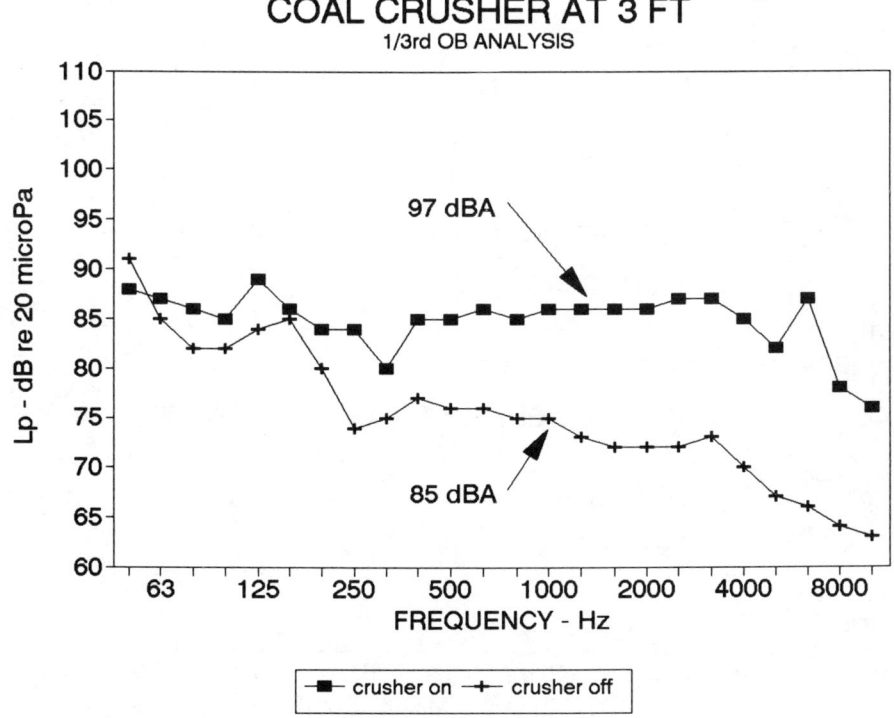

FIGURE 4.2   Comparison of coal crusher turned on and off.

of the noise priorities in the plant that had been previously estimated. It also clearly demonstrated that if the steam noise was controlled the noise levels would only be reduced by 3 dBA, since the steam noise was equal to the pumps and motors, and that the drive motor noise would have to be considered for a complete solution. Thus, by chance a clearer picture began to emerge about the noise sources in this plant.

As the plant is revisited different things will emerge that were not noticed previously. Make note of these to help obtain a clearer picture about the noise sources and their relative influence.

Vibration is another type of measurement that can be made to help identify sources. Comparison of the similarities between these spectra will assist in identifying noise sources. This is done by placing the accelerometer at various locations and obtaining a detailed spectrum. Look for the peaks in the vibration spectrum and compare these to the peaks in the airborne noise spectra. Figure 4.3 illustrates this point with a simplified spectrum of noise and vibration. Keep in mind that even though this is a complicated process, the goal is to help visualize the noise sources. In some cases octave band spectra are too broad to be of help, so a narrower band spectrum is recommended for noise and vibration.

Using the information in Table 4.1 will also help in relating the peak frequencies to something that is rotating. Noise sources can be identified many times by calculating the fundamental frequency of a rotating item. An example will help explain this concept. Figure 4.4 shows a partial one-twelfth octave band spectrum of an exhaust fan on the roof of an industrial plant and in the community. The problem was to identify which of the five fans on the roof were causing a problem in the adjacent community or if all of them were a problem. The details of this example are found in Chapter 10. The blade-passing frequency of the fan is calculated as follows:

$$BPF = rpm/60 * \text{no. of blades, in Hz}$$

This calculation determines how many times

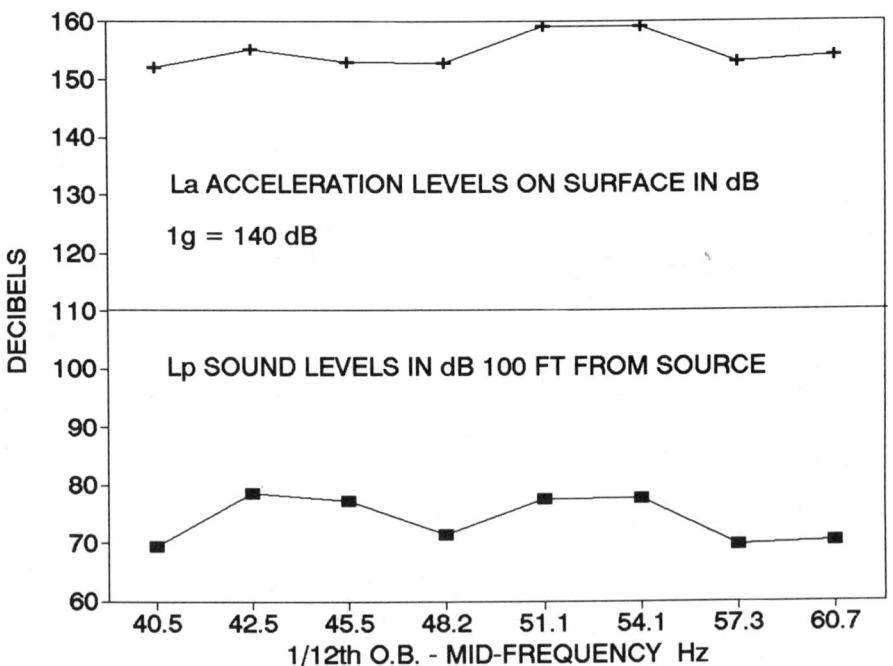

FIGURE 4.3   Vibration and noise spectrum comparison.

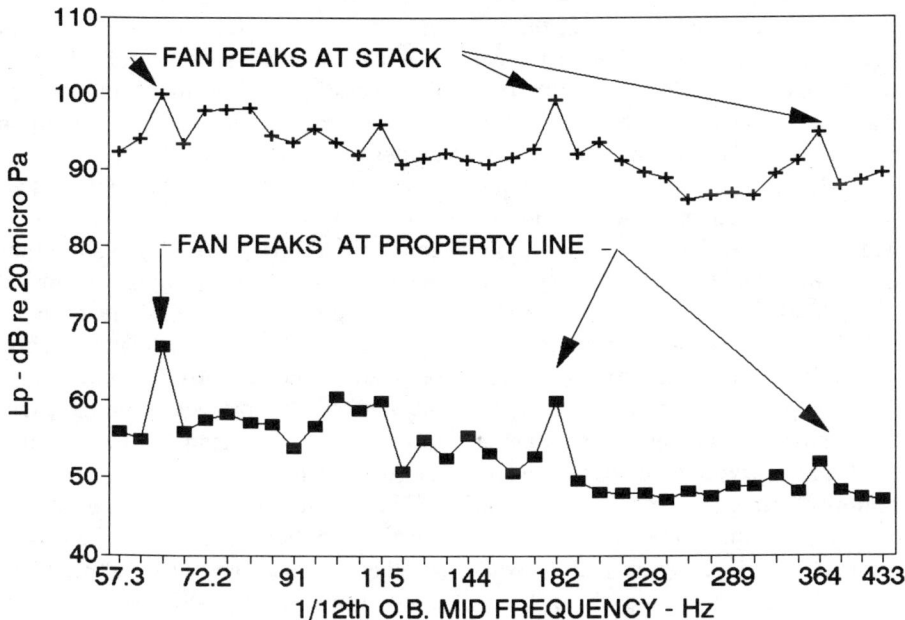

FIGURE 4.4   One-twelfth octave band spectrum at exhaust fan discharge on the roof and in the community.

per second a fan blade passes the cutoff in the casing. If the answer is compared to the peak frequency in the measured spectrum, the fan can be identified in the community. In Fig. 4.4 both spectra are shown—at the fan discharge and in the community—for comparison. In this manner the fan peak is easily identified. Keep in mind that this is a simple example, but if the concepts are understood they can be applied to more difficult examples with confidence.

The noise radiation pattern is also important to understand. Since sound is generated by a vibrating object it is helpful to know if the object that is vibrating is the source of noise or if the radiated noise is coming to the object or surface from another location. Comparison of physical dimensions of various objects will help to visualize this.

For example, a massive steel member will not radiate noise as efficiently as a lighter-weight panel that might be attached to it. If large steel members are to radiate noise significantly, the vibration amplitude must be large.

Most sources do not have that kind of energy, therefore, the light panels will radiate more effectively. An acoustically small source such as a hot gas line from a refrigeration compressor has a lot of energy imparted by the compressor. It might not radiate efficiently in free space, but if it is firmly attached to a structural member which in turn has sheet metal panels attached, the noise from the compressor can easily be felt and heard. Look at, listen to, and feel the object. Some common sense will help in visualizing the problem.

Setting Priorities

After sources have been identified, some method must be used to determine which source should be treated; or, put another way, how should the money be spent to obtain the most noise reduction for the least cost? While this can be a complicated issue, there are some simple concepts that can be used to start understanding this process. In Chapter 1 a comprehensive method was presented that set forth the priorities for a plant. First, however, let's look at a simple example to clarify the concept.

Figure 4.5 shows noise sources that have been identified in terms of dBA. The only two sources that need to be considered are the 90- and 96-dBA sources. The others will not add to the overall levels (see methods for adding decibels illustrated in Chapter 6). It is important to divide the problem into its smaller, more manageable, parts. Then it will not seem so overwhelming. Take one area or machine at a time, or whatever is reasonable to manage. If an orderly scheme can be developed for addressing the equipment in one area, then the other areas within the plant can also be done in the same manner. This will dovetail very nicely into the priority rating scheme discussed in Chapter 1. The program can be stopped at the time the desired noise level is reached by evaluating the noise reduction from each step.

Compatible Solutions

Noise control recommendations, concepts, or ideas that are considered for a particular situation may not always work over a 10- to 15-year time period; or they may not pass the rigid standards for sanitary standards, if required. No matter what noise control solution is under consideration it must be compatible with the environment in which it will be located. For example, in a food-processing plant all materials must be cleanable and must not absorb anything or provide an environment where bacteria, molds, and fungi will grow. This leaves out many acoustical absorbing materials unless they are suitably protected and have the ability to be cleaned by steam, high-pressure water, and/or solvents.

If the materials are going to be outside, they must be protected from the effects of the weather. A noise enclosure that must be opened many times a day should have heavy-duty

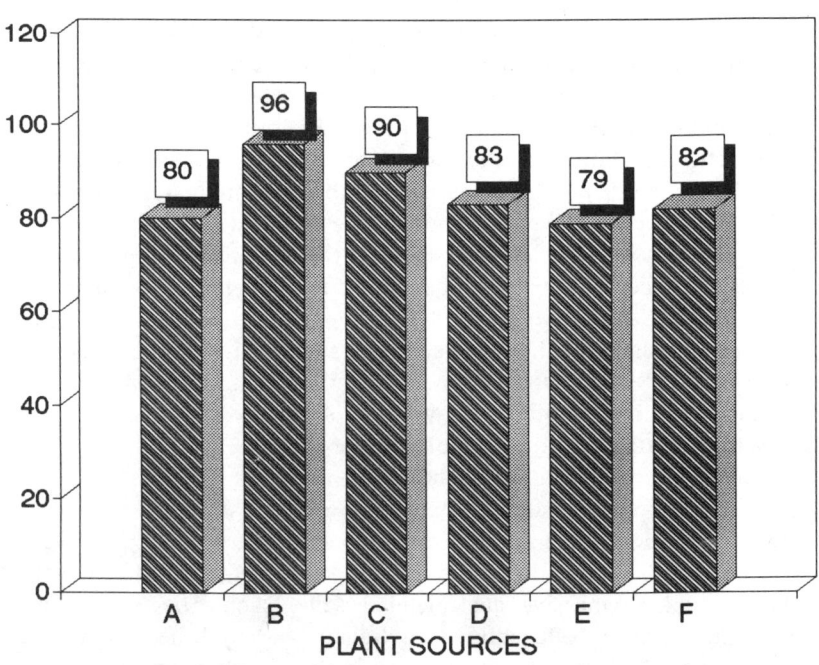

COMBINED SOURCES = 97 dBA

FIGURE 4.5   Various noise sources identified in a plant.

hardware to be compatible with installation. Think about the end result and how it will be ten years in the future. The same approach for noise control should be used in selecting materials that are expected to have a long life in industrial surroundings.

Absorption Versus Attenuation

In many cases the difference between a material's ability to absorb sound and to attenuate is misunderstood. A material that has the ability to absorb sound is usually soft and light without structural properties. It does not have the ability to contain sound alone unless applied to a more dense material. The properties of attenuating materials are: they are heavy, dense and reflective; and they may carry a structural load.

The general term *acoustical material* can be misleading because it is used to mean both absorption and attenuation. Thus, it is important to make the distinction between the two terms when they are used. Proper selection of the correct material for the type of noise source is important, as well as a knowledge of their acoustical properties and noise source characteristics. The two materials are combined in composites such as acoustical foams bonded to a loaded vinyl or lead. See Fig. 4.6 for a photograph of

FIGURE 4.6   Acoustical foam and loaded vinyl composite. (Courtesy of Kinetics Noise Control.)

this type of material. Thus, standard acoustical enclosures are also a combination of absorption and attenuation.

Absorptive material placed in an environment to reduce reflected noise buildup makes a compatible use of the material in conjunction with other noise control solutions. A concrete block enclosure will attenuate sound very well and may be used in conjunction with other noise control solutions. Keep in mind how the materials will be used, their life expectancy, and the environment. There may be other constraints that will not allow all noise control methods and materials to be used.

Noise Control Constraints

Not only must the materials be compatible with the environment, but they must fit in with other constraints such as production, maintenance, and physical. For example, a noise control solution will not stay in place very long if production and maintenance are not considered in the design. Noise control items that are in the way, not easy to remove and replace, or to work around will not be left in place very long.

For example, in a metal-working operation with cold headers, noise control enclosures were designed for a short operator, ease of maintenance, electric motor cooling, oil mist extraction, and for a long life. After 15 years the enclosure is still in place and working, because these constraints were considered in the prototype and final design models. This is illustrated in Fig. 4.7.

Another constraint could be the space that is available for the noise control device. This includes the room around a machine as well as the space to have access and maintain the equipment on which the noise control equipment is mounted or located. If the solution is working and the noise is reduced, the operator may find a way to work around the inconvenience.

An extreme example of working around a physical constraint can be illustrated with this case history. In a large manufacturing plant with over 100 punch presses, the overall ap-

FIGURE 4.7   Cold header enclosure. (Reprinted by permission from *Sound and Vibration* magazine.)

proach for noise control for small-size presses was to use acoustical panels in the press frame. A mock-up was constructed on a press with heavy cardboard before a metal prototype was built and installed. The mock-up was made of heavy cardboard and lined with acoustical foam. It was fitted on the press and taped in place. The press, a 25-ton automatic, was placed in operation. The noise was reduced from 102 dBA to 87 dBA with this treatment. The operator was so pleased with the results that she was able to continue to operate for two days with all of this material in place. It was *very* inconvenient to run the press, but as she said, with the noise level reduced this much she would work around it for a long time. When the material was removed she was very distressed. With the information gained from this mock-up a metal replica was built and installed. If there is an incentive, physical constraint barriers can be worked around. This is described in more detail in Chapter 10.

Noise Control Limitations

The constraints discussed above deal with ability to work with the installation and its effects on the results. Another limitation that can affect the noise reduction result is flanking paths. They are divided into two categories: airborne and structure-borne.

Examples of airborne paths are cracks;

clearances around openings in walls, air inlets, and outlets; and holes of any kind. For an enclosure to be the most effective it must be airtight. When a laboratory test is run on a material to measure the sound transmission loss, the material is sealed in a heavy wall. Thus, the installation will not be as effective since the laboratory conditions cannot be duplicated in the field. Figure 4.8 illustrates how a panel or material of any type can be degraded by openings. Note that as the percent of crack increases, the effectiveness decreases. A very massive enclosure can be installed, but if the percent openings are significant the results will not be as desirable.

An example of how cracks can degrade performance is to consider a door with weatherstripping seals. On a cold day if the seals are not tight cold air infiltration can easily be felt. The same is true with acoustical seals. If the seal is not tight it will leak and degrade the acoustical performance just as the weather seal lets cold air into the house.

Other airborne flanking paths might be lightweight materials incorporated into the design. For example, if a lightweight sheet metal airconditioning duct is located in a massive wall, noise can flank around the wall and enter the space through the duct. This is illustrated in Fig. 4.9. Also illustrated in this figure is an example of structure-borne noise flanking.

For any noise control design, consideration must be given to how the equipment is attached to the structure. The illustration in Fig. 4.10 shows several common structure-borne flanking paths. The noise generated from these paths can be difficult to find because vibration may travel great distances from the point of attachment to the point where noise is radiated and becomes a problem.

Good design dictates that equipment be vibration isolated along with the items attached to the vibrating equipment that can come in contact with the enclosure. These items include piping and duct work as well as anything that would transmit vibration to the enclosure. In Fig. 4.11 a resilient pipe and duct penetration is shown. If these components are rigidly attached to the wall they pass through, a major

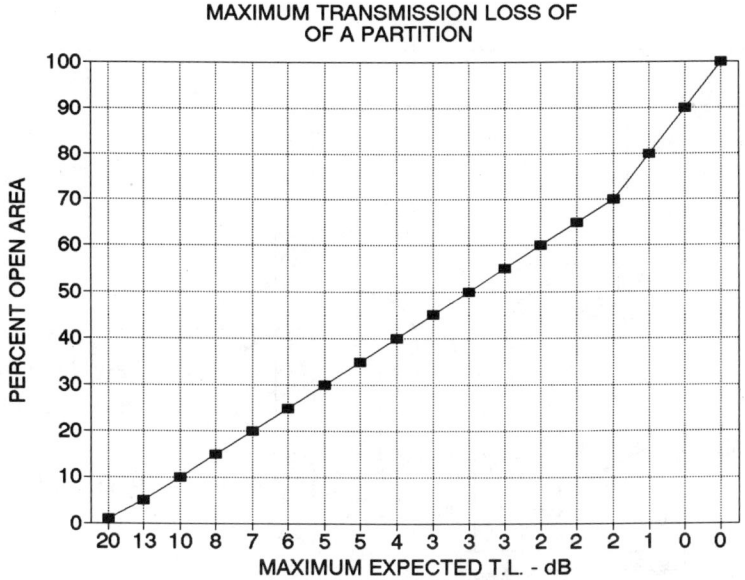

FIGURE 4.8   Effects of cracks on potential noise reduction.   (After NIOSH, *Noise Control Manual*, 1975.)

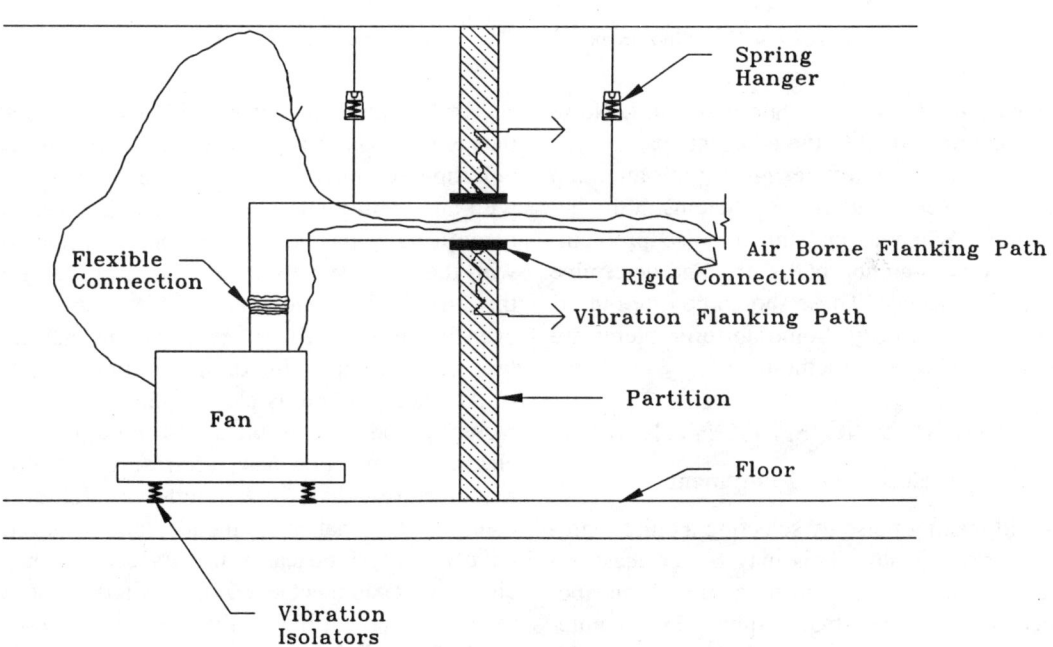

FIGURE 4.9   Structure-borne and airborne flanking paths for fan installation.

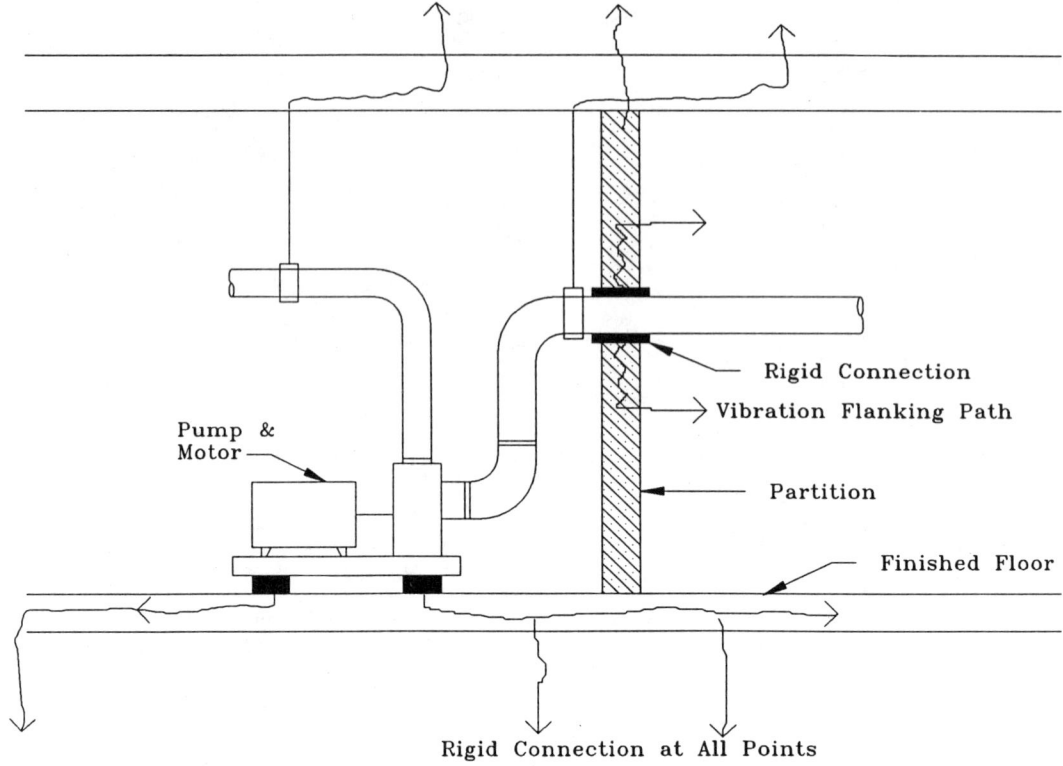

FIGURE 4.10   Structure-borne flanking paths for a pump installation.

structure-borne flanking path will result. Noise will be transmitted to the adjacent space.

Good noise control design requires attention to detail to eliminate these problems in the beginning. When diagnosing a noise problem, identify the flanking paths first and determine their significance. These should be listed in order of their priority. Some noise problems are easily fixed by this method.

SOURCE NOISE CONTROL

Select Quiet Equipment

Avoid making noise by selecting a quieter process or equipment. This may be the least expensive approach in the long run. A cheaper piece of equipment may require more maintenance; and if a noise control package must be installed as well, more time and cost is built into the system from the beginning. Further, some equipment is naturally quieter and can be used in the design.

For example, a blower with radial blades that is used as an induced draft fan on an air pollution scrubber is much noisier than a fan with an airfoil blade on the wheel. The radial blade usually has some pure tones associated with the noise spectrum that may intrude into the surrounding community. This means that even if noise control is needed, there may be less required depending on the circumstances.

Another example is electric motors. Manufacturing and process plants use thousands of electric motors. For many years several motor manufacturers have been building high-efficiency motors that are guaranteed not to exceed 85 dBA. In some cases motors can be purchased that can meet an 80-dBA criteria. These motors do have a higher first cost. Because of the higher efficiency, a payout can be developed for this approach. In the long run this approach will pay for itself compared to a standard motor. The standard motor will require a device to reduce noise. This must be planned

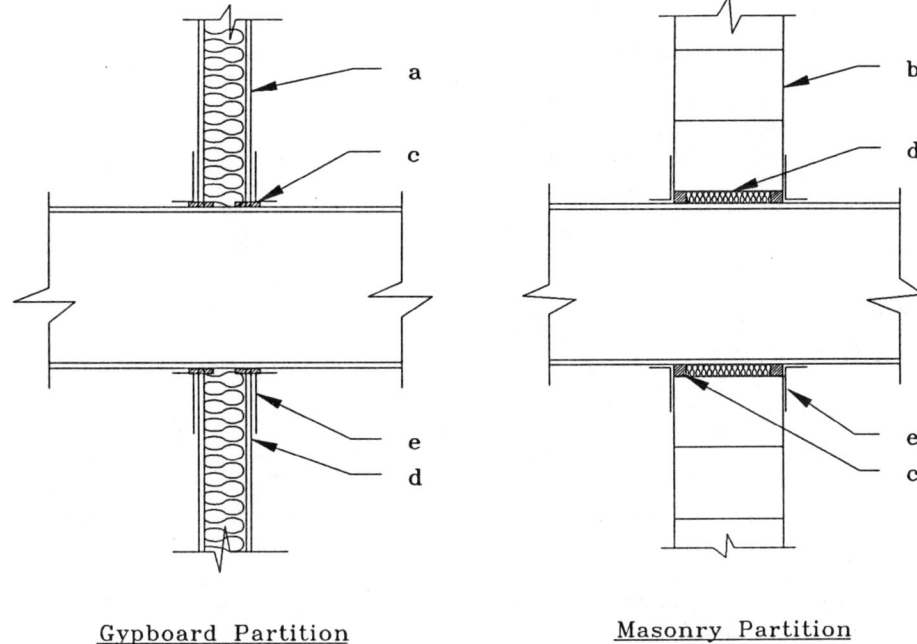

Gypboard Partition   Masonry Partition

a.   Gypboard Wall Construction

b.   Masonry Wall Construction

c.   Acoustical Caulk

d.   Pack with Fiberglass Insulation (3# to 6# Density)

e.   Sheet Metal Enclosure

FIGURE 4.11   Pipe and duct resilient connection.

for in the design in terms of space and possibly airflow over the motor for cooling.

Some equipment is inherently noisy and noise control will always be required. One example is a "Roots" or lobe-type blower. These blowers will always require inlet and discharge silencers and acoustical lagging or an enclosure to reduce noise. If the manufacturer provides noise control devices as a standard with the equipment, this may be an indication of an inherent noise problem to be considered.

Choose Equipment Based on Quiet Operating Characteristics

Most equipment has an optimum operating range. Changing operating characteristics can make a major difference in noise levels and may indicate inefficient operation. Quiet operation can usually be related to efficient operation.

Changing the operating parameters can increase noise. For example, increasing the speed of a fan to have more airflow will increase the noise level. The change of fan noise level can be related to fan speed by $50 \log_{10}$ times the ratio of the revolutions per minute. There are other similar equipment operating characteristics that can be related. These relationships are summarized in Table 4.2 to help evaluate the effects of equipment operating characteristics and noise levels.

Use Proper Maintenance

Another noise source control measure is the use of proper maintenance. Inadequate lubrication

TABLE 4.2   Effect, in dB, of Operating Characteristics on Machine Noise

	Log/Ratio		Machine Noise Characteristics
10–30	$\log_{10}$	horsepower ratio	Internal combustion engines
10	$\log_{10}$	horsepower ratio	Fans
10	$\log_{10}$	pressure head ratio	Fans
50	$\log_{10}$	rotational speed (rpm) ratio	Fans
17	$\log_{10}$	horsepower ratio	Pumps
40	$\log_{10}$	speed (rpm) ratio	Pumps
80	$\log_{10}$	velocity ratio (Mach 1 and higher velocities)	Gas Flow
60	$\log_{10}$	velocity ratio (velocities less than Mach 1)	Gas Flow
30	$\log_{10}$	pressure ratio (at velocities less than Mach 1)	Gas Flow
60	$\log_{10}$	velocity ratio (without cavitation)	Liquid Flow
120	$\log_{10}$	velocity ratio (with cavitation)	Liquid Flow

Source: Yerges, Lyle P. 1978. Sound, Noise, and Vibration Control. New York: Van Nostrand Reinhold. (Reprinted by permission.)

is often the cause of bearing noise and moving parts due to friction causing structure-borne noise. Improper installation of bearings and alignment is a major cause of noise and vibration. In addition, loose parts and improper tolerances and clearances account for a major amount of noise in plants. Well-balanced and maintained equipment, operating near peak efficiency, is normally quieter. A discussion of acoustical maintenance in Chapter 2 provides more insight. Also, common sense and simplicity can play a role in thinking through this process.

Avoid Structural Resonances

In facilities where heavy equipment is used, structural resonance may not be a problem on the equipment itself. The problem may show up on items that are attached in response to excitation of impacts, sliding parts, or rubbing contacts. These types of noise sources are difficult to find. Some examples are [Diehl, 1973]:

1. Reduce the mass of striking parts.
2. Reduce the striking velocity and rate of change of velocity by keeping acceleration as constant as possible over the time period.
3. Avoid quick deceleration of parts that can cause impacts.

4. Keep the center of gravity in proper relation to the mountings to avoid eccentric motion causing vibration and excitation of sheet metal panels.
5. Avoid large sheet metal panels that can couple to the vibration in a heavier piece of equipment. Since the sheet metal panel in a machinery guard, for example, is more compliant—that is, more flexible—it will radiate the noise more easily.

Others can be added to this brief list. The concept is to think through the process and look for these things when designing or diagnosing a problem. The solution may be as simple as isolating the panel from the structure or using expanded metal for a guard in place of a solid piece of sheet metal.

Choose Proper Flow Characteristics

Flow characteristics can affect the overall result even though proper noise control devices have been installed. In Table 4.1 are some examples of these types of noise sources. The basic principle is to keep good smooth flow avoiding turbulence. This is very easy to say but may be difficult to achieve.

As an example, consider the airflow in a duct. If the flow is smooth and not turbulent, there will be little noise to radiate. On the other hand, if the flow is turbulent, there will be low-frequency noise generated by the turbulent eddies in the air stream literally hammering the duct, causing it to vibrate at its natural frequency, and passing through other flow noise. Figure 4.12 shows three elbows in an air duct system for a fan discharge. These are from an actual noise control project to illustrate the point. The initial design specified a square elbow with turning vanes. The acute-angle elbow was installed and was later replaced by the rounded elbow with turning vanes. Changing the elbow, reducing the turbulence, and smoothing the flow reduced the low-frequency noise by 12 dB. This also reduced pressure drop, increasing flow. Thus the fan rpm was

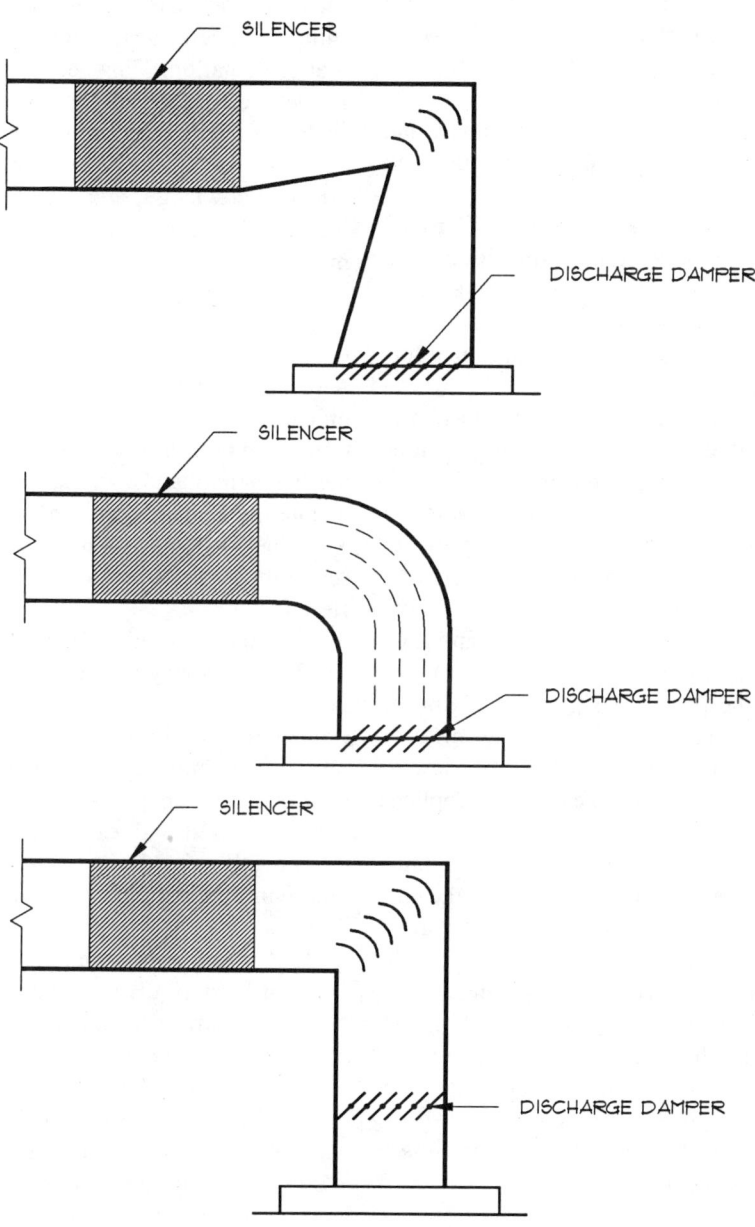

FIGURE 4.12   Discharge duct elbows.

reduced by 15 percent, further reducing noise. A case history in Chapter 10 describes this in more detail. The amount of noise reduction can be determined from the examples in Table 4.1.

In general, it is possible to reduce noise by 8 to 10 dB in the low frequencies by smoothing out the airflow. This can also be done with fluid flow. The following list briefly describes proper aerodynamic flow characteristics for good noise control design and tells what to look for when diagnosing a noise problem [Diehl, 1973]:

1. Keep velocities low in ducts, pipes, and passages. Increased flow velocities can increase noise levels.
2. Design duct work and piping to avoid turbulence. Use gradual transitions and smooth turns from one section to another to avoid turbulence. Proper design and use of turning vanes helps minimize turbulence.
3. Keep the flow inlet to equipment as large as possible and avoid any obstructions and abrupt changes in flow.
4. Provide an even flow and low velocity for the discharge from equipment to avoid turbulence. This is particularly true for fan discharge duct work where flow separation can cause turbulence and low-frequency rumble noise.
5. Know the proper design or selection of impeller vanes, fan blades, and diffuser vanes from an acoustic point of view to minimize noise for the specific application.
6. Since pumps and other similar equipment have a clearance between the impeller and casing ("cutwater") as part of the design make this as large as possible to minimize noise. Close spacing generally creates noise that can be easily transmitted by the fluid.
7. Eliminate jet velocities either inside or at the discharge of a piping system.
8. Look for and repair any leaks in a pressurized system.

All of the material in this section has been devoted to source noise control. When source control has been completed, the next step is to consider modifying the path between the source and receiver.

NOISE CONTROL ALONG THE PATH

Most noise control takes place along the path. Methods and materials include barriers, enclosures, silencers, absorption materials, and vibration isolation. This section will briefly describe these materials and give examples of each along with their limitations and longevity. There is a great deal of misapplication due to lack of knowledge, and a great deal of money is spent for short-term applications with minimum results.

Barriers

A barrier or acoustic shield can provide a significant amount of noise reduction. The reduction occurs on the side opposite the source, in the acoustic shadow zone, in much the same manner as a shadow created by a light source. The shadow zone is limited by diffraction or scattering of the sound over the top of the barrier.

Barriers are used to best advantage outside to shield freeway traffic noise and industrial noise sources. For an indoor environment, barriers will have limited results because of the reverberant buildup, or reflections, from the ceiling and other adjacent surfaces. They should be used with caution.

The following limitations apply to barriers and their applications:

1. The structure should be airtight and have at least 10 dB more noise reduction than the barrier attenuation. A lightweight wood fence might not be suitable for screening a high-noise source.
2. The distance from the barrier to the receiver must be much greater than the distance from the wall to the noise source.
3. The distance from the barrier to the source must be greater than the height of the barrier.

4. In plan view, the barrier width must be at least twice the distance from the source to the barrier or its height, whichever is smaller.

Barriers have been used as a noise control solution for many years. For example, an article in the July 1957 issue of *Noise Control* magazine illustrated the effectiveness of barriers. This article also referenced previous work done on this subject as early as 1950 [Purcell, 1957].

The illustration in Fig. 4.13 shows a barrier, noise source, and receiver in their general relationship. The approximate equation for the reduction is given below [Diehl, 1973]. This is a convenient method to use.

$$\text{Noise reduction} = 10 \log \left(\frac{20H^2}{\lambda(R)} \right)$$

where:

H = height of barrier
R = distance to receiver
λ = wavelength

As noted in this equation the wavelength is quite important. Low frequencies have long wavelengths and tend to roll over the barrier. Thus the barrier is not as effective for this type of noise source. High frequencies have shorter wavelengths and the barrier is much more effective. For middle to high frequencies the barrier is more effective. It is analogous to a light shadow. Lord et al. use the approach of a nomograph to calculate barrier effectiveness [Lord et al., 1978].

Since most effective barriers are located outside, wind and temperature conditions must be considered. The above equation assumes there are zero temperature and wind gradients. If these gradients do occur, there will be bending of the sound waves over and around the barrier. At increased temperature the sound waves will tend to bend upward, while decreasing temperature allows the sound waves to be in a more horizontal direction. This is why one can hear a train or highway noise at night over longer distances. Wind gradients also affect the sound at a barrier. The sound is higher on the downwind side and lower on the upwind side. These gradients will change and influence the shadow zone of a barrier.

BARRIER NOISE REDUCTION

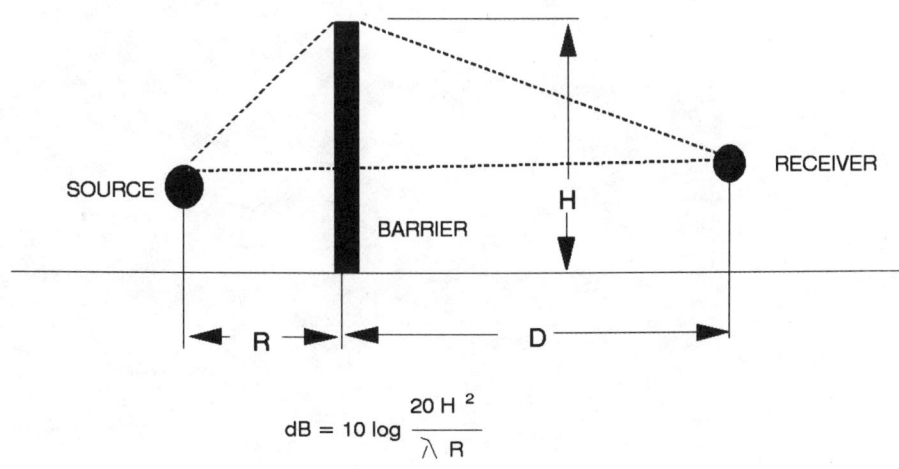

$$dB = 10 \log \frac{20 H^2}{\lambda R}$$

FIGURE 4.13   Barrier and noise source.

Material for barriers will vary depending on the economics and physical conditions. The following is a brief list of some common barrier materials:

1. Masonry of any type
2. Wood structures
3. Metal acoustical panels
4. High-density plastic modular panels
5. Loaded vinyl curtains and flexible acoustical blankets

Two examples are shown in photographs in Figs. 4.14 and 4.15. The barrier has its limitations but can provide good results. A general rule of thumb is a 5- to 10-dBA reduction based on field measurements.

Enclosures

Enclosures can cover the equipment or house personnel. Both can work to reduce employee noise exposure. If noise must be reduced by more than 15 to 20 dB, an enclosure should be considered. The attenuation of any material used for an enclosure at a particular frequency will depend on the stiffness, mass, damping, and resonances of the material. By definition an acoustical absorbing material, such as fiber-glass, is a poor attenuator or sound barrier. It is porous and allows sound to easily pass through to the other side. By contrast, heavy dense materials, such as concrete, plywood, gypsum board, or steel, are excellent attenuators. They are nonporous and good reflectors of sound energy.

The measurement that is commonly used to describe the effectiveness of a barrier to reduce sound is airborne sound transmission loss (STL). Many times it is shortened to transmission loss (TL). To determine the ability of a wall, partition, or enclosure panel to reduce sound, the TL should be measured in the laboratory. This is done using two rooms that are hard and reflective (reverberant). A sound source is placed in one room and measurements are then taken in both the source room and receiving room. This is done in accordance with ASTM Standard NoE 90-(current year). Field measurements can be made if ASTM 336-(current year) are followed.

In either case, to determine the TL one must know the properties of the receiving room. If the receiving room is hard and reflective, the sound level in it will be higher than if it is very absorptive. In the latter case the sound level will be lower in the receiving room. The TL is

FIGURE 4.14   Concrete barrier.

FIGURE 4.15   Barrier constructed of HDP die cast plastic modules.   (Reprinted by permission from Sound Fighter Systems, Shreveport, LA.)

determined so it is independent of effects of the receiving room. In the laboratory the conditions are controlled. In the field they will vary a great deal.

The sound level is measured on both sides of the wall and then corrected for the conditions in the receiving room. In many cases the TL, for the wall in question, can be obtained from material suppliers or reference books. The following equation can be used to determine the noise reduction:

$$TL = NR + 10 \log \left[\frac{S_{\text{wall}}}{(S_{\text{room}})(A_{\text{room}})} \right]$$

where:

TL = the transmission of wall in dB
NR = difference in L_p between source and receiving rooms
S_{wall} = area of common in square feet
S_{room} = area of walls, ceiling, and floor of receiving room
A_{room} = average absorption coefficient in receiving room

Use a as follows:

hard room = 0.05
normal room = 0.20
soft room = 0.40

It should be kept in mind that *TL* measured in the laboratory will be higher than measured in the field. There are many flanking paths and cracks in the field that can compromise the results (Fig. 4.9). Generally the laboratory STC data should be reduced by at least 5 points and perhaps more depending on the conditions.

ASTM E90-(current year) is the laboratory method for determining not only the TL but the sound transmission class (STC) of a material [ASTM, 1975]. This is a single-number rating system to rate walls, floors, ceilings, doors, panels, movable walls, and other products. It is calculated from the TL data based on a specific formula in the standard. The TL is measured in one-third octave bands from 125 to 4000 Hz, and then the STC can be determined. In Fig. 4.16 a typical STC curve is shown compared to a measured spectrum. The rules for calculating the STC value are:

1. The TL at any one frequency cannot be more than 8 dB below the STC curve.
2. The sum of all deficiencies below the TL cannot exceed 32 dB.

The STC is often used as a performance criterion in specifications for architectural structures and industrial materials. It is important to keep in mind that this single-number rating system is much like the EPA gas mileage rating for automobiles. The results will depend on the conditions. It should be used for comparison purposes only. As mentioned previously, the field result will be about 5 STC points below the laboratory rating.

Typical enclosures are found in Figs. 4.17, 4.18, and 4.19. These enclosures show the typical 4-inch acoustical panel for equipment, a 1-in. enclosure for equipment, and a personnel enclosure. Normal construction is to have an exterior skin of solid sheet metal ranging from 16 to 20 ga in thickness, an interior acoustical fill material of glass fiber or acoustical foam, and an interior perforated liner. In some cases the acoustical fill material will need to be protected from water, dust or other environmental contaminants. If the protective fill is 1 to 2 mils thick the acoustical absorption will not be degraded significantly.

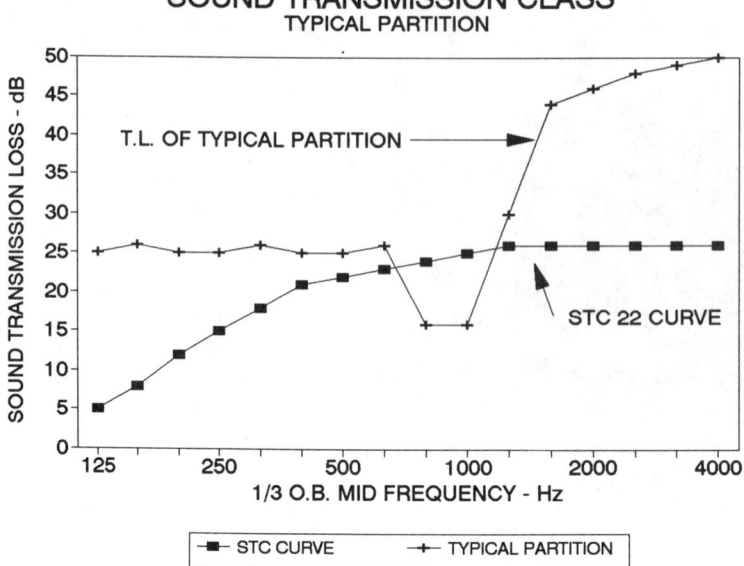

FIGURE 4.16   Typical STC curve and transmission loss spectrum.

Enclosures can also be constructed from composite materials. These are made of acoustical foam stiffened with loaded vinyl sheets to add mass. This will vary in weight from 0.5 to 1.5 lb/ft². Another material that works well for lightweight enclosures is a composite made of rubberized glass cloth, light-density fiberglass, and a loaded vinyl interior layer all quilted together to form a flexible blanket. Examples of these materials are shown in Fig. 4.20.

The typical TL and STC of these materials along with other enclosure materials are shown in Table 4.3. Standard building materials will serve well as enclosure materials. The key to a good enclosure application is to have it sealed. When ventilation is required inlet and discharge silencers can be added. Many times product flow is required and this must be accounted for in the design. Silencers can be used

FIGURE 4.17   Standard 4-in. acoustical noise enclosure.

FIGURE 4.18   Enclosure constructed of 1-in. acoustical panels.

FIGURE 4.19   Personnel room enclosure.

here as well. Examples of these are given in Chapter 10.

Acoustical Lagging

Acoustical lagging is another type of enclosure used to cover piping or duct work. It is sometimes called acoustical insulation. The materials are used to enclose the pipe containing noise from sources such as high-pressure gas flow in piping, hydraulic lines, and valves. An impervious membrane is spaced away from the surface. This air space is created by the fiberglass material. An illustration is given in Fig. 4.21, showing a cross section of the pipe or equipment wall. For piping, a preformed fiberglass pipe insulation (4 lb/ft^3) works well, because it has some stiffness and provides a suitable surface for the outer cover to be installed. It is important that the outer covering be sealed in order to work as intended. In Fig. 4.22 a typical pipe lagging is shown along with the requirements for proper sealing.

The outer cover can be one of several materials depending on the noise spectrum. The surface weight can vary from 0.5 to 2.0 lb/ft^2. The material selected will depend on the noise source and the spectrum shape. Generally a 10- to 20-dBA noise reduction can be expected if the outer cover is about 1 lb/ft^2 with 2-in. pipe insulation. Typical materials that are used for outer coverings are: 1 lb lead, loaded vinyl, 0.030 insulating aluminum, 22-ga stainless and galvanized steel, and lead aluminum laminate. Table 4.4 illustrate the noise reduction from 250 to 4000 Hz for some typical lagging materials.

FIGURE 4.20   Acoustical flexible blanket.

TABLE 4.3   Transmission Loss of Various Materials

Octave Center Frequencies (Hz)	125	250	500	1000	2000	4000
22-ga sheet metal	9	15	20	27	32	38
4" acoustical panel	23	30	42	51	59	58
2" acoustical panel	24	32	40	49	53	58
Gypsum board shaft wall	24	40	48	53	55	52
$2\frac{1}{2}$" solid wood	28	27	28	28	34	32
$1\frac{3}{4}$" solid core wood acoustical door	34	34	35	36	38	40
$2\frac{1}{2}$" solid core wood acoustical door	37	39	39	41	35	36
Metal acoustical door $1\frac{1}{2}$"	23	37	44	42	39	42
$\frac{5}{8}$" gypsum board on $2\frac{1}{2}$" metal studs	30	39	45	44	46	46
$1\frac{5}{8}$" and $2\frac{5}{8}$" gypsum board on $2\frac{1}{2}$" metal studs	22	40	52	53	55	56
6" hollow block, $\frac{5}{8}$" gypsum board on furring	28	38	47	49	52	50
8" hollow block, 1 side painted/1 side gypsum	31	35	48	57	67	68
6" solid concrete with $\frac{1}{2}$" plaster both sides	42	50	58	64	68	68
TL of $6\frac{1}{2}$" metal stud wall	34	43	52	61	61	64
2-layer/2-layer metal stud with blanket; USG-109FT	35	37	50	57	60	59
2-layer/2-layer metal stud with blanket; USG-114FT	35	45	55	57	62	55
8" solid concrete with $\frac{1}{2}$" plaster both sides	44	48	51	57	60	64
4" concrete isolated, STC 44, 53lb/ft^2	47	42	44	55	56	66
6" concrete, STC 51, 64 lb/ft^2	41	40	50	55	56	60
6" concrete floor (2.5" min.) on metal deck	34	34	38	45	55	61
4" tile below 12" air space, STC 48	37	38	43	48	53	57
6" tile below 15" air space, STC 51	38	41	47	53	57	61
$\frac{1}{2}$" plate glass	16	23	27	32	28	32
Double glazing two $\frac{1}{2}$"-$\frac{1}{2}$" air space	16	23	27	32	30	35
Lightweight roof deck 6" 20-ga metal deck	41	47	56	65	68	69

Source: NIOSH, *Compendium of Materials for Noise Control,* June 1975.

Since most noise sources of the type that require lagging have dominant noise spectrums in the mid- to high-frequency range, this approach works well. These materials do not work well for low-frequency noise sources such as fan casings or ducts with turbulent rumble. More mass is required. Material such as gypsum board, plaster, or heavy metal covering are required. A cross section of a duct or fan lagging is shown in Fig. 4.23. This can be used for forced draft fans and duct work in a power plant. For HVAC duct work, Sheetrock can be substituted for the external lagging cover.

For valves and irregular surfaces the flexible

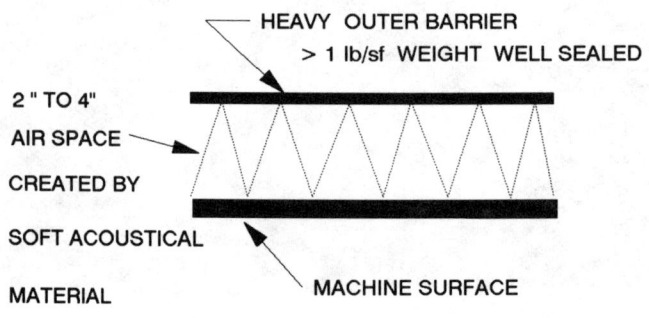

TYPICAL ACOUSTICAL LAGGING

HEAVY OUTER BARRIER
> 1 lb/sf WEIGHT WELL SEALED

2 " TO 4"
AIR SPACE
CREATED BY
SOFT ACOUSTICAL
MATERIAL

MACHINE SURFACE

FIGURE 4.21   Acoustical lagging.

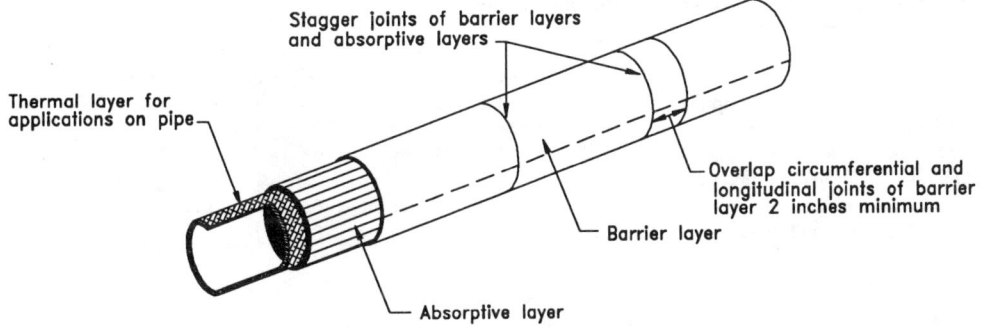

Thermal layer for applications on pipe

Stagger joints of barrier layers and absorptive layers

Overlap circumferential and longitudinal joints of barrier layer 2 inches minimum

Barrier layer

Absorptive layer

STRAIGHT RUN PIPE LAGGING

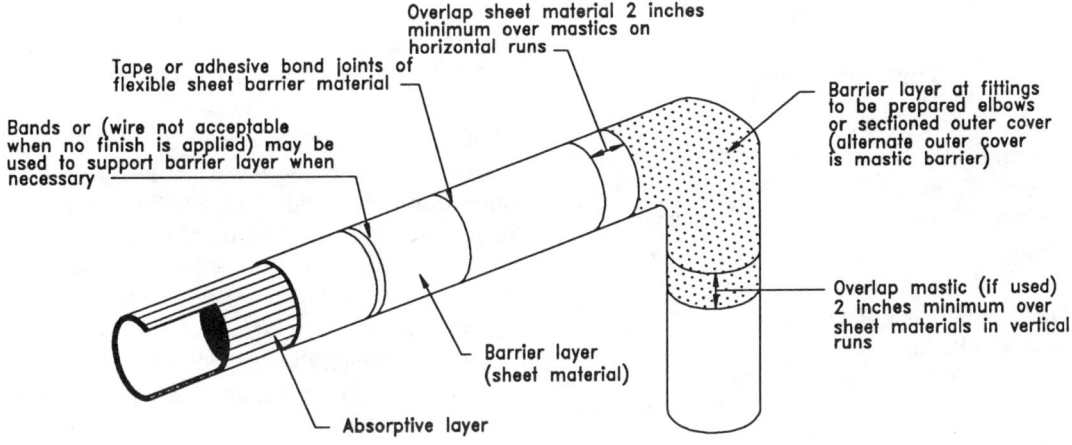

Overlap sheet material 2 inches minimum over mastics on horizontal runs

Tape or adhesive bond joints of flexible sheet barrier material

Bands or (wire not acceptable when no finish is applied) may be used to support barrier layer when necessary

Barrier layer at fittings to be prepared elbows or sectioned outer cover (alternate outer cover is mastic barrier)

Overlap mastic (if used) 2 inches minimum over sheet materials in vertical runs

Barrier layer (sheet material)

Absorptive layer

TYPICAL FITTINGS LAGGING
(For Tees and Valves Flexible Acoustical Blanket can)

FIGURE 4.22   Acoustical pipe lagging.

blanket described above works very well. The important thing to keep in mind for this approach is that the flexible blanket can be removed and replaced with ease. The acoustical lagging is a more permanent covering and must be replaced new each time it is removed. Therefore, for those surfaces or items that require maintenance the flexible blanket is the treatment of choice. The photograph in Fig. 4.24 illustrates a combination of both of these treatments on a high-pressure gas-reducing sta-

tion. The noise spectra before and after installation are shown in Fig. 4.25. This is a high-frequency-dominated spectrum. These materials work well for this application.

A number of lagging materials have been tested to determine their noise reduction capabilities [Dear, 1972; Zign, 1988]. The basic lagging material of 1 lb/ft^2 and 2-in. pipe insulation provides the best overall result for noise sources with mid- to high-frequency spectra.

TABLE 4.4   Typical Acoustical Lagging Noise Reduction (in dB)

No.	Material Description	Frequency (Hz)								
		31.5	63	125	250	500	1K	2K	4K	8K
1.	Thermal insulation only (2″)	0	0	2	3	6	10	20	25	32
2.	Add aluminum insulating metal cover	0	0	6	5	15	17	25	34	35
3.	Substitute galvanized steel cover	0	0	6	6	19	22	27	30	32
4.	2″ F.G. insulation with lead/aluminum outer cover	0	1	9	8	22	24	36	46	>50
5.	Same as 4; field tests on large pipe	0	2	2	2	3	6	10	11	17
6.	Same as 4; field tests on chiller	0	1	1	3	7	10	16	16	14
7.	4″ F.G. with aluminum insulating metal cover	1	2	10	9	24	30	38	46	>50
8.	4″ F.G. with steel insulating metal cover	1	2	11	9	24	30	38	46	>50
9.	2″ acoustical flexible blanket (lab)	0	0	4	4	8	18	16	13	15
10.	Field test 2″ acoustical flexible blanket	0	0	0	0	10	11	20	16	13

Note: Low-frequency noise reduction cannot be achieved with these types of treatments applied directly to the pipe or equipment surface. A treatment similar to that shown in Fig. 4.23 should be installed in heavy industrial applications. For HVAC application 2 layers of $\frac{5}{8}″$ Sheetrock can be attached directly to the duct.

Source: Author's private reports and technical files.

Thin-Panel Damping

Equipment may be constructed of a heavy metal structural member and have thin metal panels for covers, guards, and cosmetic finish. A thin panel attached to heavier member that is vibrating will easily resonate with the vibration input and can reradiate noise very effectively. These panels act like an audio speaker, efficiently radiating sound.

The damping technique reduces both the impact and steady-state noises at their source. The purpose of damping is to dissipate vibration energy in the structure before it can radiate as noise [Lilley, 1983]. If the vibration feels ''harsh and tingly,'' the panel may be a good candidate for vibration damping application.

Effective damping treatment depends on:

1. The panel's being in resonance, that is, being forced to vibrate at its natural frequency
2. The panel's being capable of generating the noise

All materials have a certain amount of damping, and many, like steel, aluminum, or glass, have very little internal damping. Their resonant properties make them effective sound radiators. Applying damping material to these makes it possible to control the resonances.

If the vibration causing the noise is caused from a forced, nonresonant condition, damping will have little effect. However, if the damping treatment is placed on the panel edges and/or connection points, this may be effective in reducing noise. This will be true for panels in the frequency range below critical frequency ($f_c/2$). The f_c for a thin steel panel is in the high-frequency range of about 1000 Hz. For frequencies greater than f_c, location of damping at the point of greatest displacement will have the most effective results. Lord et al. [1987] provide a more detailed treatment of this subject.

Damping materials work because they absorb the vibrational energy in shear and store it as strain energy when deformed. This retards the energy radiation through the internal friction within the material, which is known as a viscoelastic effect. Most effective damping materials are viscoelastic in nature. They can be obtained in sheet form or in a paste form troweled in place. This is illustrated in Fig. 4.26, with the application of damping material applied to the exterior of the chute.

If a solid layer of material, usually metal, is placed on top of the damping material, the results will be even more effective. This is called constrained layer damping. In Fig. 4.27 constrained layer damping material is shown be-

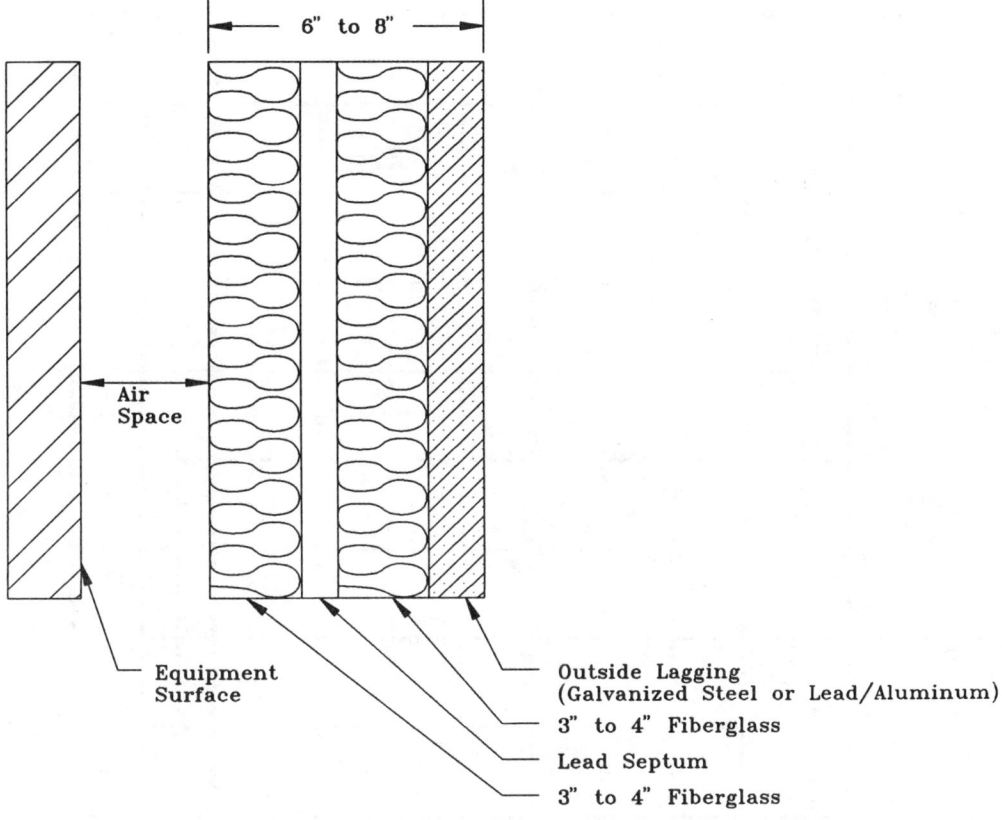

FIGURE 4.23   Ductwork acoustical lagging.

FIGURE 4.24   Acoustical lagging and flexible blanket treatment of a pressure-reducing valve station in a power plant.

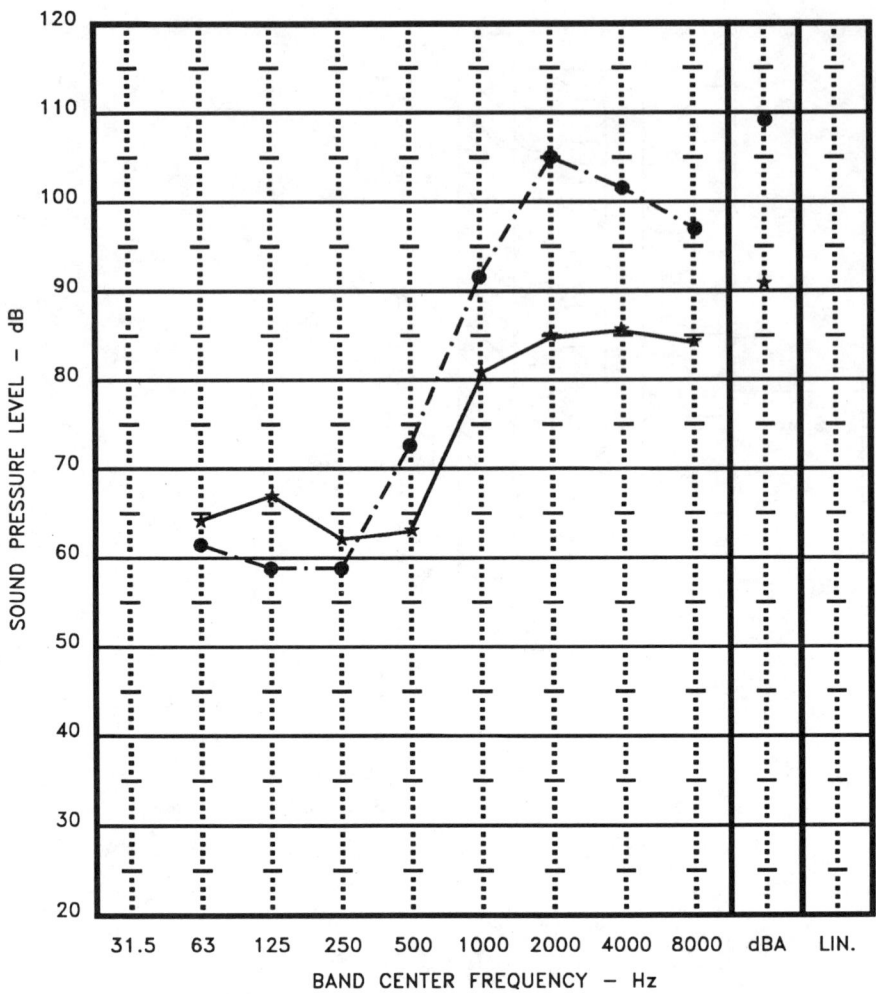

Accoustical Lagging and Flexible
Accoustical Blankets
Power Plant Gas Yard Piping

★———★   With Lagging
●–·–●   Without Treatment

FIGURE 4.25   Results of acoustical lagging treatment.

fore it is applied to a steel belt guard. The outer layer is aluminum bonded to the damping material. The composite has a pressure-sensitive adhesive. The outer constraining layer is usually about one-third the panel thickness.

The performance of damping material is measured by the "loss factor" η, which is the energy dissipated per radian of motion at the excitation frequency divided by the maximum strain energy stored in the system at the same frequency. A typical loss factor is 0.1 for satisfactory performance. Most values fall in the range of 0.1 to 0.3. If a panel is an effective radiator (i.e., in resonance) the radiated sound level can be reduced by 10 dB by increasing the loss factor by ten. Application of damping

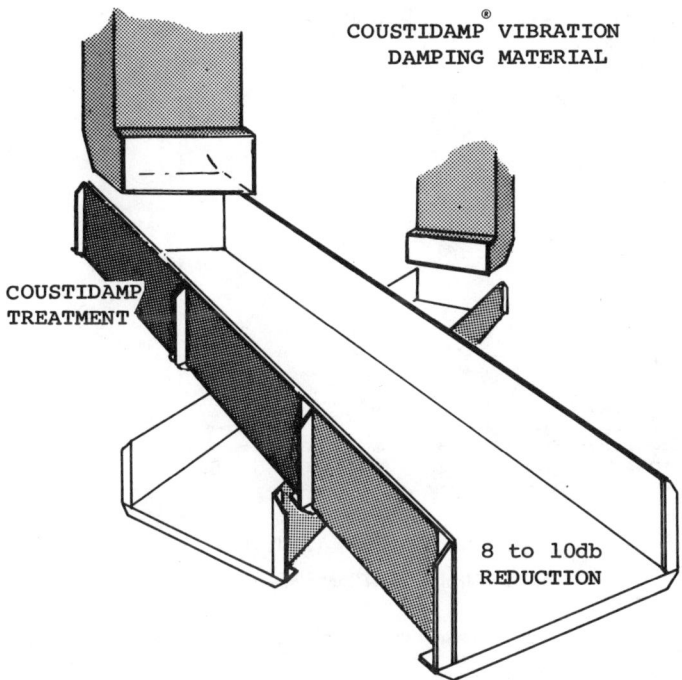

COUSTIDAMP® VIBRATION
DAMPING MATERIAL

COUSTIDAMP
TREATMENT

8 to 10db
REDUCTION

FIGURE 4.26   Damping material applied to the exterior of a chute.   (Reprinted by permission from Kinetics Noise Control, Dublin, OH.)

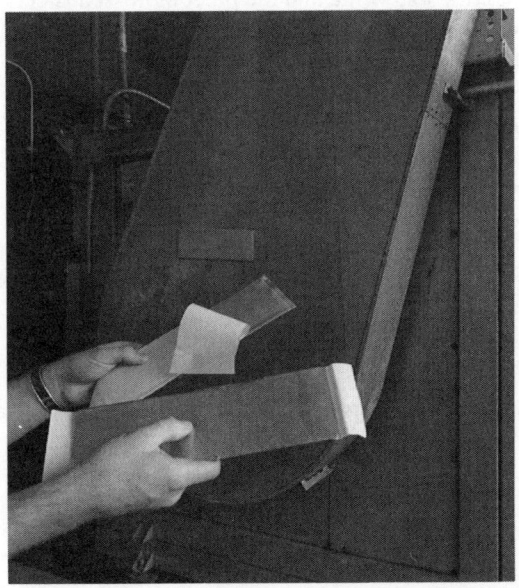

FIGURE 4.27   Constrained layer damping material ready for application to a belt guard.   (Reprinted by permission from Kinetics Noise Control, Dublin, OH.)

materials can accomplish this. Most damping material manufacturers have loss factors measured for their products.

Temperature also affects the loss factor. Most damping materials have an effective temperature range. The loss factor data will show this range. Thus the temperature must be known. Figure 4.28 shows some examples of loss factors for damping materials.

Vibration Isolation

There are numerous sources of impact vibration in the industrial environment that span a wide range of equipment. These are impacts from blasting, forge hammers, punch presses, and so on. The other type of vibration comes from rotating equipment such as pumps, compressors, fans, and engines. Vibration from this equipment or impacts can cause noise problems in adjacent areas. Excessive floor vibration can,

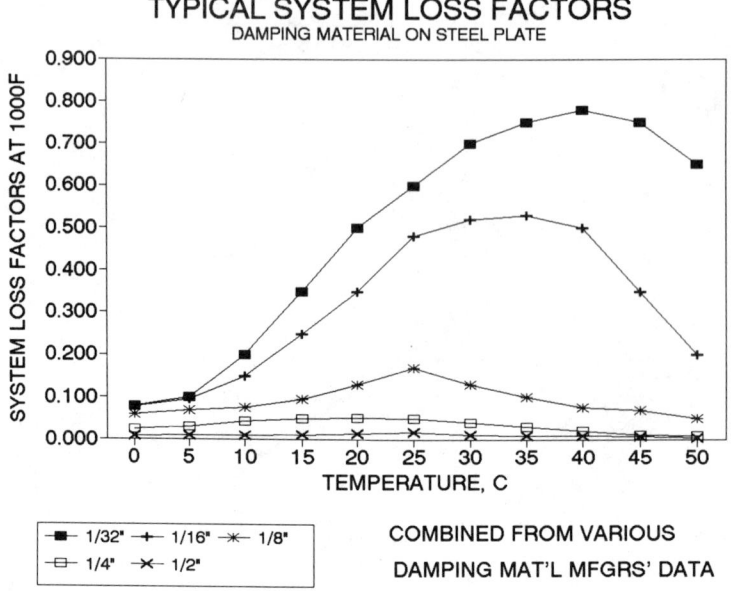

FIGURE 4.28   Typical damping loss factors versus temperature.

for example, disturb delicate optical equipment such as that found in clean rooms.

Vibration can be controlled by isolation methods and materials. This section will briefly discuss the approach and general concepts. There are many good texts on the theory of vibration and isolation, some of which are listed in the References.

Early in the twentieth century vibration isolation was accomplished with simple materials, and it was done more by tradition than science [Mason, 1966]. It was well established in the industry that cork was best under high-speed equipment such as pumps and centrifugal compressors, rubber mountings were ideal for fans and air-handling equipment, and spring mountings were the logical choice for reciprocating equipment such as compressors. These selections were so well established that the application engineer found the choice of materials limited to solving a problem, for example, choosing double rather than single deflection for rubber mountings.

The Celotex Corporation illustrated how its ceiling tile material could be used for vibration isolation [Celotex, 1937]. The vibration isolation material manufacturers had loading charts

to show the amount of material that should be used for proper installation. Cork materials were manufactured and sold in natural form and also in precompressed sheets. The natural form of cork was white and the precompressed sheets were brown. One material really worked no better than the other.

This early method of engineering served its purpose as it introduced the vibration control industry to this country. Most installations were successful because:

1. The equipment was installed in the basement or on grade.
2. The horsepower involved was small.
3. The occupants were not too demanding.
4. The building structures followed the pattern established in the early 1900s when floors were thick, spans were short, and the exterior curtain did not exist.

After World War II the air-conditioning industry really began to grow. Entire buildings were centrally air-conditioned rather than just limited spaces. Basement space, previously considered a machinery or storage area, became rentable for offices or parking. This meant that the air-conditioning equipment had to be located elsewhere in the building. Usually it was located on the roof in a penthouse equipment room over expensive lease space. In other cases, air distribution components such as air-handling units were located adjacent to office space on the intermediate floors. Buildings became much lighter weight and the floor spans were increased to provide unobstructed space for offices. Vibration problems were exaggerated by the accompanying change in structural design. The traditional method of vibration isolation failed to solve the problems that existed.

These changes ushered in the theoretical-design era based on a simplified vibration control equation. This equation states that a given percentage of the vibration can be reduced by installing the equipment, which is creating the vibration, on a system of vibration isolators. The isolation system is designed to be resonant at a frequency much lower than the disturbing frequency of the equipment (usually the RPM of the machine). When the ratio of the disturbing frequency to the natural frequency is about 3 to 1, 90 percent of the vibration is theoretically eliminated. The theoretical efficiency equation is:

$$E = 100 \left[1 - \left(\frac{1}{\left(\frac{f_d}{f_n}\right)^2 - 1} \right) \right]$$

where:

E = percentage of vibration isolated
f_d = disturbing frequency of the isolated machine
f_n = natural frequency of the isolator

The disturbing frequency should be taken as the revolutions per minute of either the equipment or the driver, whichever is lower. All equipment has some unbalance at the primary speed. This approach is conservative as any higher-frequency vibration used in the formula would result in overly optimistic values for the primary disturbance.

The formula for this natural frequency in cycles per minute is:

$$fn = 188 \sqrt{\frac{1}{d}}$$

where d is the static deflection in the resilient supports. The formula only applies to mountings having a uniform deflection (for springs) without excessive damping, that is, neoprene, rubber, or cork.

All of the vibration isolation suppliers publish an efficiency chart as shown in Fig. 4.29, expressing this relationship. Since the disturbing frequency is always known, the natural or resonant frequency can be determined by looking on the chart to obtain the proper mounting selection.

Just as a pendulum's natural frequency is a function of its length rather than its weight at the end, a vibration isolation system natural frequency is a function of the static deflection of the isolator rather than the weight of the system. For example, to isolate a compressor operating at 600 rpm at 90 percent isolation efficiency, select a mounting that has about an inch of deflection. From the previous equation the f_n would be 188 cycles per minute. Dividing 600 by 188 would result in a ratio of 3 to 1, providing nearly 90 percent efficiency. The problem was solved. The companies supplying vibration isolation materials then advocated specifications which read, "Vibration mountings shall be selected to eliminate 90 percent of the vibration transmitted by the air-conditioning equipment."

While this approach was a tremendous leap forward, it proved to be only a starting point for a number of reasons. First, the 90 percent requirement gave no consideration to the magnitude of the vibration. In approaching any problem we must remember that the vibratory force varies in proportion to the unbalanced weight, to the distance the unbalance is located from center, and to the square of the velocity or the revolutions per minute.

The vibratory force created by an 18-in. vent set running at 300 rpm is nothing compared to

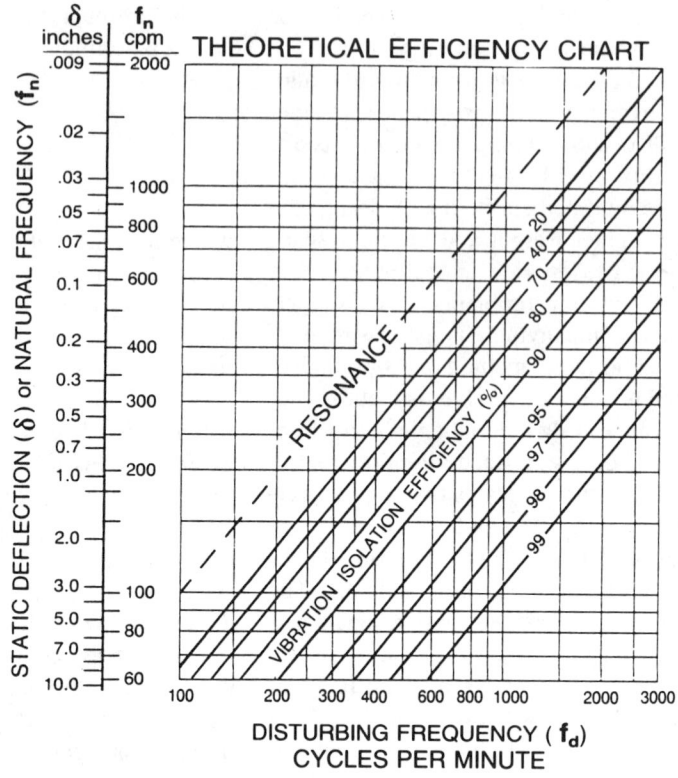

FIGURE 4.29   Vibration isolation efficiency chart. (Reproduced by permission from Mason Industries, Hauppauge, NY.)

the vibratory force from a 60-in. class-3 fan running at 1000 rpm. While a 90 percent efficiency might be an exaggerated number for the vent set, it might very well be completely inadequate for the large high-speed fan.

Second, major errors appeared because the selections were made on the false premise that assumed this theoretical equation applied to structures as we know them today. Unfortunately this is not true. The equation is based on two assumptions: the deflection in the vibration isolator is extremely large as compared to the floor deflection, and the moving mass of isolated equipment is extremely small when compared to the floor mass. While these assumptions were closer to being correct in older buildings with smaller equipment, they are seldom valid today. This is a two-mass rather than single-mass system. The floor has mass and its own spring rate, as illustrated in Fig. 4.30. This

system cannot be analyzed in terms of the simplified theoretical equation. In many cases, using the theoretical equation, an isolator will be selected that has a smaller deflection than the floor system on which it is sitting.

For example, if a 3600-rpm pump is to be isolated to eliminate 95 percent of the vibration, the deflections selected based on the theoretical chart would be about 0.05 in. Floors are designed with an allowable deflection of as much as 1/360 of the free span. If the pump is to be placed in a 20-ft bay, the floor deflection could be as much as 1/360 × 240 in., or approximately 0.666 in. If the floor had only half the maximum allowable deflection, it would still be 6.5 times less stiff than the isolator selected. The introduction of this stiff mounting will not isolate the pump. The isolators must have a deflection compared to the floor in order to be effective.

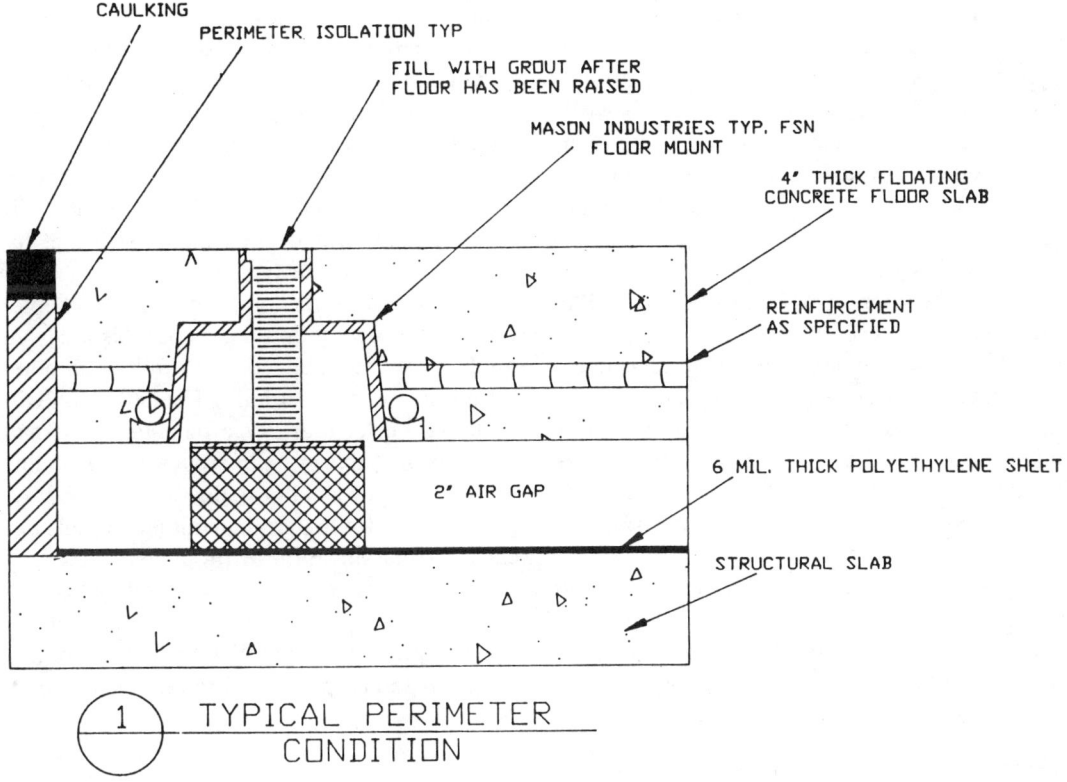

CAULKING

PERIMETER ISOLATION TYP

FILL WITH GROUT AFTER
FLOOR HAS BEEN RAISED

MASON INDUSTRIES TYP. FSN
FLOOR MOUNT

4″ THICK FLOATING
CONCRETE FLOOR SLAB

REINFORCEMENT
AS SPECIFIED

2′ AIR GAP

6 MIL. THICK POLYETHYLENE SHEET

STRUCTURAL SLAB

1 TYPICAL PERIMETER
CONDITION

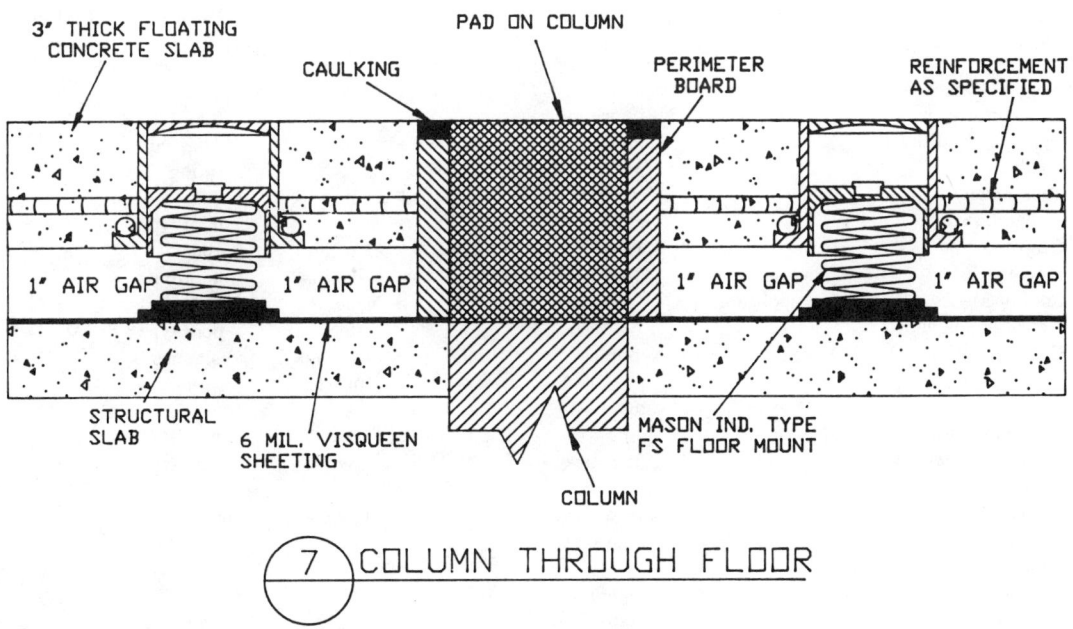

3″ THICK FLOATING
CONCRETE SLAB

PAD ON COLUMN

CAULKING

PERIMETER
BOARD

REINFORCEMENT
AS SPECIFIED

1′ AIR GAP   1′ AIR GAP

1′ AIR GAP   1′ AIR GAP

STRUCTURAL
SLAB

6 MIL. VISQUEEN
SHEETING

MASON IND. TYPE
FS FLOOR MOUNT

COLUMN

7 COLUMN THROUGH FLOOR

FIGURE 4.30   Floating floor.   (Reproduced by permission from Mason Industries, Hauppauge, NY.)

There are more complicated equations that can be used to theoretically solve any vibration application problem. This type of equation is useful, occasionally, to find out if one is working in the proper direction when a new system of isolation is developed. This type of analysis has shown that the isolator deflection should be as much as six to eight times the floor deflection for a lightweight building. It is prudent to ask about the floor or roof deflection as well as the span.

While these numbers may sound farfetched, in actual solutions of existing problems, an isolator with as much as 3 in. of static deflection might be required to solve this problem even though the theoretical charts show that a 0.05-in. deflection should have been quite satisfactory.

Isolator deflections are selected empirically based on actual field experience. These are practical values that are necessarily broad in scope. In some cases they may be more than is actually required to solve a particular problem, but a supplier must deal in broad terms.

On large critical jobs it is often important to rely on the impartial professional judgment of a qualified acoustical consultant. A specific job analysis can be made to determine the exact requirements for each piece of equipment. This determination will depend on its size and location in a particular type of structure.

The ASHRAE Systems Guide has a chapter on sound and vibration that is used as a design guide for isolator deflection, as well as all other sound and vibration criteria. The deflection will be based on location, type of equipment, and floor span. Table 4.5 illustrates some of this information for isolator deflections.

This need for higher-deflection materials has caused reclassification of products: Pad materials, whether neoprene, cork, combinations of cork and neoprene, fiberglass, sisal fibers (white natural plant material), felt, lead, or any other material provide limited deflections.

TABLE 4.5   Vibration Isolation Selection Guide

Equipment Type	At Grade			Above Grade (30-ft span)		
	Base Type	Isolator Type	Deflection (in.)	Base Type	Isolator Type	Deflection (in.)
Reciprocating compressor	Concrete inertia	Spring floor type	0.75	Concrete inertia	Spring floor type	1.5
Hermetic centrifugal chillers	None	Neoprene pad	0.25	None	Restrained spring	1.5
Tank-mounted air compressor	None	Neoprene pad	0.75	None	Spring floor type	1.5
Centrifugal pump— 10 hp	Concrete inertia	Spring floor type	0.75	Concrete intertia	Spring floor type	1.5
Rooftop air-conditioning unit	Not applicable			None or steel rail	Spring floor type	1.5
Cooling tower—500 rpm	None	Neoprene pad or mount	0.25	Steel rails or base	Restrained spring	1.5
Axial fan, 24-in. diameter over 500 rpm	Steel or concrete	Spring floor type	1.0	Concrete inertia	Spring floor type	1.0
Centrifugal fan, 50 hp and over 500 rpm	Steel or concrete	Spring floor type	1.0	Concrete inertia	Spring floor type	1.0
Packaged air handler over 15 hp	None	Neoprene floor mount	0.25	None	Spring floor type	1.0

Source: Reproduced by permission, the American Society of Heating, Refrigeration and Air Conditioning Engineers, from the 1991 ASHRAE Handbook—HVAC Applications.

These deflections are normally 10 to 20 percent of the pad's thickness. Therefore, pads are good for high-frequency noise breaks, and since their deflections are almost always small in comparison to upper-floor deflections, their use should generally be confined to basement areas, noncritical jobs, or situations where job costs must be kept to an absolute minimum regardless of performance.

Neoprene mountings and hangers fall into the 0.20- to 0.50-in. deflection range. They do provide sufficient static deflection to offer protection under small high-speed equipment such as close-coupled pumps up to 3 hp, vent sets, and small heating ventilating units, where the unbalanced forces are so small that in all probability only a noise break and minor vibration relief need be provided. Neoprene hangers are sufficiently effective for isolation of steam lines where there is seldom any real vibration but only high-frequency whistles and noises of that nature. While neoprene mountings and hangers do have their place as noted, there is a tendency for specifications to call for spring mountings throughout because the dollar difference between the neoprene mountings and small spring mountings is a very small percentage of the overall job cost. Neoprene hanger elements are often used in series with springs because the neoprene element will do a far more efficient job of eliminating the high-frequency noises than the spring alone.

Steel spring mountings are by far the most widely used commodity on critical jobs today. Steel springs are practical through 5 in. of static deflection and even more on specific occasions. Springs provide an easily variable design medium, and steel spring installations are as permanent as the machine itself when selections are made within proper stress values. Most modern isolators are simply steel springs that have been designed with large enough diameters to provide stability without the need for a supplementary, often detrimental, housing. They are generally manufactured with an adjustment bolt and a pad made of neoprene or some other material in series with the spring to attenuate the high frequencies. Figure 4.31 shows a variety of isolators used in the field today.

Another type of isolator is the air mount or air spring. In broad terms an air spring is merely a large bladder that is manufactured in a configuration to withstand as much as 100 lb of air pressure and provide a stable support point for the equipment. By proper shaping of the air spring, small units can be designed to provide the equivalent of 6 to 7 in. spring deflection in the steel spring series. Since the walls of the air spring are constructed of rubber, there is no possibility of the type of spring resonance sometimes found in the coils of large-deflection, large-diameter steel springs.

Air springs have the advantage of supporting a wide range of loads merely by varying air pressure, and the spring frequency is a function of the shape rather than the pressure. In general, air springs are installed with a replenishing air supply to compensate for leakage, or expansion and contraction where there is a wide temperature variation. Height control valves are provided to maintain elevation and compensate for external forces such as fan thrusts. Since these complete installations are generally more expensive than steel spring isolation methods, air springs are generally reserved for extremely critical locations as recommended by an acoustical consultant. These include sensitive optical equipment, electron microscopes, and manufacturing equipment in high-technology industries.

Supplementary steel or concrete equipment bases are most often used to keep equipment in alignment, for example, to tie a fan and motor together or provide the common base for a turbine-driven compressor. Bases may also be used to stiffen an existing base such as a cast iron pump base, provide stability for tall machines such as absorption machines, or tie a complete package together as in the case of a long, many-sectioned heating and ventilating unit.

In many cases, bases are constructed of steel rather than concrete because they can be shipped to the job as a complete welded assembly and installed without involving more than one trade. Their light weight, as compared to concrete, lessens the floor burden and reduces the need for strengthening the floor slab.

When steel bases are used, however, it is

(a)   (b)

(c)   (d)

FIGURE 4.31   Vibration isolators: (a) housed spring isolator, (b) open steel spring isolator, (c) neoprene isolator, and (d) spring hanger.   (Reproduced by permission from Mason Industries, Hauppauge, NY.)

most important that they be made sufficiently rigid to provide support for the mechanical equipment and do not resonate at the frequency of the equipment they are supporting. The best approach to this is the use of steel beams as the base members, and we would suggest a depth of framing equal to one-tenth the longest dimension of the base. Cross-framing may be used for additional stiffening, especially on large-horsepower direct-drive units.

Floating concrete bases are recommended for pumps because pump bases are designed for grouting to concrete floors. The use of a float-

ing concrete base provides a grouting surface and the extra stiffness the pump base requires. When stiffness is the only factor, a concrete depth equal to one-twelfth the longest dimension of the base is usually satisfactory. Concrete bases are also used for inertial purposes when an increase in mass is necessary to resist either the imbalance of the equipment or external forces.

The classic example of unbalanced equipment is the single- or double-cylinder vertical or horizontal slow-speed air compressor. These large-bore, long-stroke compressors operating at 350 rpm can be only partially balanced. The weight of the concrete base is calculated from the unbalanced force information supplied by the compressor manufacturer. The concrete inertia bases are five to seven times the weight of the compressor in order to bring the motion down to acceptable limits.

Concrete bases can be used to offer resistance to external forces such as fan thrust. This is particularly important when isolating high-pressure cabinet fans. The fan pressure will produce a thrust that tends to displace the fans horizontally. Concrete bases are suggested for the isolation of high-pressure fans above 6 in. of static pressure. The weight requirement falls somewhere between one and three times the weight of the equipment and must be determined in each situation.

Another piece of equipment that assists in isolating a system is flexible hoses. They should be installed horizontally and parallel to the axis of rotating equipment. This allows the hose to flex in the transverse direction. All hoses are quite stiff axially. Butyl rubber hose has a better chance of reducing pipe wall transmitted noise, but it does not reduce noise and vibration that is fluid borne.

Stainless steel or bronze metallic hoses are recommended for those services where the static pressure or temperature exceeds the capability of the rubber materials. While flexible hoses are not complete protection against pipe line noises, they do provide flexibility at the points of connection to the mechanical equip-ment. This reduces stresses to the flanges and allows the isolated equipment to move freely on the springs.

Since noises transmitted to pipe lines will travel for unpredictable distances, the best approach is to isolate the lines along the total length. This is particularly important within mechanical rooms and adjacent to sensitive areas. This type of isolation is normally provided by combination spring and neoprene hangers. In some cases the piping is supported from the floor. Standard spring isolation supports can be used.

While all connections to high-pressure ducts should be made flexible, these connections are not completely successful either. There is enough air turbulence in high-pressure systems to vibrate the ducts and cause possible transmission to the structure. Ducts are normally isolated by 1-in. deflection spring hangers that are installed up to 50 ft from the fan discharge.

Silencers

Silencers are engineered devices that are designed to attenuate sound waves propagating in a flowing medium. There are two categories of silencers: dissipative and reactive.

The dissipative silencer is a straight-through device with the passageway lined with acoustical absorbing material (Fig. 4.32). The reactive silencer has a series of chambers in parallel or series utilizing the reflection and expansion

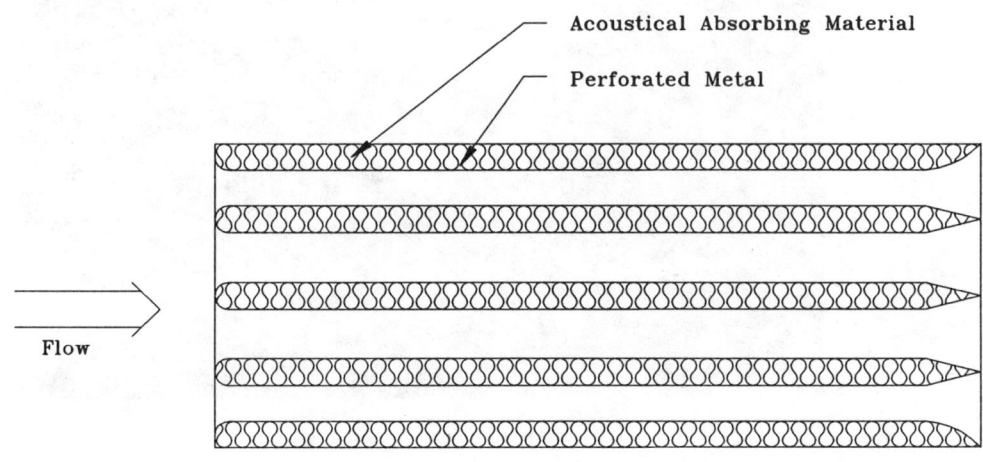

FIGURE 4.32   Dissipative-type silencer.

characteristics of the expansion chambers, side tubes, branch resonators, and tailpipes to attenuate the sound. Examples of dissipative silencer applications are on fans, gas turbines, and acoustically lined ducts, etc. An automotive engine exhaust silencer or muffler is a good example of a reactive silencer. They are also used on compressors and rotary or reciprocating equipment. This silencer is illustrated in Fig. 4.33.

Both of these types of silencers can be combined in a plenum chamber, which is an expansion chamber that is lined. This is shown in Fig. 4.34. Plenums are found in air-conditioning systems. They provide a broad range of attenuation and are very good for low-frequency attenuation.

The next type is a combination silencer, reactive and dissipative. This is also used for vent applications. When a valve opens releasing gas to the atmosphere, this type of silencer can be used to control the noise. The silencer usually has an inlet expansion chamber to help dissipate the pressure and flow followed by a lined passage to attenuate the high-frequency noise that results. This type of silencer is shown in Fig. 4.35.

Dissipative silencers can also be used on electric motor inlet cooling air fans. An example of this is shown in Fig. 4.36 on a large motor in a process plant. The construction is an acoustically lined duct with a lined acoustical target to absorb the fan noise. A noise reduction of 10 to 15 dBA can be expected.

Air noise in manufacturing plants produce levels from 100 to 120 dBA. The hoist air motor, in Fig. 4.37, has an open exhaust and is very noisy. This can easily be reduced by the application of an air exhaust silencer, shown in Fig. 4.38. Typical octave band noise spectrum noise levels with silencers installed on an open air pipe are shown in Fig. 4.39.

The performances of silencers are described in several ways depending on the method of testing or the manufacturer. The following are the common terms for describing the performance:

1. Insertion loss—difference between the sound pressure level measured at the same location before and after a silencer is installed
2. Dynamic insertion loss—the same as insertion loss except it is measured with flow in the silencer
3. Attenuation—provides information about

FIGURE 4.33   Reactive-type silencer.

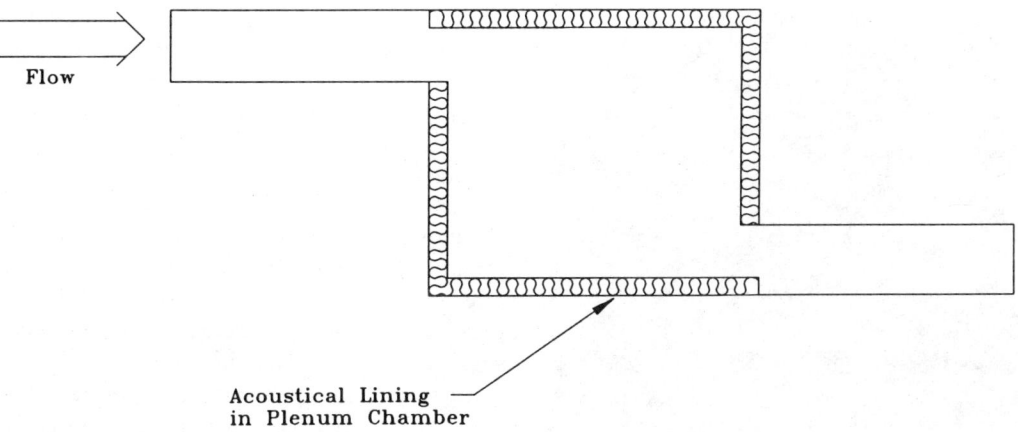

Flow

Acoustical Lining
in Plenum Chamber

FIGURE 4.34   Plenum chamber.

how the sound power decreases as the
sound wave moves through the device
4. Noise reduction—the difference between
sound pressure levels measured at the inlet and outlet of the silencer.

For more information on this subject, ref-

FIGURE 4.36   Electric motor and silencer.

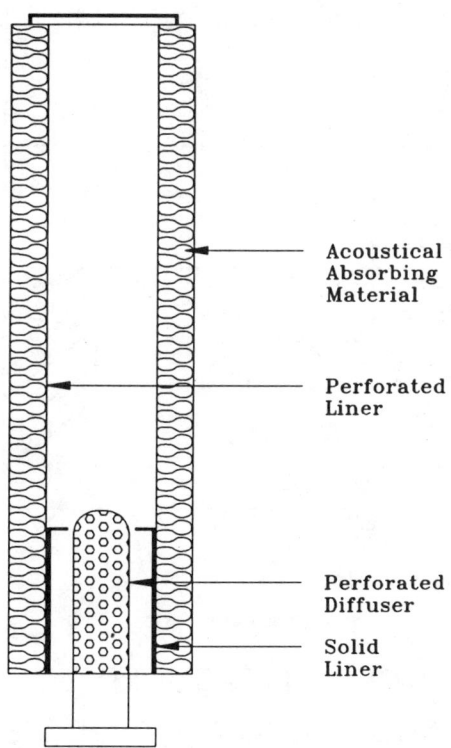

Acoustical
Absorbing
Material

Perforated
Liner

Perforated
Diffuser

Solid
Liner

FIGURE 4.35   Combination of reactive and
dissipative vent silencer.

FIGURE 4.37   Air hoist motor with exhaust opening.

FIGURE 4.38   Variety of air noise exhaust silencers.

erences such as Beranek, Lord et al., and Munjal provide a more in-depth treatment. Also, silencer manufacturers' catalogs have the required data for applying these devices.

Sound Absorption in a Space

The purpose of installing sound absorption materials in an industrial plant space is to reduce reflected noise from the ceiling and walls. This must be thought of as one of the noise control steps in the noise reduction process. The discussion in Chapter 6 about direct and reflected sound has a direct relationship to this situation. The illustration in Fig. 4.40 shows the direct

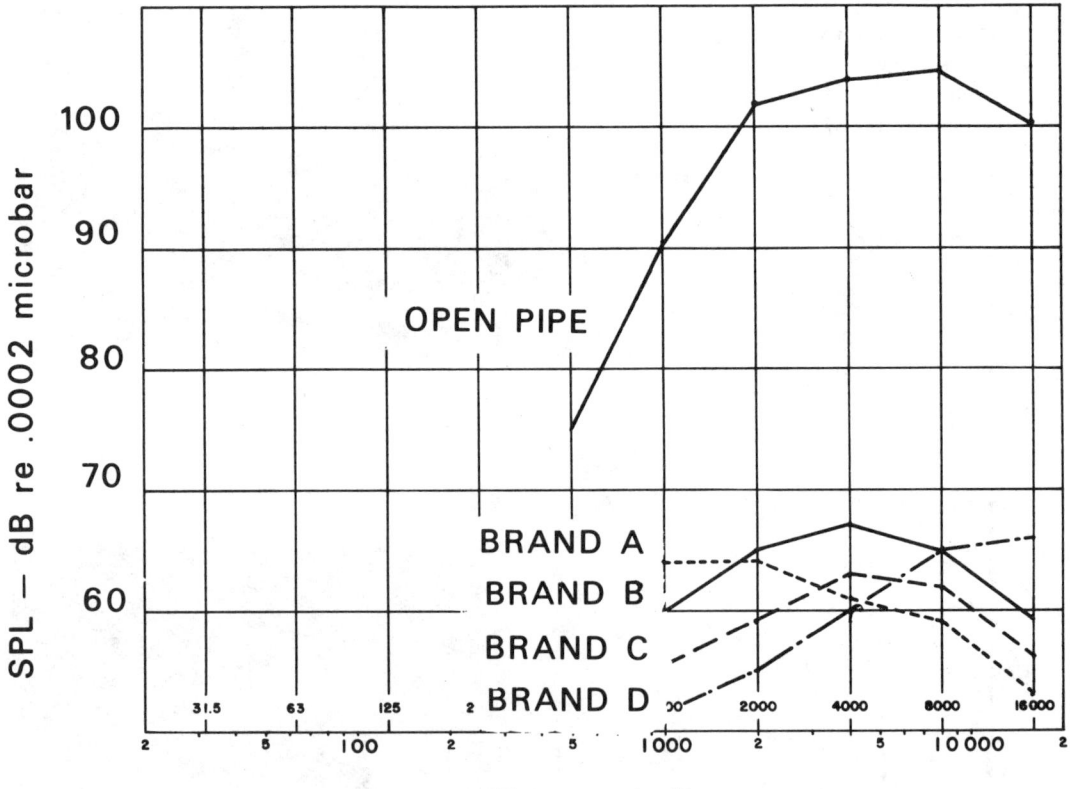

NOISE LEVEL OF ¼" NPT OPEN PIPE
AND SELECTED SILENCERS MEASURED AT
AN ANGLE OF 30° AND A DISTANCE OF 7'3".

FIGURE 4.39   Noise level results of various air exhaust silencers compared to an open pipe.

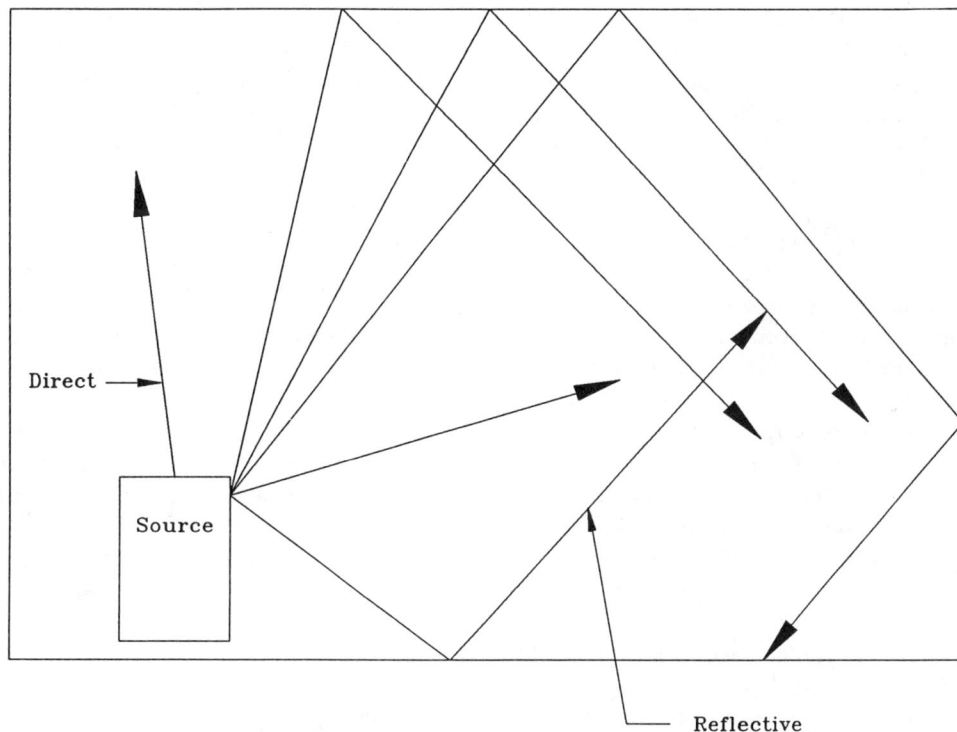

FIGURE 4.40   Direct and reflected sound from a source.

and reflected paths in a space. Since the absorbing material only affects the reflected path, it can produce limited results.

The rule of thumb to keep in mind is that absorption material will only reduce the noise level in a space by 4 to 6 dBA with a ceiling height not greater than 16 ft and a noise spectrum in the mid to high frequencies. The amount of absorbing material to achieve this result will be a coverage on the ceiling and walls of 30 to 50 percent of the surface area. For example, if the space is a gas compressor building with large engines, low-frequency source, and high ceilings the result may only be 2- to 3-dBA reduction.

The property used in determining the effectiveness of an absorption material is called sound absorption coefficient. It can be measured based on a standard test and is the ratio of the acoustic energy absorbed to the incident energy on the surface. It is measured in 125-, 250-, 500-, 1000-, 2000-, and 4000-Hz octave bands and expressed in percent or decimal

equivalent. For example, a material might have a sound absorption coefficient at 2000 Hz of 0.90, or 90 percent. This means that 90 percent of the sound will be absorbed at that frequency.

Another term used for comparison of materials is average absorption coefficient. This is the arithmetic average of 250-, 500-, 1000-, and 2000-Hz octave bands. Materials manufacturers usually provide both of these properties in their literature.

The common types of absorbing materials used in industrial noise control are as follows:

1. Functional panels—usually made of perforated metal over a metal frame that contains a 2-in. fiberglass or rock wool absorbing pad. The absorbing material may be covered with a protective film to keep out contaminants.
2. Acoustical baffles—constructed of rigid acoustical board covered with a protective film and hung in the ceiling and on the walls. They are more fragile than the

function panels but work equally well. They should be placed in locations where they cannot be damaged easily.

3. Acoustical foam—an open-cell resilient polyurethane foam having closely controlled pore size. It comes in thicknesses of $\frac{1}{2}$ to 2 in. It can be combined with loaded vinyl to provide mass and protective films to exclude contaminants.

4. Cellulose fiber—combined with a binder and can be sprayed on the ceiling and upper portions of the walls. It works best for new construction but can be used for remedial applications. The equipment in the area must be covered.

5. Acoustical plaster—a cementations-based material that is sprayed in place with controlled openings for absorption.

6. Acoustical ceiling—made from mineral fiber or glass fiber and formed into regular sizes for placement in a ceiling grid. These materials can be used in a variety of industrial applications other than office spaces.

7. Specialized acoustical materials—for use in food-processing areas and high-technology clean rooms that require materials that can be thoroughly cleaned.

Several materials manufacturers make products that fit this application. They have specialized protective films and construction that allow cleaning with steam and chemicals. There are, of course, applications where acoustical materials cannot be used.

The photograph in Fig. 4.41 illustrates some of the material discussed. In Table 4.6 are typical data for various acoustical materials as well as some other common materials found in construction.

If a more accurate determination is desired for noise reduction using acoustical materials, the following generalized equation can be used:

$$NR = 10 \log \left[\frac{\text{original absorption}}{\text{final absorption}} \right]$$

where:

NR = noise reduction

In a hard space the effect of adding a 30 to 50 percent coverage will be dramatic about 4 to 6 dBA of noise reduction. However, in a space that is much softer the effect will be small.

FIGURE 4.41   Functional acoustical absorbing baffles.

TABLE 4.6   Absorption of Various Materials

Material	Octave Band Center Frequency (Hz)					
	125	250	500	1K	2K	4K
Unpainted concrete block	0.03	0.04	0.03	0.03	0.04	0.02
Painted concrete block	0.1	0.05	0.06	0.07	0.09	0.08
Concrete floors	0.01	0.01	0.02	0.02	0.02	0.02
Resilient floors	0.03	0.03	0.03	0.03	0.03	0.02
Heavy plate glass	0.18	0.06	0.04	0.03	0.02	0.02
Standard window glass	0.03	0.02	0.02	0.012	0.07	0.04
$\frac{3}{8}''$ plywood	0.28	0.2	0.17	0.09	0.1	0.1
2″ solid wood	0.01	0.05	0.05	0.04	0.04	0.04
$\frac{1}{2}''$ Sheetrock	0.29	0.1	0.05	0.04	0.07	0.09
4″ acoustical panels	0.5	0.9	0.95	0.99	0.94	0.83
Mineral fiber ceiling tile no. 7 mtg.	0.2	0.4	0.8	0.95	0.9	0.8
1″ acoustical foam	0.16	0.35	0.45	0.84	0.94	0.85
$\frac{3}{4}''$ sprayed fibrous material	0.08	0.3	0.7	0.9	0.9	0.75
1″ fiberglass panel solid backing	0.03	0.02	0.6	0.85	0.95	0.96
2 × 4 fiberglass ceiling no. 7 mtg.	0.4	0.5	0.6	0.99	0.87	0.6

Source: NIOSH, *Compendium of Materials for Noise Control*, June 1975.

RECEIVER CONTROL

This section deals with controlling noise at the receiver of the noise, that is, employees. In some cases it may not be feasible to reduce the noise at the source or along the path. Receiver control is a method of protecting the employee by reducing the exposure to noise. This can be done with a control room as a haven from noise, hearing protectors, or administrative control of the amount of noise exposure.

Control Rooms

In many utility and process plants control rooms are used to house instruments as well as personnel. These can be fabricated from common building materials or specialized acoustical enclosure panels. In either case the amount of reduction ranges from 20 to 35 dBA depending on the levels of construction. A control room is illustrated in Fig. 4.42.

In wood-processing plants the equipment operators may be housed in a specialized space that is functionally designed for the job and to reduce noise. One such application is shown in

FIGURE 4.42   Personnel control room.

Fig. 4.43. Most enclosure manufacturers can supply a variety of personnel enclosures.

Hearing Protectors

As part of a hearing conservation program (see Chapter 5) hearing protectors are required. This is a method of controlling the noise to the re-

FIGURE 4.43   Control booth for sawmill operator.

ceiver. For those not familiar with hearing protectors Fig. 4.44 shows the variety that is available from a number of suppliers.

The manufacturer is required to test the device and issue a noise reduction rating (NRR). Since this is a laboratory test the amount of actual reduction achieved will depend on how it is worn and the fit. Lempert found that the actual amount of noise reduction reaching the ear was about half the NRR [Lempert, 1983]. The OSHA standard (Appendix B) also provides a detailed guide relating to hearing protectors. It is best to keep a variety of protectors and find the one that feels the most comfortable; that is, the one that will be worn.

Administrative Controls

If personnel can be moved about in the plant their noise exposure can be controlled. This is the basic meaning of administrative controls. In the author's 28 years of experience there have been two occasions when this worked.

A vinyl grinder operator loaded scrap vinyl for part of the day then was moved to another location where it was quiet for the remainder of the day. In some cases union work rules preclude this approach, and in other cases it may be difficult to find jobs that can be switched. This does control exposure to the employee, but at the same time the employee's audiogram should be checked on a very regular basis to be sure there is no hearing loss.

Another case involved moving the employee and the control console away from the machine to reduce the employee's noise exposure. Television cameras can be used in noisy locations for the employee to monitor as a means of controlling noise exposure. Administrative controls can take many forms, but usually require specialized circumstances to work well. The employee's hearing level must be monitored as part of the hearing conservation program.

ACTIVE NOISE CONTROL

A new technology is emerging to cover a broad range of noise control applications. This includes industrial application as well as noise control for fan systems in air-conditioning (HVAC) systems. The active noise system uses a microphone to detect noise as it propagates down a duct. This signal is fed into a digital controller than analyzes the narrow-band frequency content, then produces a signal to the speakers that is 180° out of phase with the initial signal. This significantly reduces the noise source through cancellation. As the controller is engaged it models the signal and a noise reduction can be heard. If the signal changes, as most acoustic signals do, the controller system tracks the signal producing the out-of-phase noise for cancellation. This is effective for broadband low-frequency noise sources up to about 400 Hz. It is even more effective for noise sources that have tonal components. A schematic diagram of the system is shown in Fig. 4.45.

Turbulence and airflow in air distribution systems will affect the results when a microphone is placed in the duct. In order to minimize these pressure fluctuations that are uncorrelated with the acoustic signal and contaminate

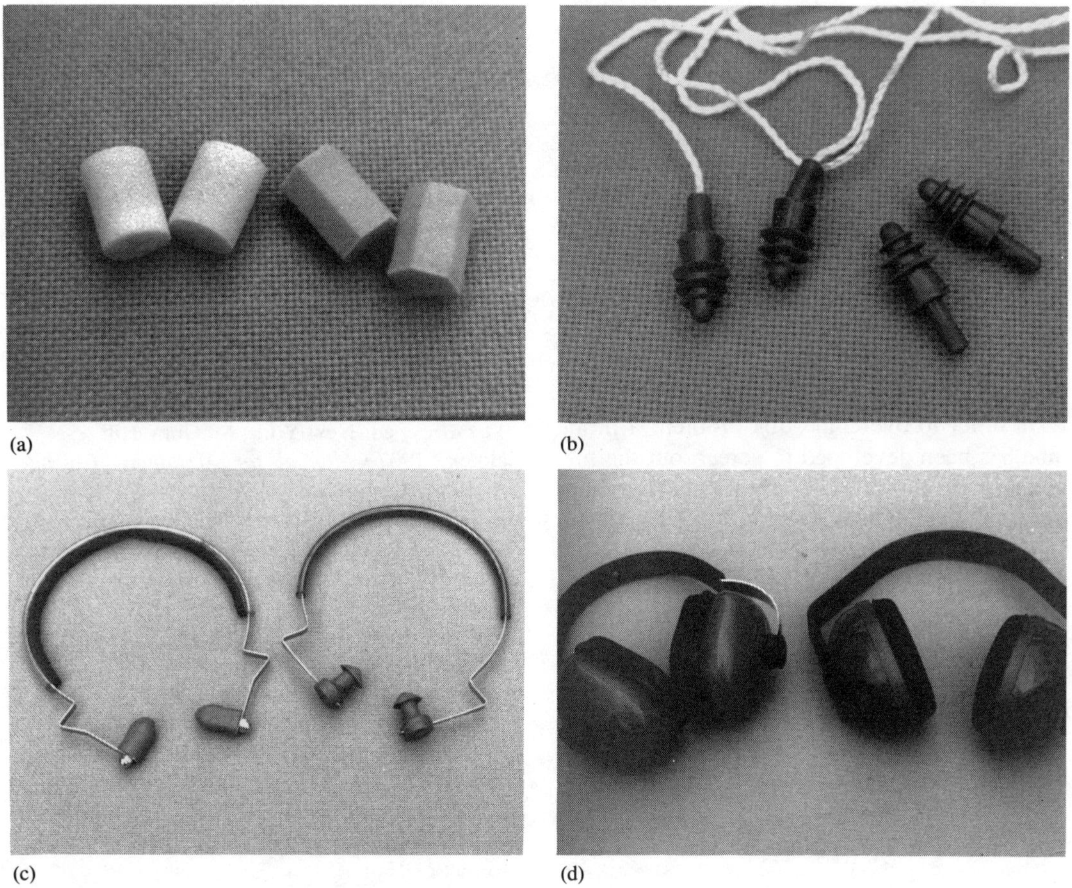

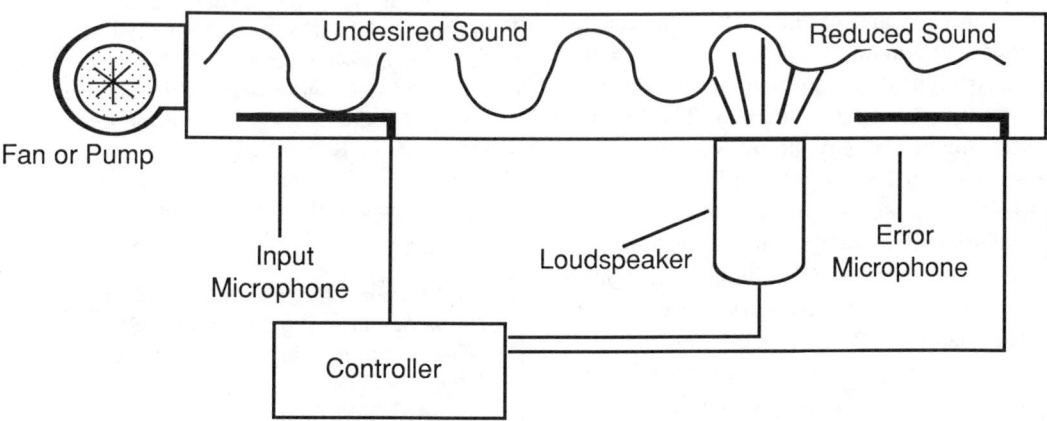

FIGURE 4.44   Variety of hearing protectors: (a) foam earplugs, (b) sized earplugs, (c) ear canal caps, and (d) earmuffs.

FIGURE 4.45   Active noise control silencing system. (Reprinted by permission from DIGISONIX, Division of Nelson Industries.)

the data, a probe tube has been developed and changes made to the digital controller [Eriksson, 1987]. Further enhancements have been made as the technology has progressed in the last few years to improve on this process.

The effects of flow turbulence produce greater pressure fluctuations than the acoustic signal on the microphone. The controller uses a correlation process between the detection and error microphone signals. The turbulent pressure fluctuations add uncorrelated noise to both signals, reducing the system's ability to effectively cancel the source [Eriksson, 1987].

In order to overcome this problem a probe tube has been developed to screen out the turbulent flow effects. This allows the turbulent pressure fluctuation to pass by, and the acoustic signal can pass through the probe tube and be detected by the microphone. It is always best to design the duct system with good aerodynamics to minimize these effects. For HVAC systems the microphone can be hidden under the internal thermal and acoustical insulation, which acts as an effective wind screen to minimize the turbulence effects.

SUMMARY

Most noise problems can be solved, or at least improved, with standard techniques and some common sense. There are times when some different approaches must be used, for instance, the problem of piping noise in a condominium: A woman calls and explains that she and her husband just bought this new place and cannot sleep. At night they can hear the water in the pipes in the wall next to the bed. The building owners are not of any help. Noise measurements reveal that the nighttime level is 30 dBA and one can hear everything—it is too quiet. The solution is to place a noise-masking unit under the bed to raise the ambient level to mask out the other noises. This works very well and they are happy. This is a simple and inexpensive solution. Fixing the pipe noise problem at the source or along the path would be quite expensive, and they would have still been able to hear other things and not have been satisfied.

Define the problem, and think about the most feasible and practical solution that is compatible with the situation.

References

ASHRAE. 1991. *Systems Guide and Data Handbook*. Chapter 52, Sound and Vibration Control. Atlanta, GA: ASHRAE.

ASTM. 1975. American Standards for Testing Materials. *Standard Recommended Procedure for Laboratory Measurement of Airborne Sound Transmission Loss of Building Partitions*. ASTM 90-75.

Beranek, Leo L. 1988. *Noise and Vibration Control*, rev. ed. New York: McGraw-Hill.

Celotex. 1937. *Manual for Architects*. Chicago: Celotex Corp.

Dear, T. A. 1972. Noise Reduction Properties of Selected Pipe Covering Configuration. *Proceedings of the Inter-Noise 72 Conference*, October, pp. 134–144.

Diehl, George M. 1973. *Machinery Acoustics*. New York: Wiley.

Eriksson, L. J., and Allie, M. C. 1987. A Digital Sound Control System for Use in Turbulent Air Flows. *Proceedings of Noise-Con*. June, pp. 365–370.

Irwin, J. D., and Graf, E. R. 1979. *Industrial Noise and Vibration Control*. Englewood Cliffs, NJ: Prentice-Hall.

Lempert, Barry L., and Edwards, Richard G. 1983. Field Investigation of Noise Reduction Afforded by Insert-Type Hearing Protectors. *American Industrial Hygiene Association Journal*, December, pp. 894–902.

Lilley, Daniel T. 1983. Understanding Damping Techniques for Noise and Vibration Control. *Plant Engineering*, April, pp. 38–40.

Lord, Harold W., Gately, William S., and Evenson, Harold A. 1987. *Noise Control for Engineers*. Malabar, FL: Robert E. Krieger.

Mariner, T. 1957. Control of Noise by Sound-Absorbent Materials. *Noise Control*, a publication of the Acoustical Society of America, July, pp. 11–19.

Mason, Norman. 1966. *Controlling Vibration Problems in Sensitive Structures* from ASHRAE Systems and reprinted in Mason Industries Catalog. Hauppauge, NY.

Munjal, M. L. 1987. *Acoustics of Ducts and Mufflers*. New York: Wiley.

Purcell, Jack B. C. 1957. Control of Airborne Sound

by Barriers. *Noise Control*, a publication of the Acoustical Society of America, July, pp. 20–26.

Thorton, Charles A., et al. 1986. *Thermal-Acoustical Blanket Insulation Field Evaluation*. Tennessee Valley Authority Document No. TVA/OC H&S/IH—86/04.

Yerges, Lyle F. 1978. *Sound, Noise and Vibration Control*. New York: Van Nostrand Reinhold.

Zigh, A., and Prasad, A. Yo. 1988. Experimental Studies on Acoustical Insulation Treatments for Machinery Noise Control. *Proceedings Noise-Con 88*, pp. 249–254.

Chapter 5

Guidelines for a Hearing Conservation Program

INTRODUCTION

The noise levels and individual employee noise exposure levels within a plant will determine the need for a hearing conservation program. The purpose is to *prevent* hearing loss. The material in this chapter will provide a general overview for a well-rounded hearing conservation program. While this material is based on the OSHA Occupational Noise Exposure Regulation, each company will need to personalize a program to suit its specific needs.

The difficulty in not having an effective hearing conservation program is the risk of hearing loss that employees may sustain. In the last few years the claims for hearing loss compensation have grown due to class action litigation brought against the employer and the companies that have equipment in the plants alleged to have caused hearing loss. Thus, the best defense against future claims is to have a continuing and effective hearing conservation program that can be monitored for individual employee hearing loss.

As the legal profession deals with more and more hearing loss cases, they are beginning to rely on the noise control engineer to interpret various types of noise data relevant to the case. These data can provide an overview of the workplace environment that refutes the emotional issues that the plaintiff might raise and refocuses a case that, at the outset, may have looked dire indeed for the defense [Fann, 1990].

Employee noise exposure is a function of the work environment and how employees interact with that environment. Their work motions must be studied and documented to accurately describe their location relative to the noise sources and the amount of time spent near those sources.

An acoustical, or noise control, engineer must be able to make accurate measurements. More importantly, these measurements must be appropriate to the case at hand. The engineer must paint a clear picture that is favorable to his or her client's case and must spend the time necessary working with attorneys to develop a verbal description that will allow individuals to vicariously experience the event and place it in proper and accurate perspective [Fann, 1990].

This additional role of the noise control engineer is essential to the case presentation. Without this, the defense counsel may find that the jury only considers indecipherable technical jargon and a plaintiff's all-too-easy-to-accept plea for justice. It's important that the judge and jury have the most accurate and nonemotional facts available to make their decision.

The major objective of any effective hearing conservation program (HCP) is to prevent permanent noise-induced hearing loss resulting from occupational noise exposures. The material in this chapter can provide the framework to help conserve employee hearing. To ensure compliance with the noise exposure regulations governing a facility, the HCP needs to contain the following basic elements [OSHA, 1974]:

1. Noise surveys
2. Education and training
3. Engineering and/or administrative controls
4. Hearing protection
5. Audiometric testing and evaluation

A draft ANSI standard, S12.13-1991, *Evaluating the Effectiveness of Hearing Conservation Programs*, provides a method for determining how a HCP is functioning for a given plant [ANSI, 1991].

The data in Table 5.1 contain noise exposure levels (NELs) that should be used in determining employee noise exposures. The NELs are also set forth in the OSHA regulation CFR1910.95, Occupational Noise Exposure. The NELs are sound pressure levels and exposure durations that represent conditions under which, it is believed, most industrial workers may be repeatedly exposed without adversely affecting their hearing. The values should be used *only* as a guide in the control of individual noise exposure. Because of individual susceptibility to noise, these exposure levels should not be regarded as fine lines between safe and harmful noise levels. It is always best to err on the conservative side regarding employee hearing loss.

The NEL values apply to the total duration of exposure per day. The exposure can result from a single event lasting for a specific time period, or due to many exposures spread out over the total workday. Examples are found at the end of this chapter.

Employees exposed to noise at or above 50 percent of the NEL (also called action level) should be included in the HCP. It is recommended that all employees with borderline noise exposures, or those with infrequent ex-

TABLE 5.1   Noise Exposure Levels (NELs) to Continuous Noise

(dBA)	Sound Level (Hr)	OR	Noise Exposure Duration per Workday (Min)
80	32.0		1920
81	27.9		1674
82	24.3		1458
83	21.1		1266
84	18.4		1104
85	16.0		960
86	13.9		834
87	12.1		726
88	10.6		636
89	9.2		552
90	8.0		480
91	7.0		420
92	6.2		372
93	5.3		318
94	4.6		276
95	4.0		240
96	3.5		210
97	3.0		180
98	2.6		156
99	2.3		138
100	2.0		120
101	1.7		102
102	1.5		90
103	1.4		84
104	1.3		78
105	1.0		60
106	0.87		52.2
107	0.76		45.6
108	0.66		39.6
109	0.57		34.2
110	0.50		30
111	0.44		26.4
112	0.38		22.4
113	0.33		19.8
114	0.29		17.4
115	0.25		15
116	0.22		13.2
117	0.19		11.4
118	0.16		9.6
119	0.14		8.4
120	0.125		7.5
121	0.110		6.6
122	0.095		5.7
123	0.082		4.9
124	0.072		4.3
125	0.063		3.8
126	0.054		3.2
127	0.047		2.8
128	0.041		2.4
129	0.036		2.1
130	0.031		1.8

TABLE 5.2   Conversion from "Percent Noise Exposure" or "Dose" to Eight-Hour Time-Weighted Average (TWA) Sound Level

% Noise Exposure	TWA (dBA)	% Noise Exposure	TWA (dBA)	% Noise Exposure	TWA (dBA)
10	73.4	92	89.4	118	91.2
15	76.3	93	89.5	119	91.3
20	78.4	94	89.6	120	91.3
25	80.0	95	89.6	125	91.6
30	81.3	96	89.7	130	91.9
35	82.4	97	89.8	135	92.2
40	83.4	98	89.9	140	92.4
45	84.2	99	89.9	145	92.7
50	85.0	100	90.0	150	92.9
55	85.7	101	90.1	155	93.2
60	86.3	102	90.1	160	93.4
65	86.9	103	90.2	165	93.6
70	87.4	104	90.3	170	93.8
75	87.9	105	90.4	175	94.0
80	88.4	106	90.4	180	94.2
81	88.5	107	90.5	185	94.4
82	88.6	108	90.6	190	94.6
83	88.7	109	90.6	195	94.8
84	88.7	110	90.7	200	95.0
85	88.8	111	90.8	210	95.4
86	88.9	112	90.8	220	95.7
87	89.0	113	90.9	230	96.0
88	89.1	114	90.9	240	96.3
89	89.2	115	91.1	250	96.6
90	89.2	116	91.1	260	96.9
91	89.3	117	91.1	270	97.2
280	97.4	540	102.2	800	105.0
290	97.7	550	102.3	810	105.1
300	97.9	560	102.4	820	105.2
310	98.2	570	102.6	830	105.3
320	98.4	580	102.7	840	105.4
330	98.6	590	102.8	850	105.4
340	98.8	600	102.9	860	105.5
350	99.0	610	103.0	870	106.6
360	99.2	620	103.2	880	106.7
370	99.4	630	103.3	890	106.8
380	99.6	640	103.4	900	106.8
390	99.8	650	103.5	910	106.9
400	100.0	660	103.6	920	106.0
410	100.2	670	103.7	930	106.1
420	100.4	680	103.8	940	106.2
430	100.5	690	103.9	950	106.2
440	100.7	700	104.0	960	106.3
450	100.8	710	104.1	970	106.4
460	101.0	720	104.2	980	106.5
470	101.2	730	104.3	990	106.5
480	101.3	740	104.4	999	106.6
490	101.5	750	104.5		
500	101.6	760	104.6		
510	101.8	770	104.7		
520	101.9	780	104.8		
530	102.0	790	104.9		

posures at or above 50 percent of the NEL, should be included in the HCP. Maintenance personnel or supervisors are some examples.

For an eight-hour workday the action level that is equal to 50 percent of the noise exposure is 85 dBA time-weighted average (TWA) noise exposure. If the work shift is different from eight hours, the TWA noise exposure in the HCP should be adjusted accordingly. Since many process and power plant operators work 12-hour shifts, the action level would be 82 dBA TWA. These are shown in more detail in Table 5.2.

NOISE SURVEYS

A preliminary noise survey is required to assess a plant or areas within a plant. This survey, which involves a walk-through of the areas in the plant, will indicate which areas or job activities require more detailed study. Noise measurements can be made using a sound level meter following the procedures found in Chapter 9. Typically, if an area has noise levels above 85 dBA (82 dBA for 12-hour shifts), then a noise dosimetry evaluation is recommended. Many times the exposure survey is the starting point. This depends to a great extent upon the experience of the individual in charge of the program and the plant's own policy and procedures.

The noise exposure objectives should be:

1. Identify all employees to be included in the HCP.
2. Provide data that will allow the proper hearing protectors to be selected.
3. Measure actual or representative noise exposures for all employees or job classifications.

A flexible sampling method for the exposure survey is required to provide the most efficient procedure to fulfill the objectives of the general survey. The survey should include:

1. Personal monitoring
2. Area sampling
3. A combination of these two procedures

Personnel Monitoring

A noise dosimeter is the most accurate method of identifying employee noise exposure. Another, but less accurate, method is to use a sound level meter and a stopwatch, then calculate the mixed exposure values. This method can work if the employee is relatively stationary or has an exposure that is simple to define. When the level of exposure is highly variable the accuracy is questionable.

Monitoring with a sound level meter requires that measurements be taken in the employee's hearing zone, about 6 to 12 in. from the ear. A sampling methodology should be developed since the sound level meters are only capable of taking instantaneous measurements. This methodology must include enough readings, taken at various times and locations, to accurately define the TWA. All the varying noise levels encountered by the employee during the shift must be accounted for when taking this type of measurement. The stopwatch is used to measure the time an employee being monitored actually spends at one location or is exposed to a specific sound level. After this information has been collected for the employee's workday, it is then used to calculate the noise exposure (see section on noise exposure calculation). Again, it should be emphasized that unless the noise exposure is simple to define this methodology may lead to erroneous exposure data and could under- or overestimate the real exposure. This method should be used when the noise levels are consistent.

For example, in a power plant study the operating personnel stayed in the lower level, mezzanine, and turbine levels during the whole shift. The noise level in these areas varied about 4 dBA. An average level could be calculated. In the coal material-handling area of the power plant the noise levels were much lower and highly intermittent. Thus, a noise dosimetry study was recommended. This was the most effective way to determine the exposure. On the other hand, all employees could be included in the hearing conservation program after the initial noise survey was completed. Each facility will have its own unique problems to evaluate and the method must be tailored to the individual facility.

The most accurate method available for the assessment of noise exposure is to use the noise dosimeter. This is an electronic device worn by the employee that automatically averages varying sound levels during a given time period.

It is not necessary to sample every employee. The use of job classifications is an efficient way to minimize the time of determining employee noise exposures. This method allows a selected number of individuals from a job classification to be evaluated. The group must engage in a similar kind of work with exposure that corresponds to the noise sources. This type of noise exposure evaluation is used to determine the noise exposure for that job classification work group.

To ensure that the data are representative of exposure—that is, that nine times out of ten the same result will be observed—more than one sample will be required. For example, if the noise is steady all the time the representative individuals should be sampled two or three times, or until the standard deviation around the mean is ± 2 dBA. If, on the other hand, the noise levels are quite variable or the employees move around the plant, then more samples will be required until this criterion has been met.

Basic statistical texts can provide the required information on various sampling methods. The variability of the noise exposure will determine the number of samples required. The purpose of the monitoring is to obtain a confidence level that at least nine times out of ten the exposure level can be predicted with a minimum number of measurements in future years.

Area Sampling

Another method of sampling is area monitoring. A sound level meter is used to determine the overall level at various locations such as column lines or center of bays within the plant. If the levels or exposure are highly variable, noise exposures should be measured with a dosimeter. Area sampling could provide a first approximation of the levels and exposure. If personal monitoring is not feasible then all employees to be included in the HCP can be identified through area sampling. While this is the least accurate method from an exposure standpoint, it is the most conservative. More employees will be included in the HCP. This may increase the cost, but it may be more prudent to pay this cost now rather than have to pay hearing loss claims in 20 years. Hearing loss claims are currently being paid in some industries.

EMPLOYEES IN THE HEARING CONSERVATION PROGRAM

The most conservative method of identifying employees to be included in the HCP is to use the highest levels rather than an average level. For this situation, all employees who work in areas with noise levels at or above 85 dBA (82 dBA for 12-hour shifts) would automatically be included in the program. The maximum area noise levels are also used to evaluate the effectiveness of hearing protectors that will reduce noise in the ear to 85 dBA or less. Methods for evaluating hearing protectors are described in the section entitled "Hearing Protection Devices."

RECORD KEEPING

All noise data of any type must be maintained for 30 years (per federal regulations). The plant medical or safety office, or outside contracting service conducting the audiograms, should be informed about the results of the noise exposure evaluation. The employees in the HCP must be informed about the results of the noise exposure study for their job classification. This should include an explanation and/or interpretation of the results. Some off-job activities have high noise exposures, so the employee and the plant authorities should be knowledgeable about all the noise sources that might contribute to any hearing loss encountered as a result of audiometric testing. This all becomes a part of the education and training program.

EDUCATION AND TRAINING

An important aspect of an effective HCP is a comprehensive employee education and training program. This is also a requirement of the OSHA regulation. Employees will be better motivated to wear their hearing protectors and to actively participate in the HCP when the reasons for having a HCP are clearly explained. Workers will understand the need to protect their hearing.

Annual training is required for all employees included in the HCP. The training must have the following basic elements:

1. The effects of noise on hearing
2. The purpose, advantages, disadvantages, and attenuation characteristics of various types of hearing protectors
3. The selection, fitting, and proper use and care of their protectors
4. The purpose of and procedures for the audiometric tests

For completeness in record keeping, attendees at training sessions should sign a form that acknowledges their attendance and the outline of the material covered. This record should be retained by the employer. In addition, the training materials used (videos, films, slide-tape programs, etc.) and a brief outline of the topics covered should always be kept on file for documentation purposes. The same training materials should not be used every year so that a fresh approach can be used to motivate the employees.

ENGINEERING AND/OR ADMINISTRATIVE CONTROLS

One effective means of reducing employee noise exposure is through the design and installation of engineering controls. This might include quieter equipment, the retrofit of existing equipment, treatment of the noise path, and/or isolation of employees.

Noise Control Engineering Methods

The following are examples of several different approaches to engineering controls:

1. Maintenance
   a. Replacement or adjustment of worn

and loose or unbalanced parts of machines

b. Lubrication of machine parts and use of cutting oils

c. Properly shaped and sharpened cutting tools

2. Substitution of machines
   a. Larger, slower machines instead of smaller, faster ones
   b. Rotating shears instead of square shears
   c. Hydraulic instead of mechanical presses
   d. Belt drives instead of gears
   e. Underwater pelletizers instead of standard pelletizers

3. Substitution of processes
   a. Compression instead of impact riveting
   b. Welding instead of riveting
   c. Hot instead of cold working
   d. Pressing instead of rolling or forging

4. The driving force of vibrating surfaces may be reduced by
   a. Reducing the forces
   b. Minimizing rotational speed
   c. Isolation

5. The response of vibrating surfaces may be reduced by
   a. Damping
   b. Additional support
   c. Increasing the stiffness of the material
   d. Increasing the mass of vibrating members
   e. Changing size to change resonant frequency

6. The sound radiation from the vibrating surfaces can be reduced by
   a. Reducing the radiating area
   b. Reducing overall size
   c. Perforating surfaces

7. Reduce sound transmission through solids by using
   a. Flexible mountings
   b. Flexible sections in pipe runs
   c. Flexible shaft couplings
   d. Expansion joint sections in ducts
   e. High transmission loss floating floors

8. Reduce sound produced by gas flow by
   a. Using intake and exhaust mufflers
   b. Using fan blades designed to reduce turbulence
   c. Using large, low-speed fans in place of smaller, high-speed fans
   d. Reducing velocity of fluid flow (air) when practical
   e. Increasing cross sections of streams
   f. Reducing the pressure
   g. Reducing air turbulence

9. Reduce noise by reducing its transmission through air by
   a. Using sound-absorptive material on walls and ceiling in work area
   b. Using sound absorption along the transmission path
   c. Completely enclosing individual machines
   d. Using acoustical absorbing barriers and baffles
   e. Confining high-noise machines to insulated rooms

10. Isolating the operator by providing a relatively soundproof booth for the operator or attendant

Engineering controls should be carefully considered and their feasibility determined. Management should take into consideration the existing technology, economic factors, benefit, maintenance, production, and practicality when evaluating the design of engineering controls.

The approach outlined in Chapter 1 provides a methodology for setting up a noise management program. An important factor to consider is the existing noise environment. This can be determined by a comprehensive survey of work exposure and area sound levels previously discussed. All of this should work together with HCP.

Any project, whether new or additions to existing facilities, should include noise at the planning stage. Purchase orders should include noise specifications (see Chapter 3) to obtain quiet equipment and systems. Vendors supplying machinery and equipment must be advised that specified low noise levels will be a factor in the selection process. Suppliers should be asked to provide information on the noise levels of their equipment. If the levels are not in accordance with the requirements, ask about the premium to achieve the noise goals specified. Noise control feasibility requires careful, objective analysis from a practical and economic standpoint.

Employees' noise exposure can also be reduced through administrative measures. This is a control, using administrative decisions, that reduces noise exposure for employees. It may work in specialized cases.

Administrative Controls

The following are some examples of different types of administrative controls:

1. Rearrange work schedules so employees will work a major portion of the shift at or near levels that will not overexpose them for the total shift.
2. Rotate employees and job scheduling to keep individual noise exposure within permissible time limits.
3. Rearrange production schedules to run portions of noisy jobs each day rather than all day to reduce noise exposure.
4. Operate occasional high-noise-producing operations at night when a minimum number of employees will be exposed.

While these controls will reduce exposure, they are not usually workable for a variety of reasons. In the past 20 years the author has had only two occasions when administrative controls really worked as designed. Cross-training, union rules, plant environment, and many other reasons may not allow this approach to function as well as it might seem.

HEARING PROTECTION DEVICES

OSHA recommends hearing protection devices (HPDs) for employees above the action level (TWA = 85 dBA). They are mandatory for all employees above 90 dBA TWA. The employer may elect, as a matter of policy, to require HPDs for all employees exposed to noise above the action level or whatever lower levels the company deems suitable to protect the employees.

There are situations when engineering and/or administrative control measures are neither feasible nor adequate, or they are in the process of being implemented. In the latter cases, HPDs are required until employee noise exposures can be reduced to a safe and acceptable level. Employers should make HPDs available to all employees exposed at or above 50 percent of the NEL (action level) at no cost to the employees.

HPDs should be replaced free of charge when necessary (and within reason) at no direct cost to the employee. Employees should have available "a variety, two or more" [OSHA, 1974] of models of HPDs when making their selection. However, the employees are not permitted to select the size of HPD they desire. Proper fitting of HPDs must be done by an appropriately trained individual (e.g., nurse, physician, or safety engineer). Employees should be trained in the use and care of their HPDs. For a hearing conservation program to be effective, workers should be supervised on the job to ensure that they continue to wear their HPDs correctly.

The use of HPDs is required whenever any of the following exist:

1. Employees with 50 percent of the NEL (85 dBA for an eight-hour workday) who have experienced a standard threshold shift (STS).
2. Employees with 100 percent of the NEL (90 dBA for an eight-hour workday).
3. All areas designated by the employer as "hearing protection required," regardless of the amount of time spent in that area.
4. A job activity that requires exposure to high noise levels that have not been measured. Some examples of this type of work are temporary construction, various maintenance projects, work using air hammers, or generator tests.

Management is responsible to ensure that all areas requiring the use of HPDs are properly designated with signs. When signs are used to designate an area, the following wording is recommended: "HIGH NOISE AREA, HEARING PROTECTION REQUIRED." These signs are available from safety equipment suppliers and are usually black letters on a yellow background.

An important factor in establishing an effective HCP is the requirement by management that HPDs not only be worn, but that they be worn properly. Management can set the proper example by wearing HPDs in designated areas even though they may not be overexposed during their short visits to the area.

Another critical element of a successful program is the proper selection of HPDs compatible for the specific noise environments where employees are located. For example, working around jet aircraft may require an earplug and an earmuff worn in series to provide the proper protection. Properly trained individuals can assess these situations and determine the proper HPD.

The basis of selecting HPDs should be, first, that it is worn; and second, that it will properly attenuate the noise to reduce an employee's noise exposure, or TWA, to 85 dBA or less. The most convenient method is to use the noise reduction rating (NRR) developed by the Environmental Protection Agency. When using the NRR, the Hearing Conservation Amendment to the Noise Standard (29 CFR 1910.95 (c)) mandates the use of one of the following methods for assessing HPD adequacy:

1. When using a dosimeter that is capable of C-weighted measurements
   a. Obtain the employee's C-weighted dose for the entire work shift and convert to a TWA, then
   b. Subtract the NRR from the C-weighted TWA to obtain the estimated A-weighted TWA under the HPD
2. When using a dosimeter that is capable of A-weighted measurements
   a. Convert the A-weighted dose to a TWA
   b. Subtract 4 dB from the NRR, then
   c. Subtract the remainder from the A-weighted TWA to obtain the estimated A-weighted TWA under the HPD
3. When using a sound level meter set to the A-weighted network
   a. Obtain the employee's A-weighted TWA
   b. Subtract 7 dB from the NRR, then
   c. Subtract the remainder from the A-weighted TWA to obtain the estimated TWA under the HPD

Note: It is important to remember that the calculated attenuation values used to determine the NRR reflect realistic values only when the HPDs are properly fitted and worn. Studies have shown that the actual noise reduction can be half of the NRR value. This can be used to determine the real effectiveness of the HPD if one wants to be conservative in the assessment.

As an alternative to using the NRR, the regulation allows employers to evaluate the adequacy of HPD attenuation by using one of the three methods developed by NIOSH.

AUDIOMETRIC TESTING

Audiometric testing will determine the overall effectiveness of the HCP. Since the objective of the HCP is to prevent occupational hearing loss, the only mechanism available for measuring its success is to conduct annual audiograms and analyze the results. Analysis of the data provides several checks for the HCP effectiveness:

1. Detecting STS in employees' hearing levels that is work related
2. Providing a record of an employee's hearing level on an annual basis
3. Strengthening the hearing protection program, by identifying weaknesses such as inadequate hearing protectors, lack of proper HPD use, and/or effective education and training of employees
4. Identifying plant areas requiring an engineering noise control study
5. Evaluating the effectiveness of engineering noise control measures by measuring

the hearing threshold of employees working near the treated equipment

6. Helping to provide justification for noise control expenditures

The following are recommendations for periodic audiometric testing (more detailed information is found in the OSHA regulation):

1. All employees included in the HCP exposed to noise at or above 50 percent of the NEL should have a baseline audiometric test. For reference see Tables 5.1 and 5.2. The baseline audiogram is a reference audiogram against which future hearing tests are compared for hearing conservation purposes.

2. All employees included in the HCP should have a periodic test at least annually after obtaining a baseline audiogram.

3. All employees should be away from workplace noise for at least 14 hours prior to their audiometric test. If the audiometric tests are to be conducted during the employees' work shift, then each employee to be tested should be instructed at the beginning of the work shift to wear hearing protection at least up to the time of the hearing test.

4. The medical professional responsible for the supervision of the audiometric testing program determines the follow-up procedures necessary whenever an individual employee shows an STS. Changes in hearing acuity that exceed an average of 10 dB or more at 2000, 3000, and 4000 Hz in either ear, relative to the baseline audiogram, are considered to be a standard threshold shift (STS).

5. When deemed practical, all employees included in the HCP should have an audiometric examination prior to leaving or retiring from the company, if their last audiometric test preceded the departure date by more than six months.

Audiometric measurements should be made with an audiometer that conforms to the requirements for wide-range, pure-tone, discrete-frequency audiometers prescribed by the American National Standard Specifications for Audiometers, ANSI S3.6-1969 (R-1973 or latest revision). If a pulsed-tone audiometer is used, the "on" time of the tone should be at least 200 msec. The instrument used should be either a manual audiometer or any other audiometer testing system of equal or greater accuracy and effectiveness.

Audiometer calibration should be checked acoustically at least annually to determine that the audiometer is within the tolerance permitted by ANSI S3.6-1969 (R-1973 or latest revision). This procedure should only be attempted by a properly trained and equipped individual. The audiometer manufacturer can provide this calibration service.

A biological calibration should be made prior to each day's use of the audiometer. This procedure consists of:

1. Testing at least one person having a known stable audiometric curve that does not exceed 10 dB hearing threshold level at any frequency and comparing the test results with the known curve

2. Registering the subject's response to distortions and/or unwanted sounds from the audiometer

If the audiometer does not pass the biological calibration (whenever the results of the "daily use" biological calibration indicate hearing level differences greater than ±5 dB at any frequency) it should be removed from service and sent out for an acoustical calibration prior to further testing. Other problems could be that the signal is distorted or there are attenuator or tone switch transients (e.g., clicks, noises, hums). Only after the problem with the audiometer is corrected to within permitted tolerances can it be put back into service.

The area designated for audiometric testing must meet the ambient noise criteria required by the standard. The sound pressure level in any octave band when measured in the au-

TABLE 5.3   Maximum Allowable Background Noise Levels

Octave Band Center Frequency (Hz)	Sound Pressure Level in Decibels (Re 0.0002 N/m²)
500	40
1000	40
2000	47
4000	57
8000	62

Source: OSHA CFR1910.95 Occupational Noise Standard Appendix D, Table D.1.

diometric booth or test room where subjects are actually tested should not exceed the values in Table 5.3. Audiometer calibration records and background noise levels in test booths or rooms should be maintained with the audiometric test results. When contract services are used, these records must be provided by them and be kept on file.

NOISE EXPOSURE CALCULATIONS

This section will provide some examples of noise exposure calculations. These are usually straightforward once the noise measurements and duration times have been recorded. The first step is to calculate the noise dose (D) in percentage of the NEL using the expression:

$$\% \ D = 100(C_1/T_1 + C_2/T_2 + \cdots + C_n/T_n)$$

Where the terms C_1 through C_n indicate the total time a worker is exposed to a specific noise level. The T_1 through T_n terms are the reference duration times for each noise level as given in Table 5.1. When the percent noise dose has been calculated, the next step is to look up the eight-hour, time-weighted, average exposure in dBA from Table 5.2. The following examples demonstrate the use of the noise dose equation.

Example 1: A group of industrial employ-

ees is exposed to continuous noise according to the following schedule:

Exposure level (dBA)	Time or duration of exposure (hr)
85	3
90	2
92	1
95	2

Problem: Is this group of employees overexposed according to noise regulations? What is their level of exposure in dBA?

Answer: Using the equation

$$D = C_1/T_1 + C_2/T_2 + \cdots + C_n/T_n$$

their dosage can be determined.

In this problem, the workers are exposed to 85 dBA for 3 hours ($C_1 = 3$), 90 dBA for 2 hours ($C_2 = 2$), 92 dBA for 1 hour ($C_3 = 1$), and 95 dBA for 2 hours ($C_4 = 2$).

From Table 5.1, it is seen that the permissible exposure time for 85 dBA is 16 hours ($T_1 = 16$), 90 dBA is 8 hours ($T_2 = 8$), 92 dBA is 6.2 hours ($T_3 = 6.2$), and 95 dBA is 4 hours ($T_4 = 4$). To calculate their noise dose, we simply substitute these numbers into the above equation and add up the fractions:

$$D = (3/16 + 2/8 + 1/6.2 + 2/4)$$
$$D = 1.0988$$

Since D exceeds unity—1—this value is in excess of the permissible levels, and thus these workers are overexposed.

To determine the level of exposure, convert 1.0988 into a percentage (% D):

$$1.0988 \times (100\%) = 109.88\%,$$

or approximately 110%

Next, using Table 5.2, look up the value for 110%: at 110% the sound level is 90.7 dBA. (*Note:* This value is often referred to as an eight-hour, time-weighted, average (TWA) noise exposure in dBA.) This TWA can be calculated using stats following equation where D = % Dose calculated or measured.

TWA noise exposure

$$= 16.61 \log_{10} (D/100) + 90, \text{dBA}$$

$$= 16.61 \log_{10} (1.1) + 90 \text{ dA}$$

$$= 16.61(0.0414) + 90 \text{ dA}$$

$$= 0.7 + 90 \text{ dA}$$

$$= 90.7 \text{ dBA}$$

Example 2: One day, an operator at a plant spends 6 hours and 20 minutes of his 8-hour shift in a soundproof control room where the sound level is 70 dBA. Because of a problem with a compressor, it is necessary for him to spend 1 hour and 40 minutes out in the plant working on the compressor. While working in the plant, he is exposed to a continuous sound level of 107 dBA.

Problem: For this operator's 8-hour workday, is he overexposed according to the workplace noise standard?

Answer: From Table 5.1, we see that no time exposure is indicated for any values less than 80 dBA. We can assume that any sound levels less than 80 dBA will contribute a negligible percentage to the total noise dose; therefore, we can ignore adding into the equation exposure times for any sound levels less than 80 dBA.

Next, we know that this operator is exposed to 107 dBA for 1 hour and 40 minutes (1 hr and 40 min = 1 + 40/60 = 1.67 hr). Thus, C_1 = 1.67 and from Table 5.1 we determine T_1 = 0.76 hours for a sound level of 107 dBA. Using the equation for noise dose, we have:

$$C_1/T_1 = 1.67/0.76$$

$$= 2.193 \quad \text{or} \quad 219.3\%$$

Obviously, this operator is overexposed despite a relatively short time spent in the plant when compared to his total workday.

From Table 5.2, we find that 219% = 95.7 dBA or, to be more precise:

TWA noise exposure

$$= 16.61 \log_{10} (219/100) + 90 \text{ dBA}$$

$$= 95.65 \text{ dBA}$$

References

ANSI. 1991. Draft American National Standard, *Evaluating the Effectiveness of Hearing Conservation Programs*. Draft ANSI S12.13-1991.

Fann, Michael. 1990. Cutting Through the Clutter. *Experts-at-Law*, September–October, pp. 60–61.

OSHA. 1983. Occupational Noise Exposure Regulation. U.S. Federal Register (39FR37773), rc:Part 29CFR1910.95.

Part II

Terminology, Criteria, Measurement, and Instrumentation

Chapter 6

Basic Terminology and Physics of Sound

INTRODUCTION

In this chapter the basic terminology of sound will be discussed to form a basis of understanding of the concepts in this field. There are many good and detailed texts on sound, acoustics, and noise that will provide the derivation of equations related to this field. This chapter is intended to provide the overall concepts to help visualize and better understand what is heard in any given environment, be it a recital hall, a busy street, or a factory. By definition *noise* is unwanted sound. Whether it is sound or noise depends on the listener. For example, the quiet sound of a dripping faucet may be a disturbing noise when you are trying to fall asleep. Likewise, the siren of a passing ambulance may be a very loud noise as you stand on the street corner, but when it is coming to your aid it is a very pleasant sound indeed. Sound may also be described as a disturbance propagating through a physical medium such as air, water, wood, or steel. As sound travels through the medium it causes it to vibrate. This vibration of the molecules in the medium creates a minute pressure change above atmospheric pressure that is sensed by the ear. This is a sound pressure and, when measured by a transducer (microphone), is called the sound pressure level (L_p).

SOUND WAVE CHARACTERISTICS

In order to better understand the generation and propagation of sound in a medium—here we will use air as the medium—visualize the following concept: noise source → noise path → noise receiver. The noise is generated at the source through some mechanism. The source vibrates and causes the air molecules to vibrate next to the source. These air molecules transmit the vibration through the air as it reaches the receiver—the ear. This transmission in turn causes the ear to vibrate by the very slight increase in sound pressure of the air above atmospheric pressure. The ear then transforms the vibrations into nerve impulses that are perceived by the brain and interpreted as sound or noise.

The sound waves spread out from a source in an idealized situation as spherical waves much like the circular ripples caused when a pebble is dropped into a still pond as shown in Fig. 6.1. If a balloon is blown up, as shown in Fig. 6.2, the force of the air is contained by the balloon's surface. This is a spherical surface that contains air pressure above atmospheric pressure. As the balloon is ruptured there is a sudden release of energy in the form of a compression wave that spreads out spherically

FIGURE 6.1   Pebble dropped into a pond.

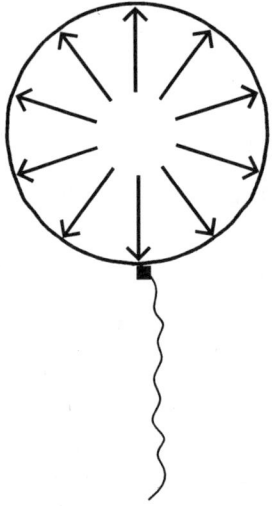

FIGURE 6.2   Balloon with interior pressure.

respectively. This is analogous to a sonic boom but at a much lower pressure.

In order to define sound further in terms of a vibrating medium we should define frequency. This is the number of vibrations or cycles per second the air particles make as they travel in both positive and negative directions from their normal position. A cross section of the sound wave in Fig. 6.4 is shown in Fig. 6.5.

The unit of frequency is Hertz (Hz). The air particles are compressing and expanding around their normal position of atmospheric pressure. The frequency is then directly related to the number of vibrations per second the air particles make. Mathematically, frequency is $f = 1/T$, where T is the time period, in seconds, required to complete a cycle of vibration. The distance a particle moves from atmospheric pressure is called the amplitude. The amplitude will then determine the intensity of the sound or its loudness. The total distance a particle moves, both positively and negatively, around its normal position is called the wavelength and is described in feet. The expression is:

$$\lambda = \frac{c}{f}$$

where:

λ = wavelength
f = frequency (Hz)
c = speed of sound (fps)

in an ever-expanding form until it reaches a boundary. This is shown diagrammatically in Fig. 6.3. The air next to the balloon, which is initially at equilibrium, is suddenly pushed outward with great force and the adjacent molecules are suddenly compressed together, and they, in turn are compressed into adjacent molecules and so forth. This is illustrated in Fig. 6.4. The compression is perceived by the ear as a "pop" when it passes by. If one were able to have the ear at several locations, or within the traveling compression wave, one would hear a series of "pops" or a continuous noise,

This discussion will help visualize the concepts of sound, frequency, period, amplitude, and wavelength. In general, the components of sound are mixed complex waves, as depicted in the graph at the bottom of Fig. 6.6. In general, the sounds that reach human ears have an infinite number of frequency components. The normal human ear is sensitive only to the components in the range from about 20 to 20,000 Hz shown in Fig. 6.7. The ear does not "hear" frequencies that fall outside this range.

The audible spectrum can be divided into various parts for analysis. This is accomplished

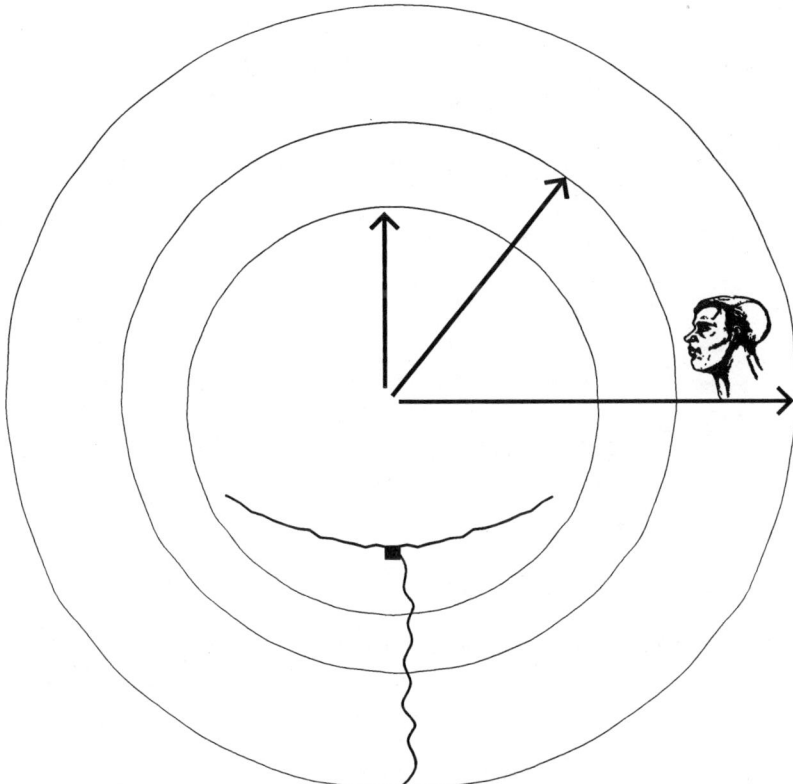

FIGURE 6.3   Balloon bursting.

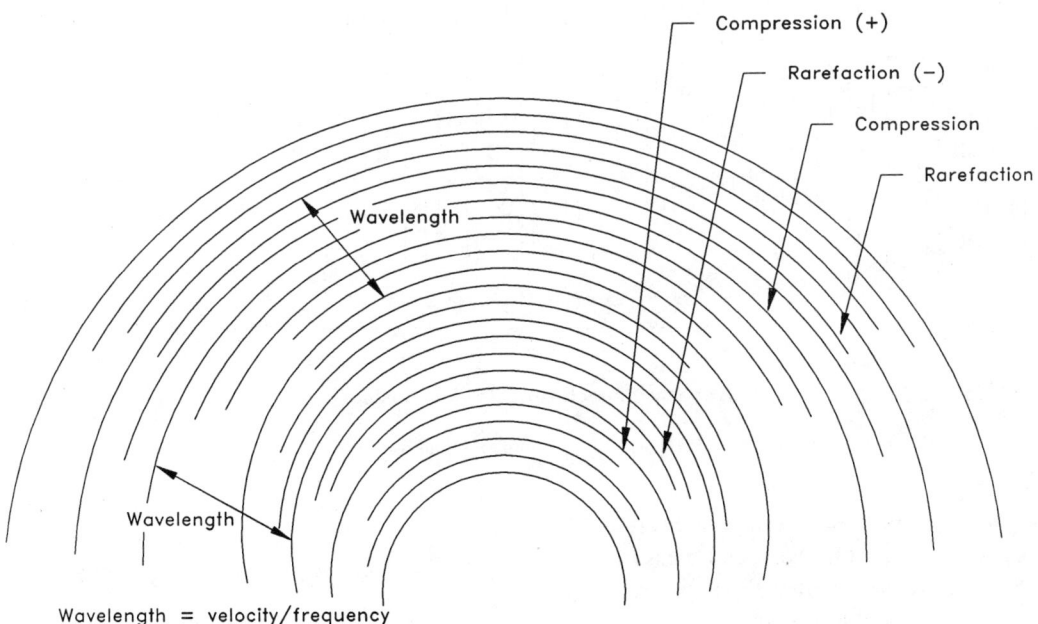

FIGURE 6.4   Sound waves traveling out from balloon bursting noise source.

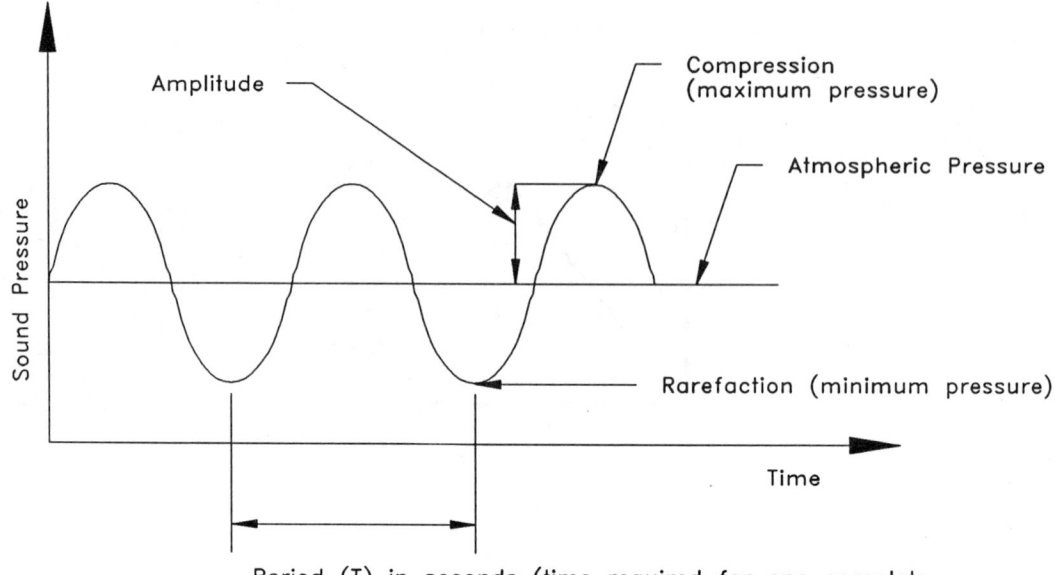

FIGURE 6.5   Cross section of a sound wave.

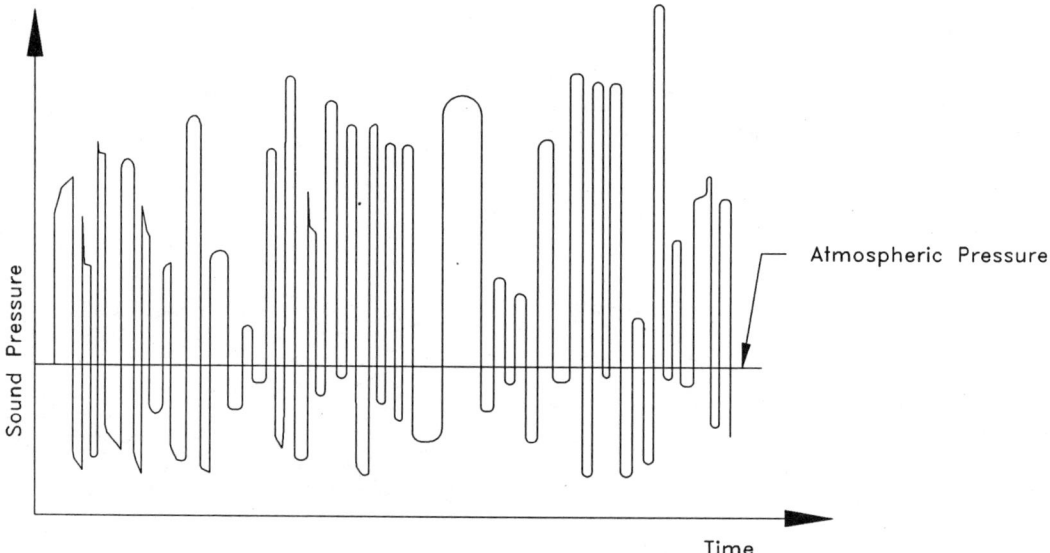

Variation in pressure caused by speech, music, or noise (complex sound)

FIGURE 6.6   Complex sound wave.

with an appropriate spectrum analyzer. The analyzer will measure the sound pressure level (L_p) for the various parts of the spectrum. For example, the audible spectrum can be divided into nine contiguous parts called octave bands. The center frequencies (Hz) of these octave bands are:

31.5   63   125   250   1,000   2,000

4,000   8,000   16,000

In the simplest form the analyzer will measure

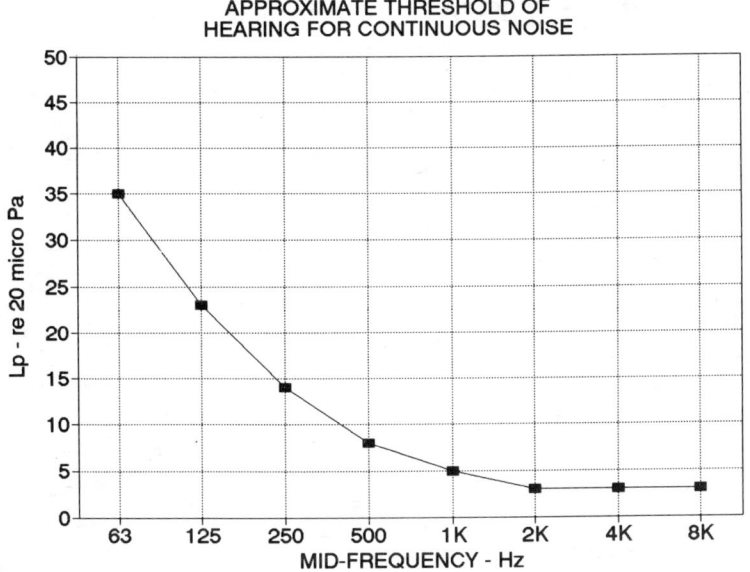

FIGURE 6.7   Threshold of hearing and feeling.   (After Noise Criteria Chart, reproduced by permission from the American Society of Heating, Refrigeration and Air Conditioning Engineers, from 1991 ASHRAE *Handbook*—HVAC Applications.)

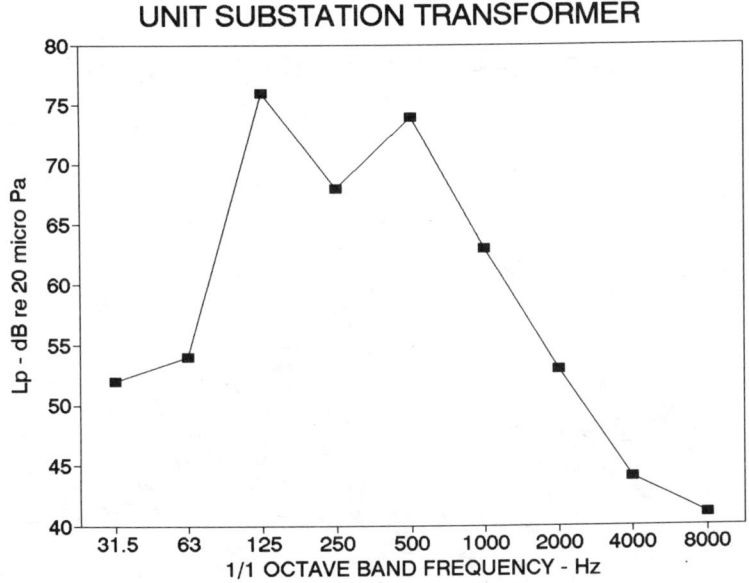

FIGURE 6.8   Electrical transformer substation—octave band spectrum.

the (L_p) in each of these octave bands. These can be plotted on a chart as shown in Fig. 6.8, which shows a spectrum of an electrical transformer substation located adjacent to occupied space in a large office building.

For more detailed information about the noise source the analyzer can divide the octave bands into three parts or one-third octave bands. The same spectrum as shown in Fig. 6.8 is plotted on Fig. 6.9 in one-third octave bands. This provides more detail about the noise source. Most standard sound level meters can

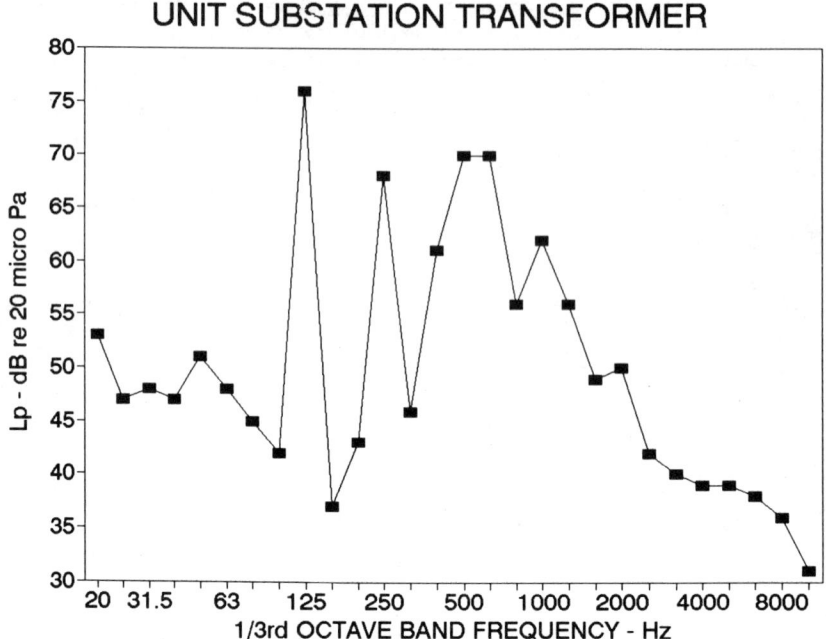

FIGURE 6.9   Electrical transformer substation—one-third octave band spectrum.

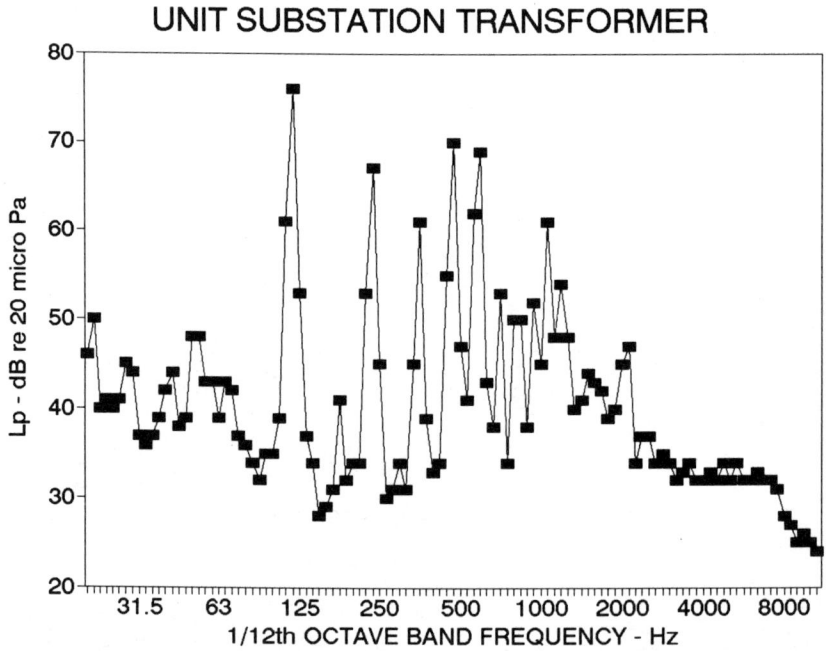

FIGURE 6.10   Electrical transformer substation—one-twelfth octave band spectrum.

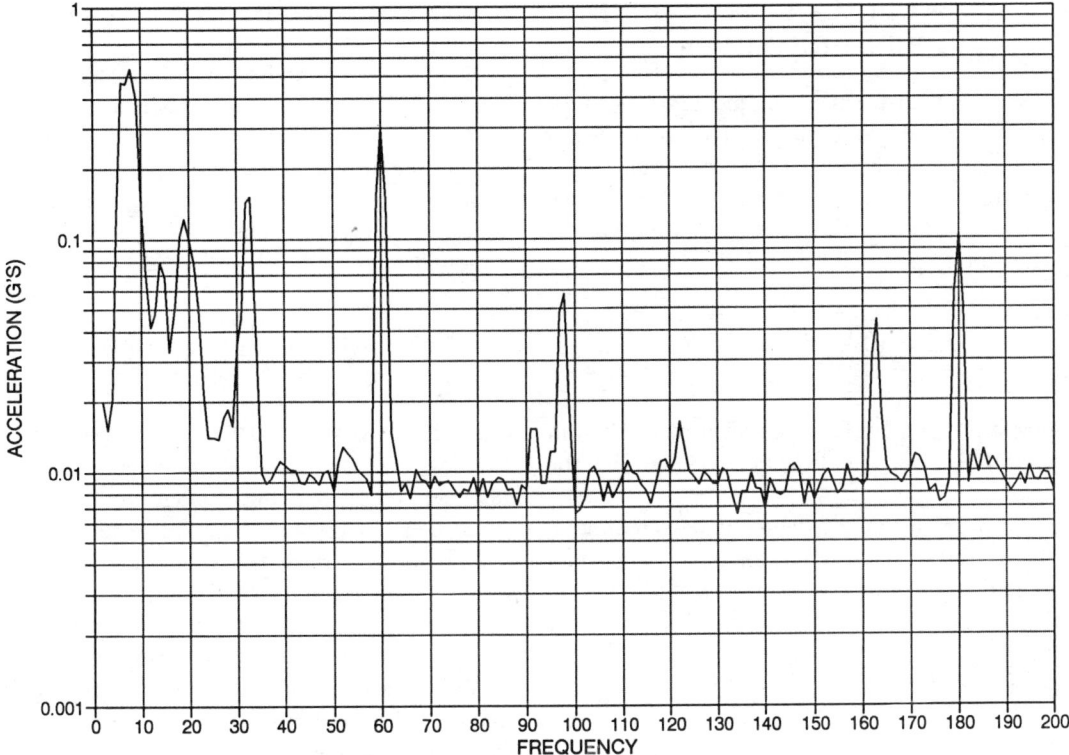

FIGURE 6.11   Electrical transformer substation—narrow-band acceleration vibration spectrum.

be purchased with the octave and one-third octave filters.

Still more information can be obtained if the octave band can be divided into smaller pieces. Specialized analyzers can be purchased that will divide the octave into 12 parts. A one-twelfth octave band spectrum of the same source is shown in Fig. 6.10. Even more detail about the spectrum can be obtained with a narrow-band spectrum. In Fig. 6.11 a vibration spectrum is illustrated with a portion of the narrow-band spectrum as measured on the partition adjacent to the substation. More detailed discussion of this equipment can be found in the references and instrument manufacturers' literature. The purpose of this discussion has been to illustrate the variety of data that can be developed to assist in solving a problem.

DECIBELS

The usual method of expressing sound pressures and sound powers is with decibel levels.

This method allows a logarithmic rather than a linear scale to be used. The range of sound pressure can vary from 3×10^{-9} pound per square inch (psi), the normal threshold of hearing, to 3×10^{-2} psi, the threshold of pain. In terms of decibels this ranges from 0 to 130 decibels (dB). The term pound per square inch (psi) is given above to illustrate a point. This might be easier to understand, initially, than the usual metric notation convention of 20 micro Pascal (Pa). There is an advantage in using small numbers to carry out the calculations rather than very large numbers as indicated above.

The decibel is well suited to the field of acoustics. It was originated in the field of telecommunications many years ago. The decibel is a logarithm to the base ten of the ratio of a quantity referenced to a base reference quantity as shown in the following equation:

$$dB = 20 \log_{10} \left(\frac{pressure}{reference\ pressure} \right)$$

SOUND POWER AND SOUND PRESSURE LEVELS

The energy that causes the air particles to vibrate is called sound power. The sound power level describes the acoustical power radiated by a given source with respect to the international reference of 10^{-12} watt. The equation that defines sound power level (L_w) is:

$$L_w = 10 \log \left(\frac{W}{W_{re}} \right) \quad \text{(dB)}$$

where W_{re} is the reference power of 10^{-12} watt and W is the sound power radiated by the source.

In order to have a better understanding of this process consider the example of a small siren that generates 0.1 watt of sound power:

$$L_w = 10 \log \left(\frac{W}{W_{re}} \right) \quad \text{(dB)}$$

$$= 10 \log \left(\frac{0.1}{10^{-12}} \right) = 10 \log (10^{11})$$

$$= 10 \times 11 = 110 \text{ dB}$$

This illustrates very clearly that a very small amount of acoustical power, 0.1 watt, results in a very loud sound source.

Even though sound power levels cannot be measured directly, as the acoustical power is emitted by a source, the air particles compress and expand around atmospheric pressure; thus, what are actually measured with a sound level meter are changes in sound pressure.

Sound pressure level is defined as:

$$L_p = 10 \log \left(\frac{P}{P_{re}} \right)^2 \quad \text{(dB)}$$

where P_{re} is 20×10^{-6} Pascal, or 20 micro Pa. This level was chosen as a reference because it has been found that the average young adult can perceive a 1000-Hz tone at this reference.

In practical terms, just what is the relationship between L_p and L_w? Before an analogy is considered, note that in the previous equation the L_p, or sound pressure level, is proportional to the logarithm of the rates of pressures squared. Therefore, both sound power level and sound pressure level are associated with sound power. This relationship can be expressed as:

$$L_p = L_w + K$$

The value K is a constant value that is a func-

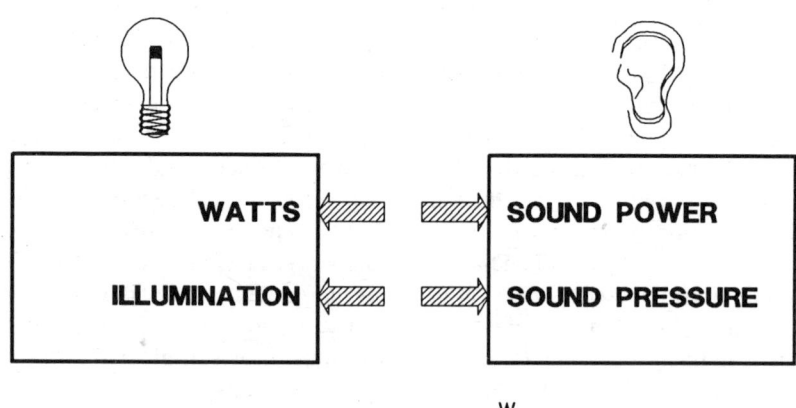

$$L_W = 10 \ \text{LOG}_{10} \ \frac{W}{W_{re}}$$

$$L_P = 10 \ \text{LOG}_{10} \ \frac{P}{P_{re}}$$

FIGURE 6.12   Analogy of sound power and sound pressure.

tion of the environment in which the sound source is located.

To visualize this, consider an ordinary 100-watt electric light bulb that is placed in a room painted with white paint. The room will seem very bright. If the same bulb is placed in a room painted flat black, what is the result? The bulb seems dimmer. The light output of the bulb did not change, but the way the light rays were reflected did. The resulting footcandles changed from room to room, but the lumen output remained constant (see Fig. 6.12). This analogy will help you understand the concept as it applies to acoustics and sound waves.

For example, a 750-hp diesel engine located in a pump room with hard reflective walls will produce a higher sound pressure level at 10 ft than if the same engine were placed on a pad out-of-doors. Inside, the sound is easily reflected by the hard walls; often it is said to be bright. In the outdoor environment the sound is not reflected, but goes out into space. If the building, depending on its size, were lined with acoustical material, part of the sound would be absorbed much like the black paint on the wall of the room described above. The engine sound power has not changed; it is the environment (K value in previous equation) that has changed and affects the resulting sound pressure level. More will be said about this later in the chapter.

DECIBEL ADDITION

Since we are immersed in a "sea" of sound (noise), it is rare when an individual is exposed to only one noise source. Similarly, when a new piece of equipment is installed, or a new facility is built, there are several noise sources that contribute to the overall or total noise level. Therefore, it is important to be able to calculate using decibel addition the impact one noise source will have on the overall noise level. The following material shows a method of decibel addition using logarithms.

Equation method—Adding two or more sound pressure levels:

$$L_p \text{ (total)} = 10 \log \left(\sum_{i=1}^{n} 10^{L_{pi}/10} \right) \quad \text{(dB)}$$

TABLE 6.1   Decibel Addition Chart

Numerical Difference Between L_1 and L_2	L_3 Amount Added to the Higher of L_1 and L_2
0–1	3
2–4	2
4–7	1
7–10	0.5
>10	0

Note: Step 1: Determine the difference between the two levels to be added (L_1 and L_2).

Step 2: Find the number L_3 corresponding to this difference in the table.

Step 3: Add the number L_3 to the highest of L_1 and L_3 to obtain the resultant level: $L_r = (L_1 \text{ or } L_2) + L_3$.

Because noise levels vary so widely, noise measurements are really estimates, and it is recommended that the decibel levels be rounded to the nearest whole number.

Table and graph method—An alternate method to add decibels is to use a chart such as Table 6.1 or the graph in Fig. 6.13. This will be slightly less accurate but should suffice for most applications. The method works as illustrated in Table 6.1. In many cases it is necessary to subtract ambient levels to obtain the actual levels of a source without the effect of the background. Keep in mind that meaningful data cannot be obtained unless the background noise or ambient is at least 3 dB below the level of the source. If it is desired to calculate the levels, use the following equation:

$$L_p = 10 \log (10^{L_{pa}/10} - 10^{L_{pb}/10}) \quad \text{(dB)}$$

Example using the table method: Use Table 6.1 to determine the overall or total level from $L_{p1} = 80$, $L_{p2} = 85$, and $L_{p3} = 87$ (dB).
Solution:

$$\left. \begin{array}{l} 85 \\ \quad\ +1 = 86 \\ 80 \\ 87 \qquad\qquad +3 \end{array} \right| = 90 \text{ dB} = L_p \text{ (total)}$$

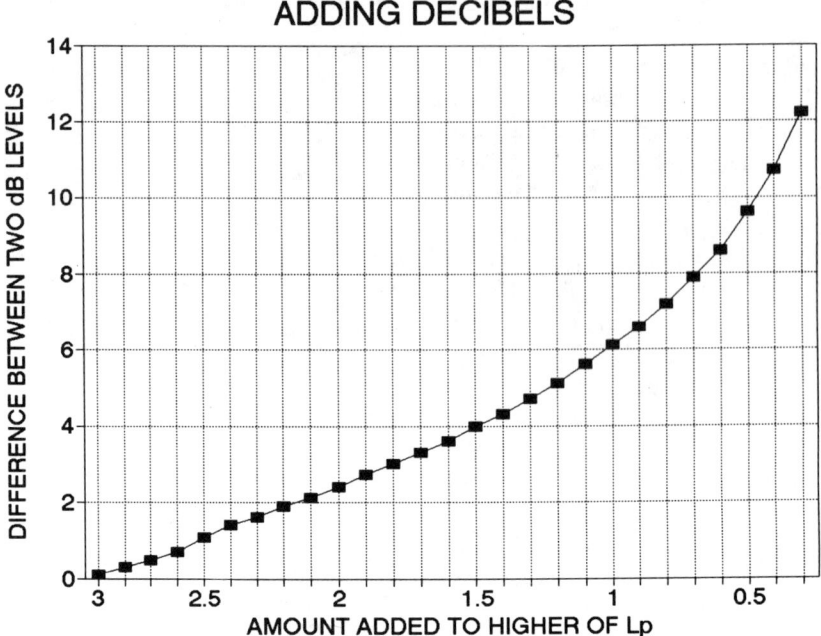

FIGURE 6.13   Adding decibels.

A-WEIGHTED SOUND LEVELS

As acoustics has evolved, various techniques have been developed to describe noise environments, the subjective response, and how the human ear might be damaged with respect to exposure to noise. Weighting curves were devised for this purpose. They were placed within the sound level meter so that they could be electronically switched and used to weight the overall level being measured. The most notable of these curves is the "A-weighting" curve. This has been found to describe the subjective or human response to noise and annoyance, and how the ear might be damaged by noise. The usual designation is dBA. In terms of level notation it is L with a subscript a, or L_a.

Table 6.2 shows the values of that are subtracted from or added to each center-frequency octave band for A weighting [ANSI, 1985]. Note the increasingly larger amount that is subtracted from the lower frequencies. The ear is less sensitive in terms of annoyance and damage. To be sure, if the level is high enough there will be annoyance or damage. The spectrum in Fig. 6.14 is an A-weighted spectrum of the spectrum originally shown in Fig. 6.8.

TABLE 6.2   Random Incidence Relative Response Level as a Function of Frequency

Frequency (Hz)	A Weighting (dB)	C Weighting (dB)
31.5	−39	−3
63	−26	−1
125	−16	0
250	−9	0
500	−3	0
1000	0	0
2000	+1	0
4000	+1	−1
8000	−1	−3

Source: This material from American National Standard S1.4 1983 *Specifications for Sound Level Meters* Copyright 1983 is reprinted with the permission of the Acoustical Society of America. Copies of this standard may be purchased from the Standards Secretariat, Acoustical Society of America, 335 East 45th Street, New York, NY 10017-3483. No further printing without prior written permission of the Acoustical Society of America.

The other values shown in the table pertain to C weighting. This has a much flatter response with very little weighting in the low frequencies. Most sound level meters have at least these two weighting networks and some have a linear; that is, there is no weighting and all frequencies are measured equally. If one has data

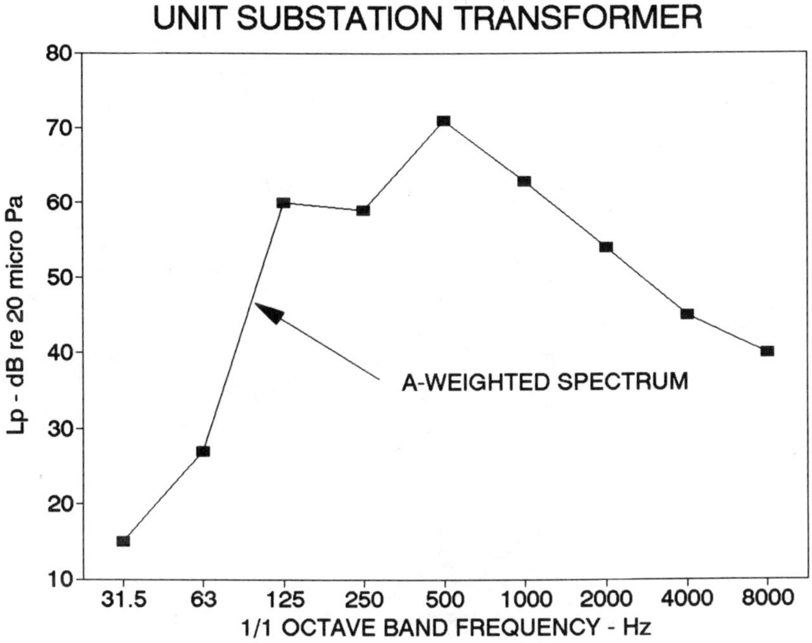

UNIT SUBSTATION TRANSFORMER

FIGURE 6.14   A-weighted octave band spectrum of Fig. 6.8.

reported from measurements in these three weightings, conclusions can be drawn about the character of the noise in broad terms, that is, low frequency, middle frequency, or high frequency. Also, knowing the type of noise source will help confirm more about the data.

DIRECTIVITY FACTOR

All noise sources will have a certain directivity factor (Q), in which the sound waves are radiated. Assume that a machine radiates a certain sound power, which in turn produces a sound pressure. As the sound pressure waves radiate away from the center of this machine, the machine, in most cases, is a point source and the sound waves radiate in a full spherical pattern. For most industrial applications, Q can be simplified and expressed as one of the patterns depicted in Fig. 6.15.

SOUND FIELDS

Sound has been defined as vibrations in air or another medium. When the vibration is created at a point, the energy generated (sound waves) will spread out through the air until it reaches a boundary. Airborne sound can be described by longitudinal waves, the particle velocity of which is in the same direction as the propagation of radiated sound energy. Two definitions can be used to describe this:

Sound source (cause)—fluctuating mechanism causing the disturbance,
Sound field (effect)—the disturbance in the air or medium

The decibel, as discussed previously, is the quantitative measure of the sound in the sound field. If a reasonable measurement of the sound in the field can be made to completely describe the source at various distances, it may then be possible to calculate the sound power being radiated. Keep in mind the previous equation relating the source sound pressure, sound power, and the environment.

The remainder of this chapter will deal with methods of relating noise measurements to source identification and sound power computation. Commonly used terminology will also be introduced describing particular types of

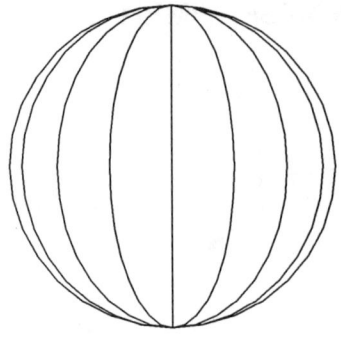

Spherical Radiation
Q = 1

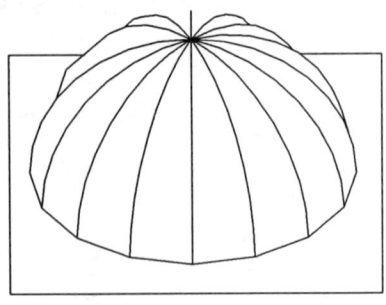

1/2 Spherical Radiation
Q = 2

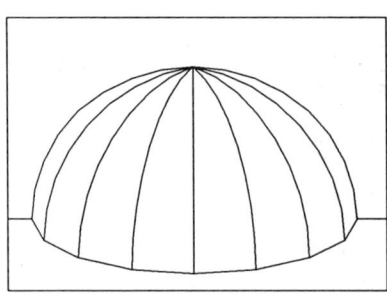

1/4 Spherical Radiation
Q = 4

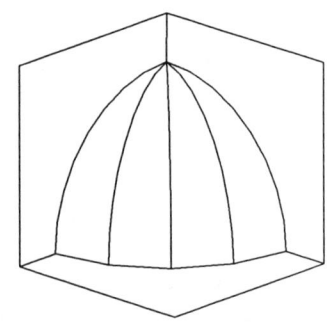

1/8 Spherical Radiation
Q = 8

Q = 1   for free field radiation, and there are no
reflecting surfaces

Q = 2   where the sound source is on the floor
in the center of a large room or on ground
level outdoors

Q = 4   when the source is near the intersection of
the floor and a wall (1/4)

Q = 8   when the source is near the intersection of
the floor and two walls (1/8)

FIGURE 6.15   Directivity.

sound fields. The concepts presented will be of importance to anyone concerned with noise by relating the various sound levels measured from a source, the location of an unknown source, or in predicting resulting sound level caused by a source located in a known environment.

SOUND FIELDS WITHIN AN ENCLOSED SPACE

When a sound source exists in an enclosed space, there are two distinct sound fields that exist at the same time—direct and reverberant.

The direct field is dependent upon the source and distance a receiver may be from the source. The reverberant field is, on the other hand, dependent on the size and reflectiveness of a room's surfaces. The reverberant field is the more complicated. For example, noise sources located outdoors only have a direct-field component since there are no bounding surfaces for reflections.

All surfaces in a space will absorb and reflect sound energy. The common term used to describe a room's absorption characteristic is the absorption coefficient (α). This term is defined as the ratio of the acoustical energy absorbed by the surface to the acoustical energy incident upon the surface when the sound field is perfectly reverberant. The absorption coefficient is also a function of frequency and is provided in octave bands from 125 Hz to 4000 Hz. The absorption coefficient is measured in accordance with ASTM standards in a laboratory reverberation room.

Since various surfaces of a room will have different areas and different absorption coefficients, the average absorption coefficient of the room should be calculated. The general equation is:

$$\overline{\alpha_i} = \frac{\sum\limits_{i=1}^{n} S_i\, \alpha_i}{\sum\limits_{i=1}^{n} S_i}$$

where:

S_i = area of the i_{th} surface of the room (m^2 or ft^2)

α_i = absorption coefficient of the i_{th} surface both α_i and α are frequency dependent

It is often advantageous to be able to determine what portion of the sound pressure level is the result of the energy from the direct field and what portion is from the reverberant field. The latter is the field that is composed of the sound energy that has gone one or more reflections from various surfaces in the space. It is also assumed that this field is diffuse, that is, well dispersed. Furthermore, it should be pointed out that only the reverberant portion of the total sound pressure can be reduced by adding acoustical absorbing materials to surfaces in the space. A cautionary note: merely reducing the reverberant energy may only solve a portion of a noise problem. The direct field, on the other hand, will always exist, even if the source was removed and placed outdoors. Thus the direct field will be dealt with separately.

In order to determine the overall sound pressure level that will exist in a space when a sound source is introduced, it is necessary to know the acoustical properties of the room. One of these properties to be defined is the room constant (R) using the following equation:

$$R = \left(\frac{(S)\overline{\alpha}}{1 - \overline{\alpha}} \right) \qquad \text{(ft}^2 \text{ or m}^2\text{)}$$

The room constant can be calculated using absorption coefficients and areas of each surface. This is done for each octave band.

An increase in the average absorption coefficient increases the room constant. This means that more reflected energy is absorbed and thus the sound levels are reduced. In other words, the higher the R, the lower the reverberation. This is true up to a point. As a rule of thumb, for a hard space with a ceiling height not over 16 to 18 ft, the overall noise can be reduced 4 to 6 dB when 30 to 50 percent of the wall and ceiling surface is covered with an appropriate absorption material such as 4- to 6-lb/ft^3 glass fiber.

Before the sound pressure level, L_p, can be predicted in a space where a new noise source is introduced, the location of the source with respect to the receiver must be known. The distance between these two locations is referred to as r (radius). In an idealized situation, when the L_p is some distance r (ft) from a small point source, the source radiates acoustical energy spherically from the source's center. In addition, the energy being measured at a radius or distance r from this sphere is the direct energy (sound). The L_p at distance r is affected by the directivity factor, Q, described earlier. Therefore, the sound pressure due to the total energy at some radius is:

$$L_p = 10 \log \left(\frac{Q}{4\pi r^2} + \frac{4}{R} \right)$$

(dB, metric units)

$$L_p = 10 \log \left(\frac{Q}{4\pi r^2} + \frac{4}{R} \right) + 10$$

(dB, English units)

The term used to express the direct-energy portion of the total energy is expressed as:

$$\text{Direct field} = \frac{Q}{4\pi r^2}$$

The energy from the reverberant field is a function of the acoustical absorption of the space based on the room constant, expressed as:

$$\text{Reverberant field} = \frac{4}{R}$$

Recall the discussion relating sound power and sound pressure where:

$$L_p = L_w + K$$

The constant or factor K is the effect of the space on the sound source. Thus:

$$K = 10 \log \left(\frac{Q}{4\pi r^2} + \frac{4}{R} \right)$$

So, the overall L_p is:

$$L_p = L_w + 10 \log \left(\frac{Q}{4\pi r^2} + \frac{4}{R} \right) + 10 \quad \text{(dB)}$$

Note that the additional 10 dB is the result of converting from the metric system to the English system of units. This is a fundamental equation in room acoustics.

When a noise source is located in the center of a large room or outdoors, the term $4/R$ in the above equation for the reverberant field ap-

proaches zero. Therefore, the equation becomes:

$$L_p = L_w + 10 \log \left(\frac{Q}{4\pi r^2} \right) + 10 \quad \text{(dB)}$$

which can be rewritten as:

$$L_p = L_w + 10 \log Q - 10 \log (4\pi r^2) + 10$$

(dB)

This expression describes the acoustic field generated by any noise source in a free-field condition. It demonstrates that the sound pressure decreases as a function of distance or radius squared. For every doubling of distance from a common reference point, the sound pressure level decreases 6 dB. Conversely, for every halving of distance the sound pressure level increases by 6 dB. This factor can be useful when attempting to locate a new piece of equipment in a plant without exceeding a predetermined criteria such as a boundary limit noise level.

The condition for the free field only exists when the direction of particle velocity is normal to the face of wave propagation. Hence, all the sound waves are traveling radially outward from the acoustic center, and at a certain distance these sound waves are no longer considered spherical but planar.

A near field exists when the particle velocity is not in the direction of wave propagation. This is analogous to turbulence, in this case, a turbulent acoustic field with the wave amplitude varying in a random fashion. The boundaries of a near field are not precise and the only way to accurately determine them is to measure the L_p. If the L_p does not decrease, or decreases very slowly moving away from the source, the near field likely exists. However, when the L_p decreases in a regular fashion (i.e., 6 dB/ doubling of distance), a region called the far field has begun. In the far field, the particle velocity is primarily in the direction of wave propagation, and the acoustic intensity is related to the mean-square sound pressure. This condition of near field far field occurs indepen-

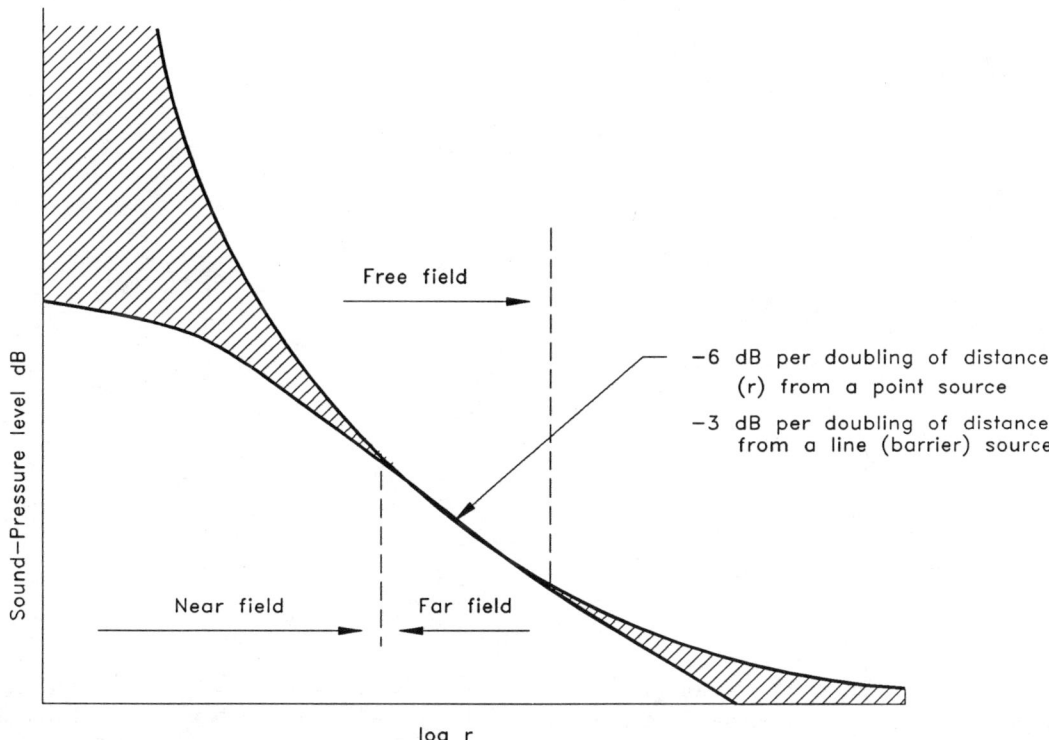

FIGURE 6.16   Variation in sound pressure level with distance from a source.   (Reprinted by permission from D.D.R., Inc.)

dent of surrounding conditions. This will happen regardless of whether the source is placed indoors or outdoors.

The reverberant field, described previously, is said to exist for a sound source located in an environment where the sound waves are reflected back toward the source. The reverberant field is different from the near field. The particle velocity is in the direction of wave propagation, but the wave itself has undergone a reflection at the boundary. When many reflections are present and seem to come from every direction the sound pressure level, as measured or heard, can be quite uniform.

A good way to visualize this is to think about an engine, compressor, or motor located in a small building with hard reflective walls. Noise measurements will reveal that the noise levels will be somewhat uniform throughout the area. There will be higher levels closer to the source, but as the microphone is moved away the noise level tends to stabilize within 2 to 3 dB depending on the size of the space. There will not

be a free field existing since the noise level does not fall off at the prescribed rate of 6 dB per doubling of distance.

These three fields are shown in Fig. 6.16 [Reynolds, 1981]. As the distance from the source increases, the highly variable near field becomes a predictable far field, which becomes a less variable, but predictable, reverberant field. Before measurements are made, the receiver position in the sound field must be known. A great deal can be determined if the measurement field conditions are known; conversely, mistakes can be made if these field conditions are unknown or misunderstood.

References

American National Standards Institute. ANSI S1.4-1983. *American National Standards for Sound Level Meters.* Amended 1985. New York: Secretariat Acoustical Society of America.

Reynolds, D. D. 1981. *Engineering Principles of Acoustics: Noise and Vibration Control.* Boston: Allyn & Bacon.

Chapter 7

Noise Criteria

INTRODUCTION

Consideration of any noise control problem involves not only measurement of the noise sources but determination of the desired noise level that must be achieved when the project is complete. The level that is desired is usually determined by criteria established by a regulating agency. In the industrial setting noise levels are usually compared to some hearing damage risk criteria. These criteria are currently under the regulations established by the Department of Labor and the Occupational Safety and Health Administration (OSHA) in CFR1910.95, Occupational Noise Exposure. A copy is included in Appendix B for reference.

Other criteria might deal with the maximum noise level that an industrial plant can produce at its property line. This level might be established by a city or county ordinance or as part of a zoning requirement. These standards are usually termed environmental criteria if they involve the workplace and the community.

Other criteria that deal with noise generated on the interior of a building, but not related to the workplace or community, are used to quantify communication ability in this environment. Two of these are Preferred Speech Interference Levels (PSIL) and Noise Criteria (NC). These criteria could be used for office settings, laboratories, clean rooms, control rooms, and similar space.

The Occupational Safety and Health Act (OSHA) of 1970 and the Noise Control Act of 1972 under the Environmental Protection Agency (EPA) are the two major pieces of federal legislation that really define noise as a concern on a national level. While the former provides for workplace protection, the latter provides for the division of powers for federal, state, and local governments. The EPA has the primary responsibility for safeguarding the sound levels in the community by analyzing and reducing the sources. The state and local governments then are responsible for enforcement of the noise regulations by establishing the levels permitted in the environment. Without these laws, the whole field of noise control might not be as large as it is today.

The remainder of this chapter will discuss the various criteria for both noise and vibration and explain how they can be applied within the context of noise control evaluation.

THE WORKPLACE ENVIRONMENT

Since the purpose of this book is related to the industrial environment, all the other types of noise sources in the environment such as aircraft, traffic, railroad, and similar noise sources will be left to other texts and the references at the end of this chapter. The workplace noise environment is composed of three basic elements: manufacturing (in a generic sense) facilities, office areas, and the community around the facility. Work must be done by all employees, and this involves communication within a variety of locations within the facility; areas may be extremely noisy, moderately noisy, or quiet, such as an office. At the same time, the facility should not be annoying to the community. The individual concerned with plant noise may be able to relate equally to all of these and in particular their unique criteria. The criteria for each of these three areas will be discussed in the following sections.

Manufacturing Environment—Employee Noise Exposure Criteria

The basic criteria for the manufacturing area are found in the Department of Labor (DOL) OSHA CFR1910.95, Occupational Noise Exposure. They are designed to protect employees from hearing damage while on the job and were part of the larger OSHA Act passed in 1970. The noise exposure portion of the OSHA Act was adopted from the 1969 amended Walsh-Healy Public Contracts Act, which established allowable maximum noise levels on a national basis. All businesses that had government contracts in excess of $10,000 were covered. OSHA is applicable to all businesses that affect interstate commerce and one or more employees. Thus it is far reaching.

In general the Occupational Noise Standard was designed to protect employees who are exposed to noise of a sufficient level and duration that could cause permanent hearing damage. The standard does three things: First, it sets the maximum noise levels for employees exposure; second, it sets the duration (time allowed) of noise exposure at given levels; and third, it sets forth in very general terms what actions the employer must take to comply if the levels are found to be in violation of the standard.

The three basic steps that the employer must take are as follows:

1. The noise levels must be measured to determine if a hazard exists,
2. If a hazard exists, feasible engineering or administrative control must be developed and used to reduce the noise exposure. While these are being developed and installed, personal protective equipment must be used to lower the employee noise exposure.
3. If the controls fail to lower the employee noise exposure sufficiently, then an effective hearing conservation program must be put in place. Chapter 5 provides a detailed discussion of this process and the requirements.

In recent years, and after the initial flurry of noise control activity in the 1970s, a lot of companies have used only the hearing conservation program as their line of defense for protecting employees' hearing. This has been, in part, due to the lack of enforcement by OSHA and also due to the perceived economic consequences if noise control methods and materials were installed. On the other hand, a lot of large companies have always had safety and medical programs designed to afford a high degree of protection to their workers. Noise was one part of this overall program. Appendix D has a generic hearing conservation program that can be used as a guide or modified to suit individual company needs.

The OSHA noise standard is based on a daily noise of 100 percent. An exposure to a level of 90 dBA for an 8-hour shift results in a dose of 100 percent. The maximum noise levels and duration of exposure are given in Table 7.1. This is based on a doubling of 5 dBA for each halving of time. More detailed information is provided in Appendix B, OSHA Noise Standard.

The equation that allows the noise exposure dose D to be calculated is given as follows:

$$D = C_1/T_1 + C_2/T_2 + \cdots C_n/T_n$$

where:

C = time of exposure at given level
T = allowable time of exposure at given level

TABLE 7.1   Maximum Allowable Noise Levels, OSHA 29CFR1910.95

dBA	Allowable Time (hr)
90	8
95	4
100	2
105	1
110	$\frac{1}{2}$
115	$\frac{1}{4}$

Source: U.S. Department of Labor, 29CFR1910.95, Occupational Noise Standard.

As an example, the following values were determined for an employee from a noise level survey using a sound level meter measuring in dBA:

dBA	Actual time (C)	Allowable time (T)	C/T
92	2.0	6.0	0.33
95	2.0	4.0	0.50
98	1.0	2.6	0.38
105	0.5	1.0	0.50
80	0.5	32.0	0.02
90	8.0	8.0	0.025
Totals	8.0 hr	% dose D	1.98

The % dose D is calculated from the above equation and the summation converted to a percentage by multiplying by 100. This value, the time-weighted average (TWA) exposure, can either be looked up in a table as found in Appendix B or calculated from the following equation:

$$TWA = 16.61 \log_{10} (\% \ dose/100)$$
$$+ 90 = dBA$$

For this example the result is:

$$TWA = 16.61 \log_{10} (198/100) + 90$$
$$TWA = 16.61 \log_{10} (1.98) + 90$$
$$TWA = 16.61(0.297) + 90$$
$$TWA = 4.97 + 90$$
$$TWA = 94.97 \text{ or } 95 \text{ dBA}$$

The results mean that this employee has been overexposed to noise according to the OSHA criteria and that some action is required. This example, of course, is idealized since the noise levels are not that steady for those periods of time. However, by examining the results of the tabulation above it can be seen that if the noise levels of 98 and 105 dBA could be reduced for this employee the percent dose would be 0.755, or 75.5 percent, which is equal to a TWA of 87.9 dBA. This level exceeds the action level of 85 dBA for hearing conservation but does not require engineering or administrative control action.

Other texts and references may express the TWA or equivalent level as $L_{eq}(OSHA)$ or L_{OSHA}. This means that the dBA level represented by this calculated constant level would result in the same hearing damage risk as the time-varying sound for the employee in the example above.

If the company, for whatever reason, elects to reduce noise and not have a hearing conservation program, then the TWA must be reduced to 85 dBA or less. The OSHA standard then establishes a maximum TWA of 85 dBA or a noise exposure of 50 percent for this condition. This is shown in Table 5.2.

Looking further at the example above, if the noise sources that create the levels of 92, 95, 98, and 105 dBA can be reduced the new percent dose would be 0.27, or 27 percent. This is slightly over a TWA of 80 dBA and below the action level of 50 percent, or 85 dBA. Thus no hearing conservation program is required. In real terms, however, a lot of work is required from the point at which a problem is defined to the point at which one can say, "The levels have been reduced and no further action is required."

From this example we have determined a criteria from the OSHA Standard for Engineering and Administrative Controls and also criteria for a Hearing Conservation Program. This example also illustrates how an examination of the noise sources that comprise the total exposure can be used to only reduce those sources that are required. A careful reading of this standard will be helpful in understanding all its ramifications and how to apply it to a specific situation. There are many options between noise exposure and engineering control that can be taken to have a viable program that is cost-effective.

Criteria for Communication in the Manufacturing Environment

Everyone has experienced the inability to communicate at one time or another in the presence of high background noise, whether it was at a football game or on the factory floor. In any manufacturing environment communication is an important issue. Instructions must be given and received, safety warnings heard, telephones answered, and conversations of a social nature conducted. Of course, there are other types of communications that must be carried out, but these examples give you the idea.

In general the noise level in this type of environment will be somewhat high and the ability to communicate may be difficult in the presence of a high background noise level [Webster, 1969]. When it is necessary to evaluate this type of situation the Speech Interference Level (SIL) can be used. This was developed to assist in such evaluations where no speech sounds were reflected back to the listener. With the new octave band convention used, the designation was renamed Preferred Speech Interference Levels (PSIL). By definition the PSIL is the arithmetic average of the 500-, 1000-, and 2000-Hz octave bands. This was also developed as a simplified substitute for the articulation index, which is a much more complicated rating scheme requiring more detailed noise data. In cases where octave band data are not available, the PSIL can be estimated from the following equation:

$$PSIL = L_A - 7 \quad (dB)$$

The only caution for using this equation is that if the noise spectrum in a facility or other environment is low frequency in nature, then the PSIL should be used as the criteria. If there is any doubt about this, the PSIL should be calculated and compared to the dBA value. In general if the difference is 7 to 8 or more, then the PSIL will be the best judge of speech interference levels. Figure 7.1 provides the best graphic of PSIL and dBA for rating speech communication ability from speech interference levels. Note that the difference in the values for PSIL and dBA on the x-axis is 8 dB.

Telephone communication, a very important activity, also uses PSIL criteria. The data below can be used for establishing criteria for this purpose:

PSIL range	Degrees of difficulty
43–63	Satisfactory to slightly difficult
63–83	Difficult to unsatisfactory

The problem of communication in a plant is well exemplified by a company that switched to "just-in-time" manufacturing. This change required a completely different office setup for manufacturing engineering. The engineering staff was placed in small office cubicles scattered around the manufacturing floor. During this process the major product line was being converted from metal-working manufacturing to printed circuit board manufacturing, an inherently quieter process. Nonetheless, there were very loud complaints from engineering personnel coming from a relatively quiet office environment to a noisier manufacturing environment.

The plant management did not want to have everyone in a closed office since that would defeat the purpose of the open environment to aid in communication and problem solving between engineering and production personnel. A study was done to determine the problems associated with speech interference and telephone communication for these office cubicles. The results are shown in Table 7.2. Note that data were taken in octave bands and the PSIL calculated was then compared to the dBA in order to determine the most appropriate criteria. The result was a recommendation to rearrange some offices away from noisy equipment, provide noise control on some equipment, and use a complete office enclosure in some locations where other means of controlling noise were not feasible.

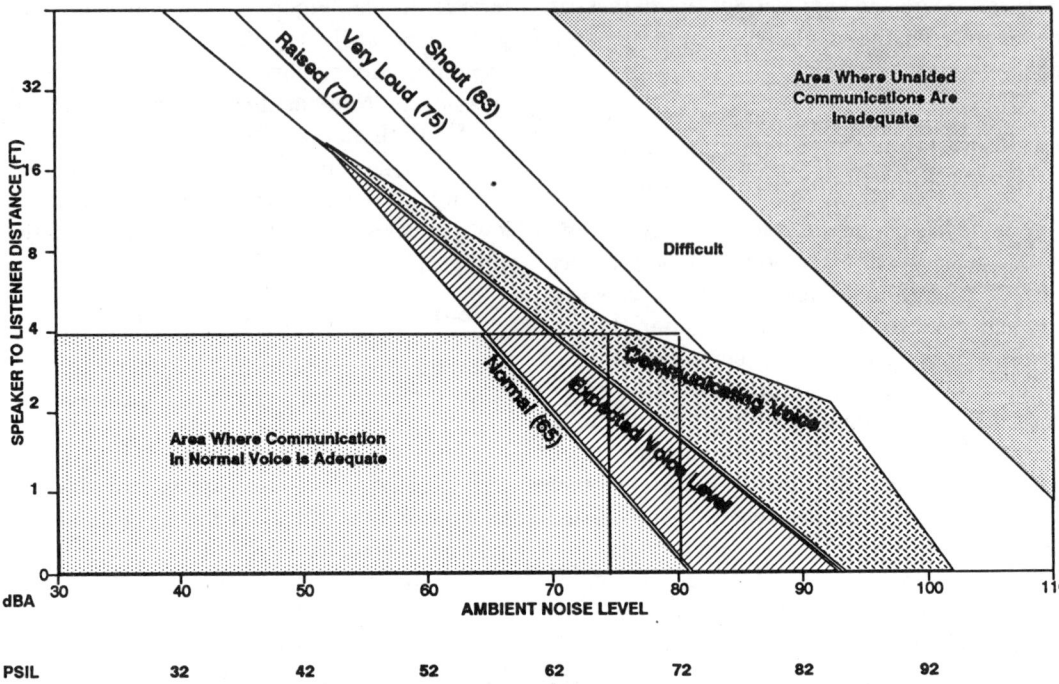

FIGURE 7.1   Speech interference levels.   (Reprinted by permission from Dr. John C. Webster.)

In summary, communication in various environments is extremely important, and the speech interference criteria provide a recognized method for evaluation in a variety of situations.

Criteria for Office Environments

Within a manufacturing facility there are offices that house the management and adminis-

TABLE 7.2   Plant Office Location Noise Levels

Location	PSIL	dBA
A	76	85
B	75	83
C	72	82
D	71	79
E	69	77
F	68	75
G	67	74
H	65	72
I	64	71

trative staff. Offices are found in the manufacturing areas as well. Even though these have been discussed in the previous section, the criteria described in this section will be of help in evaluating these offices in another way.

The purpose of this type of criteria is to evaluate intrusive noise sources in the various types of space found in an office environment, or any other space that is shown in Table 7.3. Intrusive noise is usually from heating, ventilating, and air-conditioning (HVAC) sources. Other sources of intrusive noise are:

1. Manufacturing noise
2. Traffic or aircraft noise
3. Noise from adjacent offices

The American Society of Heating, Refrigeration and Ventilating Engineers (ASHRAE) has developed a detailed chapter (Chapter 42) in its *Systems Guide and Data Handbook* devoted to sound and vibration as related to the

TABLE 7.3   Criteria for Acceptable HVAC Noise Levels in Unoccupied Rooms

Occupancy	Preferred	Alternate
Private residences	RC25-30(N)	NC25-30
Apartments	RC30-35(N)	NC30-35
Hotels/motels	RC30-35(N)	NC30-35
Individual rooms or suites	RC30-35(N)	NC30-35
Meeting/banquet rooms	RC35-40(N)	NC35-40
Halls, corridors, lobbies	RC40-45(N)	NC40-45
Service/support areas	RC25-30(N)	NC25-30
Offices		
Executive	RC25-30(N)	NC25-30
Conference rooms	RC25-30(N)	NC25-30
Private	RC30-35(N)	NC30-35
Open-plan areas	RC35-40(N)	NC35-40
Business machine/computers	RC40-45(N)	NC40-45
Public circulation	RC40-45(N)	NC40-45
Hospitals and clinics		
Private rooms	RC25-30(N)	NC25-30
Wards	RC30-35(N)	NC30-35
Operating rooms	RC25-30(N)	NC25-30
Laboratories	RC35-40(N)	NC35-40
Corridors	RC30-35(N)	NC30-35
Public areas	RC35-40(N)	NC35-40
Churches	RC30-35(N)	NC30-35
Schools		
Lecture and classrooms	RC25-30(N)	NC25-30
Open-plan classrooms	RC35-40(N)	NC35-40
Libraries	RC35-40(N)	NC35-40
Courtrooms	RC35-40(N)	NC35-40
Legitimate theaters	RC25-30(N)	NC25-30
Movie theaters	RC30-35(N)	NC30-35
Restaurants	RC25-30(N)	NC25-30
Concert and recital halls	RC15-20(N)	NC15-20
Recording studios	RC15-20(N)	NC15-20
TV studios	RC20-25(N)	NC20-25

Source: Reprinted by permission. The American Society of Heating Refrigeration and Air Conditioning Engineers, from the 1991 ASHRAE "HANDBOOK," HVAC Applications.

systems found in occupied structures. One of the major sources for information of this nature as well as acceptable criteria for occupied spaces is found in their handbook. The *noise criteria (NC)* is a tabulation of various occupied spaces with the acceptable NC values. This has been in use for many years [Beranek, 1960]. These values are shown in Table 7.3 along with *room criteria (RC)*. The RC provides a more balanced and bland-sounding spectrum [ASHRAE, 1991]. It is neither too rumbly sounding in the low frequencies nor too hissy in the high frequencies. These are shown

graphically in Figs. 7.2 and 7.3. The RC curves have been published in the ASHRAE guide for several years and are becoming more prominent in HVAC design applications.

Another criterion, suggested by Beranek, extends the range down to 16- and 31.5-Hz frequency bands. This method predicts the probability of complaints for a low-frequency and high-frequency imbalanced spectrum. This method can also be applied to automobile and airplane passenger compartments to predicate acceptability [Beranek, 1989]. These criteria are called the *balanced noise criterion curves (NCB)* and are shown graphically in Fig. 7.4. These can be compared to the NC and RC curves in Figs. 7.2 and 7.3.

Many offices in manufacturing plants now have the need for very quiet environments for board rooms, teleconference rooms, presentation rooms, and training rooms. These environments pose a real challenge for the noise control engineer in places where there may be high noise from manufacturing activities.

The noise criteria were not an original development by ASHRAE. They were adopted by ASHRAE and included in its reference guides. In the late 1940s and early 1950s, when offices were first air-conditioned, there was a concern about speech interference from noise from air-conditioning fan and air distribution systems [Beranek, 1954]. The noise criteria in the ASHRAE Guide has wide acceptance and is used on a daily basis. This type of criteria is known as a consensus standard.

To determine the NC value for a space, noise data must be measured in octave bands with a sound level analyzer. The data are plotted on the chart and the NC level determined from the location of the highest point on the chart. This will provide a good estimate of the NC level that can then be compared to the criteria table. Because sound level measurements are estimates, many times it is best to consider the resulting NC values estimates. The levels can vary depending upon the source(s), so a series of measurements should be made and plotted to obtain a range for the space.

An example of typical data is plotted in Fig. 7.5 for training facilities located in the office

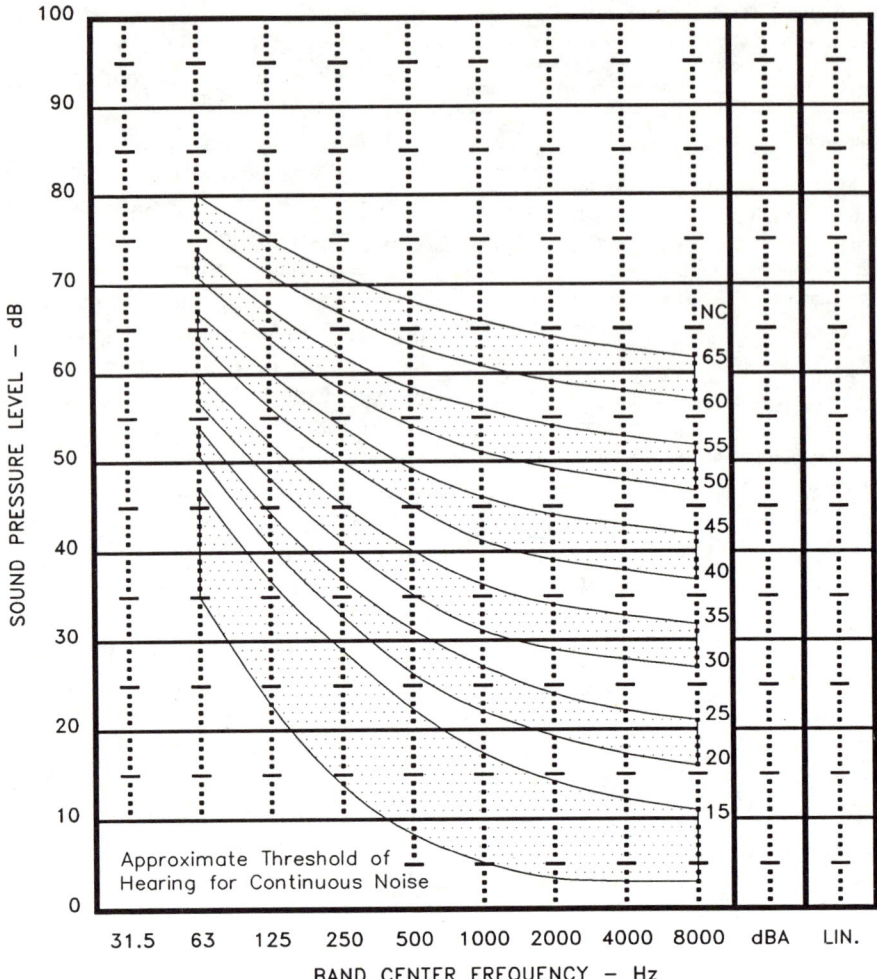

FIGURE 7.2   Noise criteria (NC) curves.   (Reprinted by permission from Dr. Leo Beranek, *Noise Reduction*, 1960.)

spaces for a manufacturing plant. The training rooms are located next to a process line and under a space that houses heavy processing equipment. Note that the results are less than satisfactory for the criteria for this type of environment. A typical criterion for this type of space is NC35. Since it is located next to and under process equipment, some judgment can be used to determine a criteria range. The individual charged with the design can also use some judgment not only in selecting the criteria but in interpreting the measured results.

The NC values are also used for design goals for the office spaces. Not only is the HVAC system noise control design required, but building construction systems, such as partitions, roofs, exterior walls, and doors, are also required to impede intrusive noise. The ASHRAE guide is a general reference for HVAC noise and vibration control design. Other references must be used for selecting the proper elements of building construction to meet the NC values required for the use spaces. Building material suppliers are also good sources of information for these data. Suppliers have references and test data for their materials that can

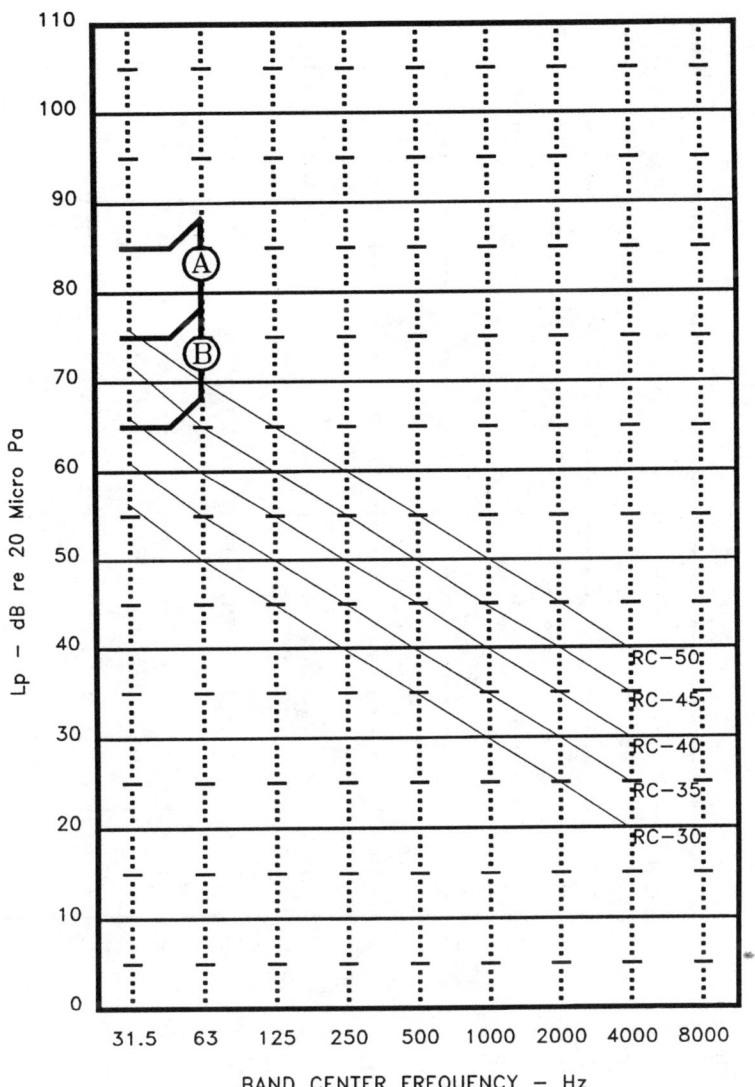

FIGURE 7.3   Room criteria (RC) curves.   (Reprinted by permission, The American Society of Heating, Refrigeration and Air Conditioning Engineers, from the 1991 ASHRAE *Handbook—*HVAC Applications.)

be used in the evaluation. Many textbooks on noise control have tabulated data for this purpose.

In conclusion, the design of an office environment in a manufacturing facility is an important issue since good speech communication is required in many spaces. The NC values can provide a guide and the criteria for this purpose depending on the use of the space.

THE COMMUNITY ENVIRONMENT

There are several parameters of sound that affect people and their response. These are the level of sound, the frequency content, and the variation of these two with time. A complete picture of the impact of various sounds on the community would require a continuous record-

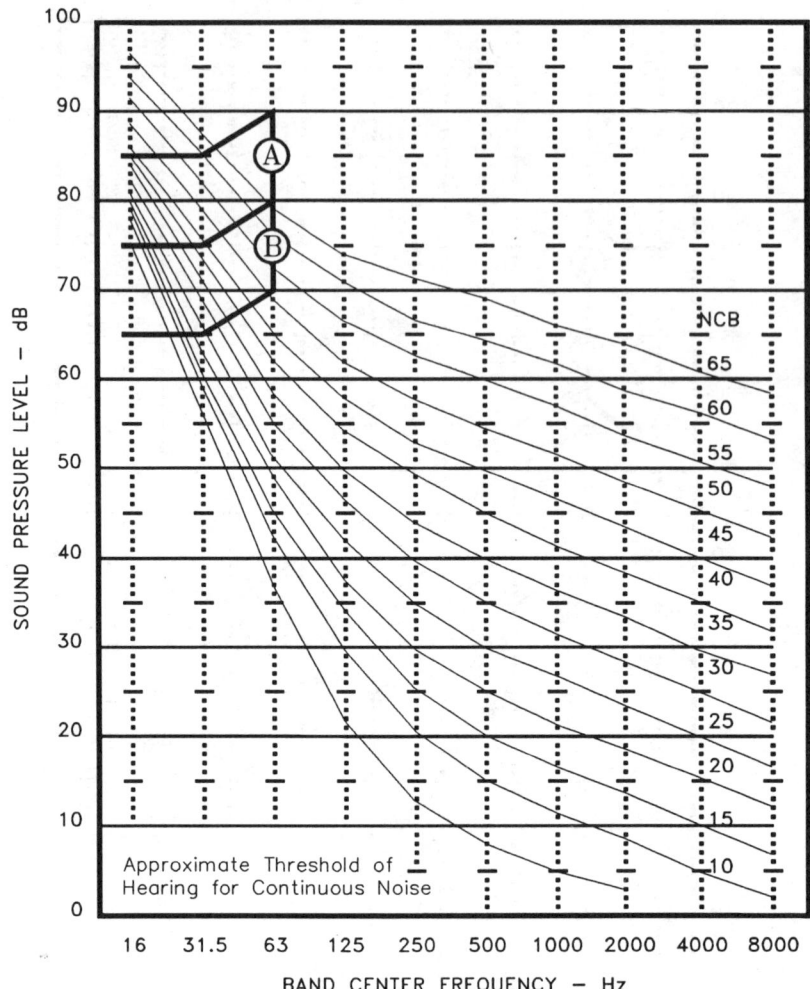

FIGURE 7.4   NCB noise criteria curves.   (Reprinted by permission from *Noise Control Engineering Journal.*)

ing of these parameters. While there are some specialized instruments that will provide this information, plant personnel do not usually have access to this or even a more simplified octave band analyzer. Thus a simplified rating scheme that has been widely studied and accepted is the A-weighted sound level (dBA). This has been found to correlate reasonably well with a community's response to noise. The dBA combines the overall frequency content of the noise source in a weighting network in a sound level meter that has a response much like the human ear.

A manufacturing or process plant will usually be located next to other property that may be sensitive to noise. Even if the plant is out in the country away from urban noise, the plant noise may be more disruptive due to the lower ambient sound level. Thus the plant, especially at night, may be heard quite easily. A plant located in the city may also be disruptive to neighbors depending upon the proximity and plant noise level. Criteria for noise flowing from the plant to the community can be established based on Environmental Protection Agency (EPA) or Housing and Urban Devel-

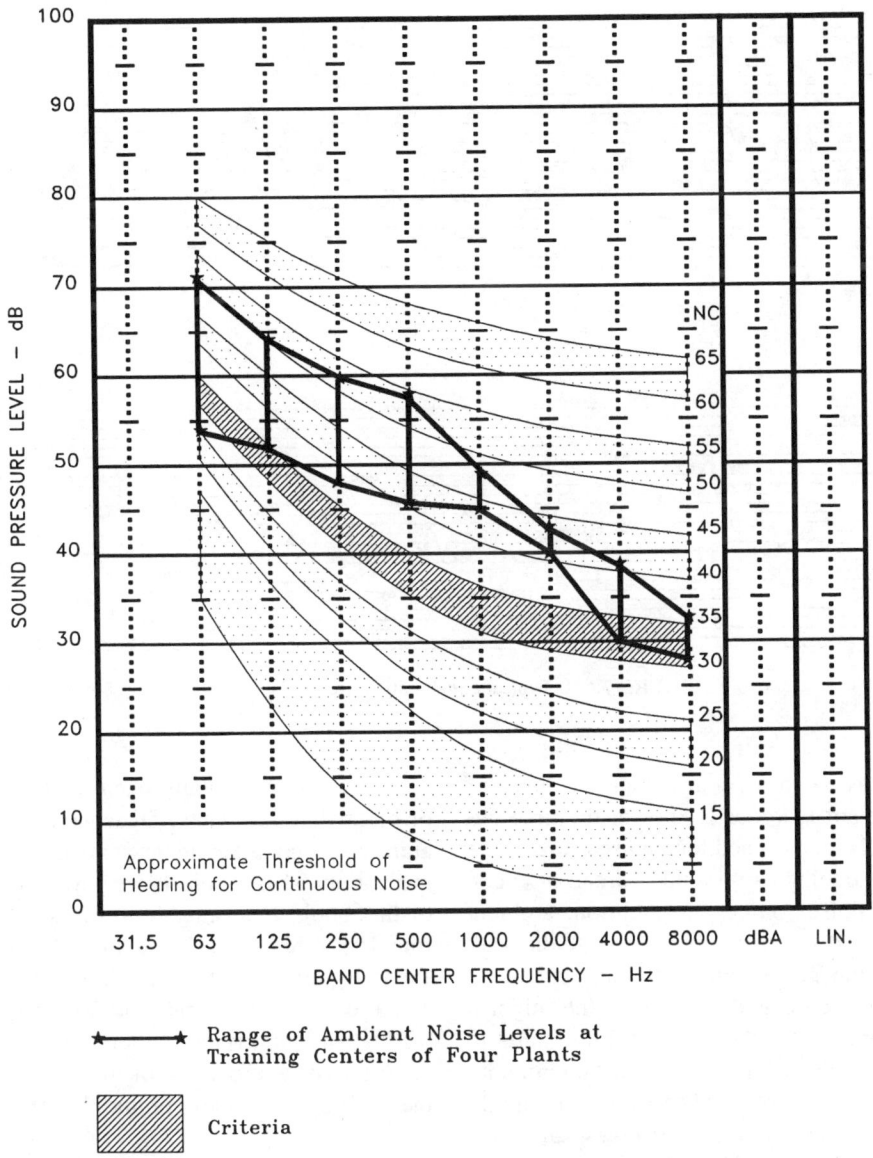

SOUND PRESSURE LEVEL – dB

BAND CENTER FREQUENCY – Hz

Approximate Threshold of
Hearing for Continuous Noise

★————★ Range of Ambient Noise Levels at
Training Centers of Four Plants

Criteria

FIGURE 7.5   Range of noise levels in four corporate training facilities.

opment (HUD) guidelines. The city, county, or state where the plant is located may have property line noise ordinances.

A plant to be located next to a residential community may have to comply with property line noise limits as part of a zoning ordinance. Finally, a community may sue a company that operates a plant alleged to create a noise nuisance. The criteria will then be established by the court. There are several methods of establishing noise criteria for plant noise as it relates to the community.

The temporal pattern of noise can be evaluated with a strip recorder coupled to a sound level meter. This will provide a tracing of sound level in dBA versus time. For example, Fig. 7.6 illustrates strip chart tracing from a community noise evaluation. The various sources

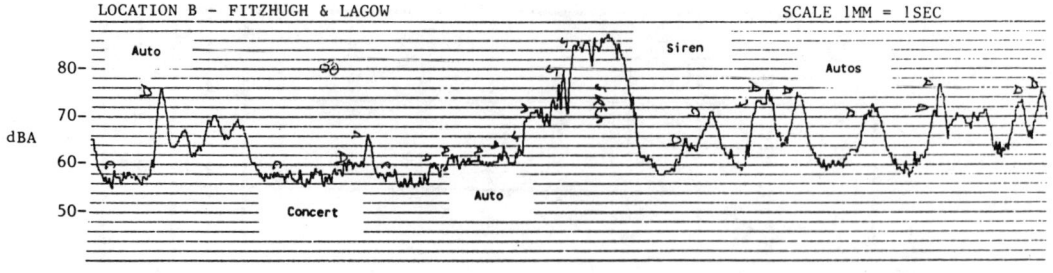

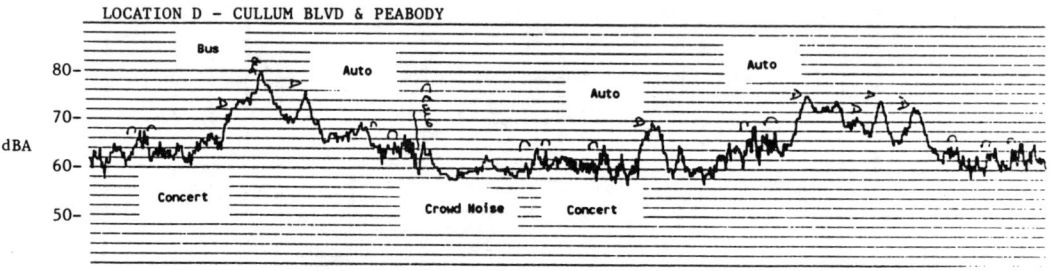

FIGURE 7.6   Typical community noise levels in dBA.

can be noted as they occur. When the speed of the strip chart recorder is known, the time duration can be determined for a noise source. The strip chart tracing can provide a trend or cyclic nature of noise sources. In addition, ambient noise can be distinguished compared to the peak values of the noise sources. If one watches a strip chart tracing as the day turns into night it can be noted when the ambient noise starts to trend downward with time. In most community environments between 11 P.M. and midnight the ambient level starts to decrease and will tend to level off in the early morning hours. Thus noise sources can be more easily distinguished.

For evaluation purposes the noise level time history can be divided into three segments:

Day—7 A.M. to 7 P.M.
Evening—7 P.M. to 10 P.M.
Night—10 P.M. to 7 A.M.

A review of data in each of these periods is important to learn how the ambient and noise sources change in level with time. Most plant noise complaints usually occur in the evening or night time segments. Distance and temperature also play a significant part in the overall evaluation. These subjects are beyond the scope of this book, but valuable information can be found in other texts listed in the references. What is important to note, however, are the noise descriptors. This notation provides a method of simplifying the temporal detail and allows a characterization of the noise environment. The four values commonly used are:

L_{max} = maximum level occurring in the measurement period

L_{min} = minimum level occurring in the measurement period

L_n = percent of time level is exceeded in the measurement period; i.e., L_{50} is sound level exceeded 50 percent of the time, L_{10} is the sound level exceeded 10 percent of the time

L_{eq} = equivalent level, that constant level that has the same sound energy as the time-varying sound for the measurement period

There are instruments, available for purchase or rent, that can measure all of these data. These noise monitors can be placed at the property line for a period of time and the results printed out. The important thing to understand is that a single dBA sound level reading will not begin to describe the community noise impact that your plant might have—more information is required.

With the methods established by the EPA the time periods are divided into two segments:

Day—7 A.M. to 10 P.M.
Night—10 P.M. to 7 A.M.

EPA uses the L_{eq} descriptor as a means of providing a day/night average in dBA. This is called L_{dn} and represents an average over a 24-hour period with a 10-dB nighttime penalty for the 10 P.M. to 7 A.M. period. It is useful for predicting the long-term effect of environmental noise such as airports. The L_{eq} descriptor is combined in an equation for the two time periods and weighted for the nighttime. The EPA Levels Document (as it is usually called) provides a very detailed discussion of all the ramifications for this as well as a variety of criteria [EPA, 1974]. Figure 7.7 is adapted from this document. It shows the threat and severity of community action and should be used only as a guide. If local ordinances exist they must be followed.

It should be noted that since L_{dn} is a day/night average, peak levels or pure tones from a plant may be the major source for a noise complaint. Therefore, a good subjective description of the noise being heard by the complainants is very important. These can then be accounted for in the L_{dn} criteria established.

Another set of criteria for the community environment is provided by HUD. This is an A-weighted single-number rating scale to assess the exterior noise environment. For this rating system a subjective evaluation is pro-

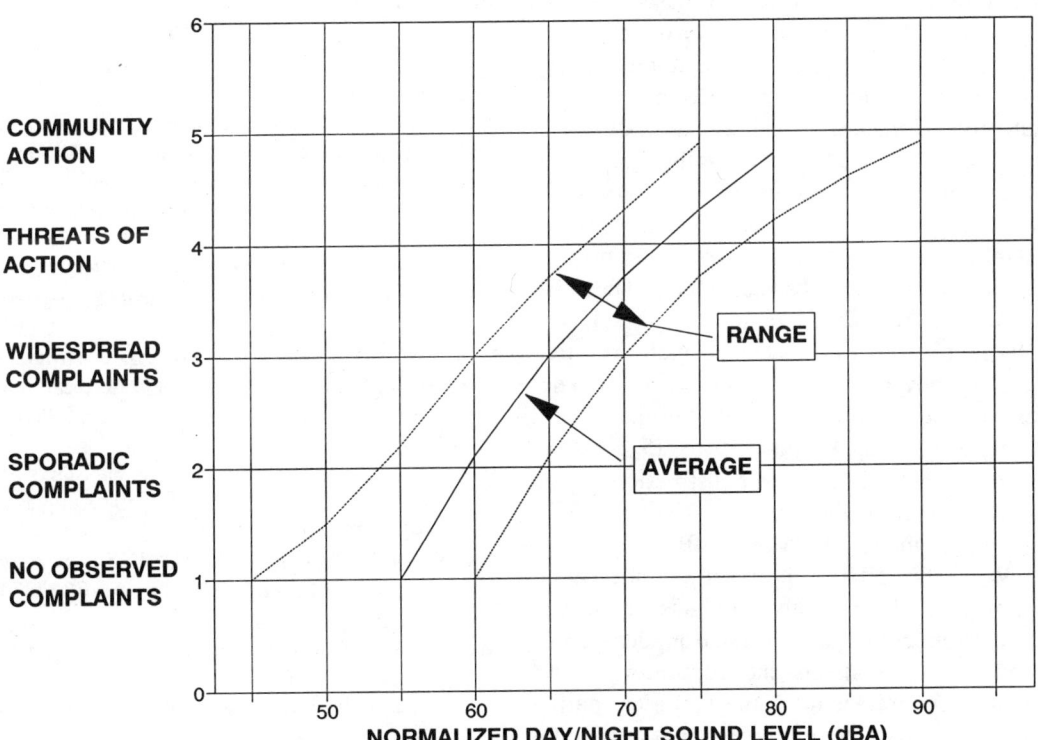

FIGURE 7.7   Community noise data adjusted for conditions of exposure.   (EPA Levels Document, March 1974.)

TABLE 7.4   Site Acceptability Standards of New Construction

	Day/Night Average Sound Level (db)	Special Approvals and Requirements
Acceptable	Not exceeding 65 dB[a]	None
Normally unacceptable	Above 65 dB but not exceeding 75 dB	Special approvals[b] Environmental review[c] Attenuation[d]
Unacceptable	Above 75 dB	Special approvals[b] Environmental review[c] Attenuation[e]

[a]Acceptable threshold may be shifted to 70 dB in special circumstances pursuant to Section 51.501(a).

[b]See Section 51.501(b) for requirements.

[c]See Section 51.501(b) for requirements.

[d]5 dB additional attenuation required for sites above 65 dB but not exceeding 70 dB, and 10 dB additional attenuation required for sites above 70 dB but not exceeding 75 dB. (See Section 51.501(a).)

[e]Attenuation measures to be submitted to Assistant Secretary for CPD for approval on a case-by-case basis.

Note: All levels in dBA.

Source: U.S. Department of Housing and Urban Development, Environmental Criteria and Standards 24CFR51.

vided that will help evaluate the complaint potential of intrusive noise. This is based not only on the dBA level but the number of intrusions in a given time period above the ambient level. Intrusive noise sources considered by HUD are automobile, truck, railroad, and aircraft. HUD provides a method of determining how these sources impact the community by allowing the noise levels to be predicated. It also provides methods of calculating noise reduction primarily using barriers and distance effects. The noise criteria for HUD are found in Table 7.4 [HUD, 1984]. This provides site acceptability standards for new construction sites. In order to determine the potential levels on the inside of a dwelling one can subtract 15 to 25 dBA depending upon the type of construction, wood frame to brick. If the windows are open the reduction might be only 5 or 10 dBA.

While the HUD criteria might not be directly applicable to industrial plants, in the way they are structured, they provide an idea of how intrusive noise affects the community. This could be important for some facilities, particularly those having railroad and truck traffic for incoming raw materials and transportation of finished products.

References

ASHRAE. 1991. *ASHRAE Systems Guide and Data Handbook.* Chapter 42, Sound and Vibration Control. Atlanta: ASHRAE.

Beranek, Leo L., ed. 1960. *Noise Reduction.* New York: McGraw-Hill.

Beranek, Leo L. 1989. Application of NCB Noise Criterion Curves. *Noise Control Engineering Journal,* pp. 45–52.

EPA. 1974. Information on Levels of Environmental Noise Requisite to Protect Public Health and Welfare with an Adequate Margin of Safety. 550/9-74-004.

HUD. 1984. Environmental Criteria and Standards. 24CFR51 44FR40860, July 12, 1979, amended by 49FR880, January 6, 1984.

OSHA. 1983. Occupational Noise Exposure Regulation. U.S. Federal Register (39FR37773), rc:Part 29CFR1910.95.

Webster, John C. 1969. Effects of Noise on Speech Intelligibility. In Proceedings of *Noise as a Public Health Hazard.* ASHA Report Number 4, pp. 49–73.

Chapter 8

Instrumentation for Noise Measurements

INTRODUCTION

Instrumentation for noise measurements is an extremely broad subject. This discussion, therefore, is confined to the most important principles involved and the most frequently used instrumentation for noise control.

Noise-measuring instruments play a basic role in noise control. With instruments, measurements are made to identify noise sources, determine the magnitude of the noise at a particular distance from the source, and measure the frequency characteristics of a noise source. Measurements provide data for comparison with various standards. The data are also used for specific design of noise control techniques.

INSTRUMENTATION

Figure 8.1 is a block diagram of a typical sound level meter which consists of four basic components:

1. Microphone or transducer—converts the changes in sound pressure level into variations in electrical current by means of a pressure-sensitive diaphragm
2. Amplifier—increases the minute electric current signals to a usable level
3. A-weighting network or other electronic filters—analyze the electric signal

4. Meter—displays analog or digital values of the electric signal in decibels (dB)

Sound level meters come in many different shapes and sizes, depending on the use and manufacturer. There are four broad classifications of meters and each classification has its own stringent specification [Peterson, 1980]. The four classifications are:

1. General purpose
2. Survey
3. Precision
4. Special purpose

The two classifications that are normally used are the survey meter, classified as type 2, and the precision sound level meter, classified as type 1. The type 2 meter has an accuracy of ± 2 dB at most frequencies and is designed as a general-purpose sound level meter. The type 1 is built to more stringent specifications and has a much higher accuracy of ± 0.5 dB at most frequencies and is a precision sound level meter. Usually, the type 2 meter has just two or three weighting networks. An octave or one-third octave and air filter set can be added. The type 1 meter may have the various filter sets built into the system.

The International Standards Organization

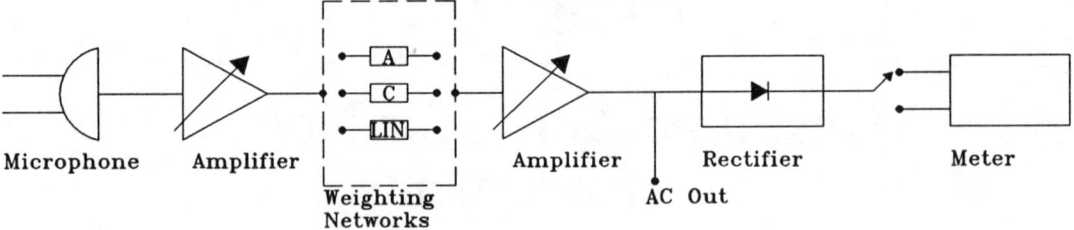

FIGURE 8.1   Block diagram of sound level meter.

(ISO) and the American National Standards Institute (ANSI) have developed standards for various sound level meter classifications. Selection of a sound level meter should take into consideration the applicable standard. For example, sound level meters should meet as a minimum ANSI S1.4A-1985 Type S2 American National Standard for Sound Level Meters. There are numerous manufacturers offering a variety of instruments for sale.

The operation of a sound level meter generally provides for two settings, fast and slow. The fast needle setting gives a true indication of the noise level averaged over a very short period on the order of one-eighth of a second. Due to the quick response one can have a good indication of the time-varying quality of an acoustic signal. However, the response may be too quick to allow an accurate measurement. The slow setting averages the signal over a longer period of time, about a second. This allows a single measurement to be more reliable. Most measurements or specifications state that a "slow" meter response should be used. Regardless, it is a good practice to note which meter speed was used.

Almost all sound level meters have an A-, B-, and C-weighting network. Originally, these networks were developed to approximate the loudness level sensitivity of the human ear. Over the years, the A- and C- weighting networks have been used extensively while the B network has received little usage [Lord, 1987].

The values in Table 6.2 are the same A- and C-weighting corrections used in the sound level meters built to ANSI standards [ANSI, 1983]. The A-weighting network reduces the low-frequency sound levels below 500 Hz and slightly increases the high-frequency sounds between 1000 and 4000 Hz. This is how the human ear responds to sound and explains why high-frequency noises are usually more annoying than low-frequency noises. The C-weighting, on the other hand, reduces sounds below 100 Hz and above 5000 Hz. Unlike the A-weighting, the frequency region between 100 and 5000 Hz is linear; that is, all frequencies are given equal weight.

For a given acoustic signal, the A- and C-weighting networks can be used to determine some basic frequency content without listening to the sound. When the A-weighted sound level (dBA) and the C-weighted (dBC) are compared, the following rule of thumb is useful:

If dBC − dBA ≥ 2, the noise is basically low frequency.

If dBC − dBA = 0, the noise is almost broadband with equal sound level frequency.

If dBC − dBA ≤ −2, the noise contains high-frequency components.

More detailed information is desirable relative to the frequency content and time-varying qualities of the noise. But as a first approach, if this is all the information that is available, some basic judgments can be made when comparing dBC − dBA.

The following items are parameters that may physically influence the noise measurements. These items are always there, and the person conducting the noise survey should be aware of them. Each sound level meter has its own distinctive operating characteristics. The first step is to read the manual and become familiar with how this particular instrument operates.

MICROPHONES

All microphones, because of their physical presence, exhibit some form of directionality at high frequencies. The size of the microphone determines the frequency at which the directionality begins. The smaller the physical dimensions of the microphone, the higher the frequency. For measurements near or above this critical frequency, errors of 2 dB or more may result. For a typical one-inch nominal microphone, the directivity effects do not occur below 10,000 Hz. As a rule of thumb, the critical frequency (f) is equal to:

$$f = \frac{15,000 \text{ Hz}}{r}$$

where r is the radius of the microphone in inches.

The general type of microphones are [Irwin and Graf, 1979]:

Condenser—a single diaphragm is connected to one side of a capacitor which changes in response to the change in sound pressure.

Ceramic—a piezoelectric crystal is located behind a diaphragm; when the sound pressure changes it strains the crystal, creating an electrical signal that is used by the instrument.

Dynamic—the diaphragm is connected to a coil that moves in response to the sound pressure. The electrical signal from the moving coil is used by the sound level meter.

Figure 8.2 illustrates a response of a typical condenser microphone from the author's equipment.

Some microphones are designed for 0° (frontal) incidence; that is, the microphone is to be "pointed" at the sound source. For the condition where the sound environment is very reflective, such as inside a building, a random incident corrector attached to the microphone may be used to improve the frequency response of the microphone. Some instruments have a switch built in that allows random or frontal incidence.

REFLECTIONS

For most industrial situations, the noise source will not usually be in an acoustically free field, and reflections from surrounding surfaces may occur. When this occurs, caution must be exercised in interpreting the measurements. Many measurements may need to be made to accurately describe the sound field.

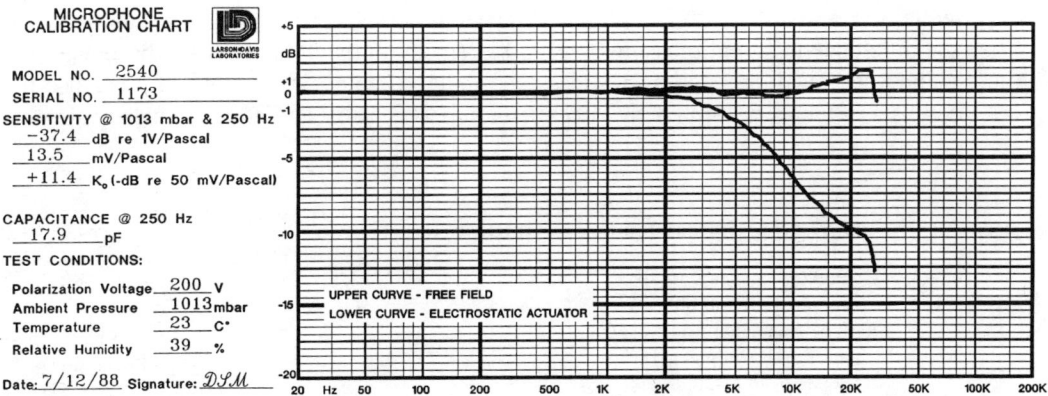

FIGURE 8.2   Typical frequency response curve of a microphone.   (Courtesy of Larson-Davis Laboratories.)

Another form of reflection can occur from the operator back to the measuring instrument. The microphone should be extended as far from the sound level meter as possible. In addition, the sound level meter should be extended at least one arm's length away. At low frequencies (<250 Hz), where the wavelength approximates the operator's physical dimensions, false readings may result because of reflections. Therefore, a tripod is recommended to obtain the most accurate data.

BACKGROUND NOISE

The noise level measured in industrial settings can be significantly altered by background or ambient noise levels. The most desirable situation is to be able to measure the noise level of a machine without anything else operating. But when this is not possible, a technique such as logarithmic subtraction can be employed to determine the noise level of the machine. The noise level is found from:

$$dB(\text{machine}) = 10 \log (10^{L_{p1}/10} - 10^{L_{p2}/10})$$

where:

L_{P1} = sound level of machine and background
L_{p2} = sound level of background only
dB = sound level of machine

The values found in Table 8.1 will provide an easy method of determining the background noise corrections.

When the difference in sound pressure level with and without the machine operating is less than 3 dB, this technique is limited by other factors and a true level of the source cannot be determined.

MICROPHONICS

The transmission of mechanical vibrations from the sound level meter housing to the microphone, or the internal electronic components, is called microphonics. This situation occurs only in high-intensity sound fields greater than 130 dB and can influence the output of the meter. Vibration isolation of the meter usually alleviates this problem; a better approach is to place the microphone on a tripod with a cable to connect to the sound level meter. This allows the operator to be in a less noisy environment to concentrate on the measurement data.

Another source of extraneous noise can be from the microphone cable. This source is usually in the form of low-frequency noise. It can easily be detected due to its transient nature. This noise can be minimized by keeping the cable in a stationary position, using special low-noise cables in critical applications, or modifying the cable dielectric.

R. F. PICKUP

Radio frequency signals may be picked up when long microphone or headphone cables are used and the sound level meter amplifier gain is set high [Beranek, 1960]. This situation can occur when low-level ambient community noise levels are being measured and the sound level meter is turned into a rather expensive radio. To remedy the situation, either disconnect the headphone cables or electronically balance the microphone cables. Radio frequency signals

TABLE 8.1   Background Noise Corrections

Difference between L_p with equipment on and off	0	1	2	3	4	5–9	>10
L_p to be subtracted from total L_p (all equipment operating) to obtain true L_p of equipment in question	>10	7	4	3	2	1	0

can also be picked up on a ground loop if the instruments are plugged into an electrical power outlet.

EXTRANEOUS PICKUP

When measurements are taken close to high-voltage electrical power transmission lines, care must be exercised in using shielded cables and protecting tape-recording devices. This can result in pure tones or static giving false readings at 60 cycle (Hz) and harmonics. The influence of this can be minimized by proper orientation and shielding of the sound level meter with respect to the magnetic field.

WIND NOISE

Whenever measurements are made in the presence of wind, noise can be generated across the microphone. Air blowing over a microphone creates a pressure differential and turbulence. The turbulence generates noise that increases with wind speed. The addition of a windscreen moves the pressure gradient and turbulence further away from the microphone, allowing a smooth flow of air and minimizing the influence of turbulence.

Whenever sound measurements are made out-of-doors, a windscreen should *always* be used over the microphone. It is good practice to use a windscreen all the time since it will protect the microphone in any situation, even with indoor measurements. If the microphone gets dropped the windscreen will cushion the impact. The steps outlined below can be used to determine the contribution of wind noise. Take measurements of the noise source with and without the windscreen, then:

1. If there is no change, assume that wind noise has no influence
2. If the change in L_p is equal to that shown, assume that wind noise is dominant
3. If the change in L_p is less than that shown, assume that wind noise was measured without the windscreen and the noise source was measured with the windscreen

For outdoor measurements standards usually recommend that data not be taken if winds are greater than 7 mph. This is specified in ANSI S1.13-1971(R1986) American National Standard for Measurement of Sound Pressure Levels. Experience has shown that accurate data can be obtained measuring dBA only in winds 10–15 mph. If linear and low-frequency data are to be obtained, the standards should be followed.

Other applications for measuring sound in high-velocity airstreams is for ventilating ducts. Since the speed of airflow is high, nose cones or other special protectors are required to minimize wind noise at the microphone. This will provide accurate data. Standard windscreens will not be effective for this application.

BAROMETRIC PRESSURE

Daily changes in atmospheric pressure at any one location have a modest influence on sound level measurements. A much more significant influence occurs in changes of elevation. The standard barometric pressure at sea level may be 760 mm Hg, whereas at an elevation of 1000 meters, the pressure may only be 675 mm Hg. This change in atmospheric pressure would alter all sound level readings by 1 dB. This means that during the calibration procedure, the gain control of the sound level meter should be adjusted to read 1 dB less than the value normally obtained at sea level. For other elevations, appropriate correction factors should be applied.

TEMPERATURE

Typically, the most temperature sensitive elements in a sound-measuring system are the microphone and batteries. As the ambient temperature decreases, the useful life of the batteries also decreases. This is especially true in subzero temperatures where normal eight-hour battery life can be shortened to two hours or less. Some manufacturers have available special insulated battery cases designed to minimize temperature effects. This case can be extremely useful when prolonged outdoor measurements are required.

The microphone system may also be sensitive to changes in temperature. Rochelle salt microphones are extremely sensitive to temperature. In fact, damage to the microphone may result if exposed to very high ambient temperatures. The condenser and ceramic piezoelectric microphones, which are in much more common usage than the Rochelle salt microphone, are virtually insensitive to temperature variations. For condenser microphones, there is usually less than a 0.5-dB calibration drift over the temperature range of $-40°$ to $150°C$ ($-40°$ to $302°F$) and $-40°$ to $60°C$ ($-40°$ to $140°F$) for ceramic piezoelectric.

HUMIDITY

The condenser electret microphones, unlike the ceramic, can be sensitive to humidity under severe conditions. If moisture has entered and condensed in the space between the cartridge electrode and the diaphragm, a crackling or popping noise may occur. This can be observed in the earphone output of the sound level meter and as erratic needle behavior on the meter. This excessive moisture condition can be remedied by placing the microphone under a high-intensity light bulb for five to ten minutes. Moisture can condense in the microphone if one moves from a cool air-conditioned space to a hot, humid outdoor environment. It might be prudent to keep the microphone and sound level meter outdoors under constant temperature and humidity conditions.

ADVANCED MEASUREMENTS

In many cases, after the normal noise survey is completed and results are reported, there may be a need to conduct some further investigations of noise sources to determine the feasibility of engineering noise control. While the details of this type of investigation are beyond the scope of this material, it is important to mention a few items so that when working with a consultant or noise control engineer the terms will be familiar.

As mentioned previously, the standard octave measurements can be divided into three parts to reveal more about the noise source.

Likewise there are some spectrum analyzers that have the capability of dividing the octave band into 12 parts. Such analyzers provide even more detail about the noise source and allow one to draw conclusions about the noise source based on speed, flow rates, and other operating parameters. In addition there are many analyzers that provide measurement options in the narrow-band mode, allowing even finer detail to be viewed. With the advent of computer-based instruments many other types of analysis can be made. For example, with a two-channel spectrum analyzer it is possible to look at two locations at once and compare the two within the instrument electronics or software.

One type of analysis that can be made is coherence. In order to evaluate the contribution from structure-borne and airborne noise paths, a two-channel instrument is necessary so that each of these contributions can be determined. Coherence is a measurement that is useful for determining these energy paths and contributions. With the use of two transducers, energy (vibration, acoustic, etc.) is measured and compared. The dependence of the energy measured at one transducer (accelerometer or microphone) on the energy measured at the other is compared (0 being no dependence, 1 being absolute dependence) for each frequency in the analysis range. With this measurement, it is possible to determine if the vibration or acoustic energy measured at one location contributes at another location or is composed of many sources.

For example, a high coherence value (0.8 to 1.0) for wall vibration compared to the measured sound pressure level at the source location could indicate that most noise experienced at this location is due to the contribution from the structure-borne path. A low coherence value would indicate that the wall is not a contributor but is from another source or multiple sources. Coherence measurements must usually be made at several locations to thoroughly evaluate the problem.

Other measurements such as the transfer function can also be made during the same set of measurements. The transfer function would indicate the amount of energy that is traveling or being attenuated from one point of measurement to the other point of measurement. This is helpful in evaluating the compliance of a structure, for example, how well the energy transfer is accomplished or how easy it is for the structure to sustain the energy as it is damped out naturally. Other terms such as cross spectrum, autocorrelation, and phase may also be used and can be measured with this type of analyzer.

Another type of measurement that is becoming more useful is sound intensity. This can provide a method of determining the sound power level of equipment using a specialized analyzer. The basic system consists of a two-channel spectrum analyzer having intensity measurement capability and an intensity probe. The probe consists of a two-microphone system. The microphones are either facing each other or in parallel. They must be phase matched to provide the accuracy required.

Since the sound intensity is related to the sound power level by the area of the source, the sound power level of a source can be obtained directly with calculations performed by the analyzer. While it is not possible to measure the sound power directly, this provides a valid method of obtaining the sound power level without having to take the equipment to a laboratory and obtain it in an anechoic room or reverberation room. Large and small equipment can be measured in situ.

Sound intensity provides a method for determining power flow from a piece of equipment and finding out which components are the noisiest. This method allows diagnostic work to be done for noise control purposes. The various sources on a piece of equipment can be listed in order of priority. In addition the various pieces of equipment can be listed in order of priority. Thus sound intensity provides another tool in the noise control plan for best managing a project. This measurement method requires specialized instrumentation and very accurate techniques. Some equipment is available for rent or lease and can be mastered.

References

ANSI. 1983. American National Standards Institute. *ANSI S1.4 1983 Specifications for Sound Level Meters*. Amended 1985. New York: Secretariat Acoustical Society of America.

Beranek, Leo L., ed. 1960. *Noise Reduction*. New York: McGraw-Hill.

Irwin, J. D., and Graf, E. R. 1979. *Industrial Noise and Vibration Control*. Englewood Cliffs, NJ: Prentice Hall.

Lord, Harold W., Gately, William S., and Evenson, Harold A. 1987. *Noise Control for Engineers*. Malabar, FL: Robert E. Krieger.

Peterson, Arnold P. G., ed. 1980. *Handbook of Noise Measurement*. Concord, MA: General Radio Company.

Chapter 9

Noise Measurement Procedures

INTRODUCTION

The goal is to make valid measurements. In order to achieve this goal, it is helpful to recognize that the results of a measurement are determined by a number of factors, among which are the following:

1. The phenomenon being measured
2. The effect of the measurement process on the phenomenon being measured
3. The environmental conditions
4. The characteristics of the type of transducers and instruments being used
5. The way the instruments are used
6. The observer or surveyor

Although many useful measurements are made by people with little background in acoustics, the chances of making valid measurements are increased as the understanding of these factors becomes more thorough. Thus, firm knowledge of the fundamentals of acoustics, transducers, instruments, and measurement techniques is helpful in making noise measurements.

An important rule in any measurement task is to look at the results and determine if they are reasonable. If not, look into the possible sources of error. These sources can be simple things like poor connections, plugs in the wrong place, no power, low batteries, controls set incorrectly, damaged equipment, stray grounds, and magnetic fields.

Recognizing the accuracy limitations of sound measurements is also important in order to be reasonable in the approach to a measurement problem. Thus, consistency to 0.1 dB or better is attainable in only a few laboratory calibration procedures in acoustics and not in general acoustical measurements. Field calibrations of sound level meters at one frequency with a calibrator may be consistent to 0.5 dB or slightly better. A repeatability of ± 1 dB is difficult in general measurements, even under carefully controlled conditions. In general a ± 2-dB repeatability is expected in normal field measurements [Bruel and Kjaer, 1971].

It is useful to think of a measurement result simply as an estimate of the sound level. It is an estimate not only because of various uncertainties in the measurement, but also because sound levels being evaluated are not always stable. These uncertainties can be due to such items as:

1. Changes in equipment being measured
2. Instruments being used
3. Nature of sound source
4. Experience of surveyor

MEASUREMENT TECHNIQUES

There is more to noise measurement than just pointing an instrument at a source and reading the numbers. There are standard techniques or procedures that should be followed whenever noise measurements are conducted. The American National Standards Institute (ANSI) has several standards that should be followed and referenced when taking measurements. Three of these are:

1. ANSI S1.2-1962 American National Standard for Physical Measurement of Sound (partially revised by S1.13-1971 and by S1.2-1972)
2. ANSI S1.4-1985 American National Specification for Sound Level Meters
3. ANSI S3.15 Method for Measurement of Community Noise

Others might apply for typical situations. For example, when noise from gas turbines is measured ANSI B133-1972 should be followed. This standard also references some of the basic standards indicated above. There are also international standards published by the International Organization for Standardization (ISO) in Geneva, Switzerland. They are somewhat compatible with the American ANSI standards. They can usually be obtained through ANSI in New York City.

When more sophisticated acoustical information is needed, such as noise intensity or coherence, more elaborate measurement techniques are required. A more complete list of standards is found in Appendix C. The information presented in this chapter provides a basic foundation for conducting general sound surveys.

Preparation beforehand can save frustration and worthless data recording during the measurement period. To ensure the accuracy of the data, a calibrator must be used before and after each measurement period. This assures that the instrument is still in calibration. Instruments that have a digital display do not drift as much as the older technology having an analog meter display.

A calibrator is a single- or multiple-tone device that produces a known sound pressure level when it is placed over the microphone. The type of calibrator will depend on the instrument manufacturer. It is important to document information to identify all measuring and calibrating instruments and conditions of use. As a minimum, the following recommended procedure should be used when preparing for and conducting a sound survey [Peterson, 1980]:

1. For best use of an instrument, it is recommended that the manufacturer's instruction manual be thoroughly reviewed.
2. Check the battery level and replace the battery if required.
3. If there is any damage to the carrying case(s) or instrument casing, the entire instrument should be thoroughly checked for possible damage to the microphone or electronics. This step is very important whenever an instrument is used by more than one person.
4. Calibrate in accordance with the manufacturer's instruction manual before the survey. Record these reading(s) on an appropriate data sheet. Also, record the equipment manufacturer, model number(s), and serial number(s) of the calibrator and all instruments used for the survey.
5. Sound level measurements should be obtained with the meter set to the appropriate weighting scales and "slow" response.
6. It is recommended that the operating conditions of the machinery be documented. This documentation should include production rate, any unusual conditions, distance the measurements are made from each machine, locations of adjacent machines. A diagram of equipment layout and measurement position is extremely important.
7. Perform a calibration check at the end of the survey period, and record the reading (without adjustment) on the same data sheet. The manufacturer's instruction manual should provide the calibration procedure the particular instrument re-

quires to ensure optimum accuracy of all instruments. If the postsurvey calibration check should be outside the acceptable range provided in the instruction manual, then the sample results may be in error and the survey should be repeated.

STEADY, FLUCTUATING, AND NONSTEADY SOUND LEVELS

When noise measurements are taken, two distinct types of noise can usually be heard: steady and fluctuating noises. The type of noise really depends on the type of industry. For example, in process and power plants, the noise is usually steady. However, for many manufacturing plants there may be steady noise as a background but with fluctuating or nonsteady noise above and below the steady noise. Listen to the noise in the environment in which you find yourself and distinguish these two different types of noises.

A steady noise is defined as noise that does not vary more than 3 dB [Lord et al., 1987] during a typical observation of 10 sec with a slow meter setting. If octave band measurements are being made and a discrete tone is heard, this tone can be distinguished in the octave band data if the noise level is 5 dB or much higher than the adjacent bands. Thus, it is recommended that the meter be moved around to determine if this could be a result of a standing wave and obtain an average level for the area. This is why a sound level measurement should be considered an estimate. A certain amount of judgment is required to determine what a reasonable level would be for a given situation.

Steady noise that fluctuates less than 3 dB can be accommodated by observing the instrument display over a 10-sec period and recording the central tendency of the needle. This is called fluctuating levels. For fluctuations greater than 3 dB but less than 10 dB, a slightly more involved procedure is required. Observe both the "average" needle tendency and the range (maximum to minimum) of values in a 5-sec period. Additional measurements of the "average" should be recorded until the number of data points is greater than the observed range in decibels of the meter fluctuation; that

is, if the range is 10 dB take 10 or more measurements and average them. The arithmetic mean of the "average" readings should be calculated on an energy (logarithmic average) basis and used as the noise level for that particular position. If the instrument has an equivalent level function this can be used to provide the result.

Nonsteady noises (those levels that vary more than 10 dB) often must be carefully measured with a sound level meter. The maximum and minimum levels must be documented as a minimum to bracket the range of levels. In addition, a histogram can be made very easily over a 5- to 10-min (length will depend upon the type and range of levels) period. Simply list the levels in 1-dB increments in the left-hand column of a page, then take a moment every 5 or 10 sec and mark an X by the level. Over a 5- to 10-min period a large number of measurements can be observed and a visual image will emerge as the average sound level. The data can be manipulated statistically if desired to obtain a mean and standard deviation.

If a noise dosimeter is available, it can be placed at the location and left to operate for a half hour to an hour and the results read out. Some noise dosimeters can be programmed to give all the data necessary about the statistics of the noise environment at that location.

When discrete tones are generated (and heard), patterns of standing waves or large variations in sound pressure may develop. The sound pressure level difference can vary 10 dB or more. This difference may be determined when the sound level meter is moved around and large variations occur. To determine the average noise level, move the microphone through the maximum and minimum levels and record the central tendency of the needle or digital display. If the instrument is able to measure the equivalent level, place the instrument in that mode and move the microphones around until the reading is steady and record the result.

PROCEDURES FOR DEVELOPING A NOISE CONTOUR MAP

Noise contour maps are often useful because they provide a visual image of the sound field over an area. They can be used to help define hearing hazard areas where the use of hearing protection devices is required. Although noise contours are a useful tool for visualizing the noise zones, it should be remembered that they represent approximations of the noise levels. Consequently, noise contours should be used as general guidelines for directivity of noise sources, implementation of hearing conservation programs, and posting of warning signs. There are two techniques that can be used for obtaining a noise contour map. They are the *direct method* and the *grid method*.

Direct Method

The direct method entails the tracing of equal contour lines representing a defined sound level on a scale equipment or area layout drawing. Each contour is drawn progressively as mea-

surements are made, as illustrated in Fig. 9.1. A color coding or line symbol for each contour or spacing between contours is helpful in interpreting the results. The following procedure for the direct contour method is recommended:

1. Walk through the plant or unit while observing sound level measurement (SLM) readings, guiding your path to maintain a constant reading (example: 85 dBA) using the A-weighted and "slow" response settings on the meter. Follow each path to the edge of the survey area or until it closes on itself forming a loop or circular pattern around a noise source or sources.
2. Trace the path followed on a plant layout or area map. This can most readily be done by a second person following the surveyor.

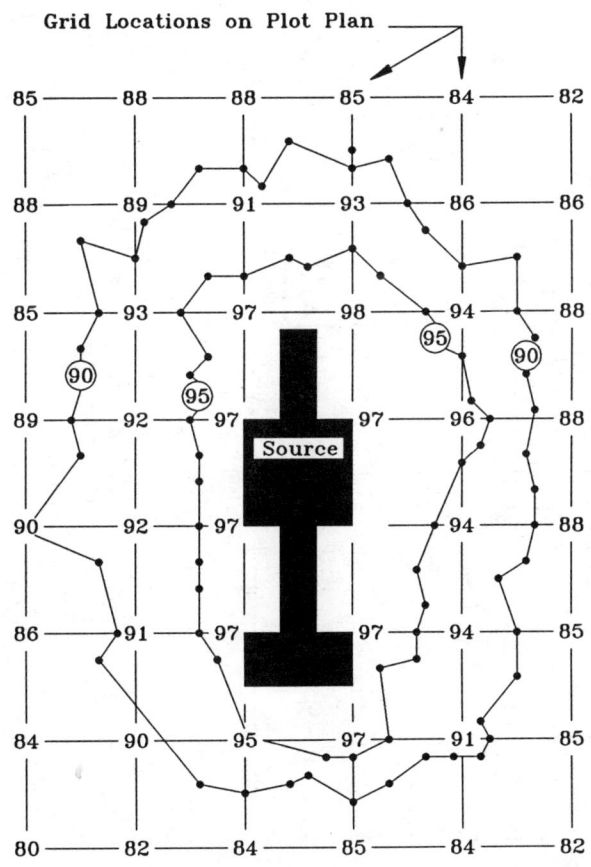

FIGURE 9.1   Hand-drawn noise contours for 90- and 95-dBA levels.

3. Repeat the process for as many contours as desired or needed. In finding a starting point for the higher-level contours, it is important to approach all potential noise sources in the area to find any closed contours that might not be apparent on a straight walk-through or boundary traverse.

The number of contours required depends on the evaluation to be performed and the degree of analysis needed in each zone. As a minimum, seven contours forming eight hearing-hazard zones are recommended. The zones should be bounded by contours of 85, 90, 95, 100, 105, 110, and 115 dBA. Usually the higher levels, such a 105, 110, and 115 dBA,

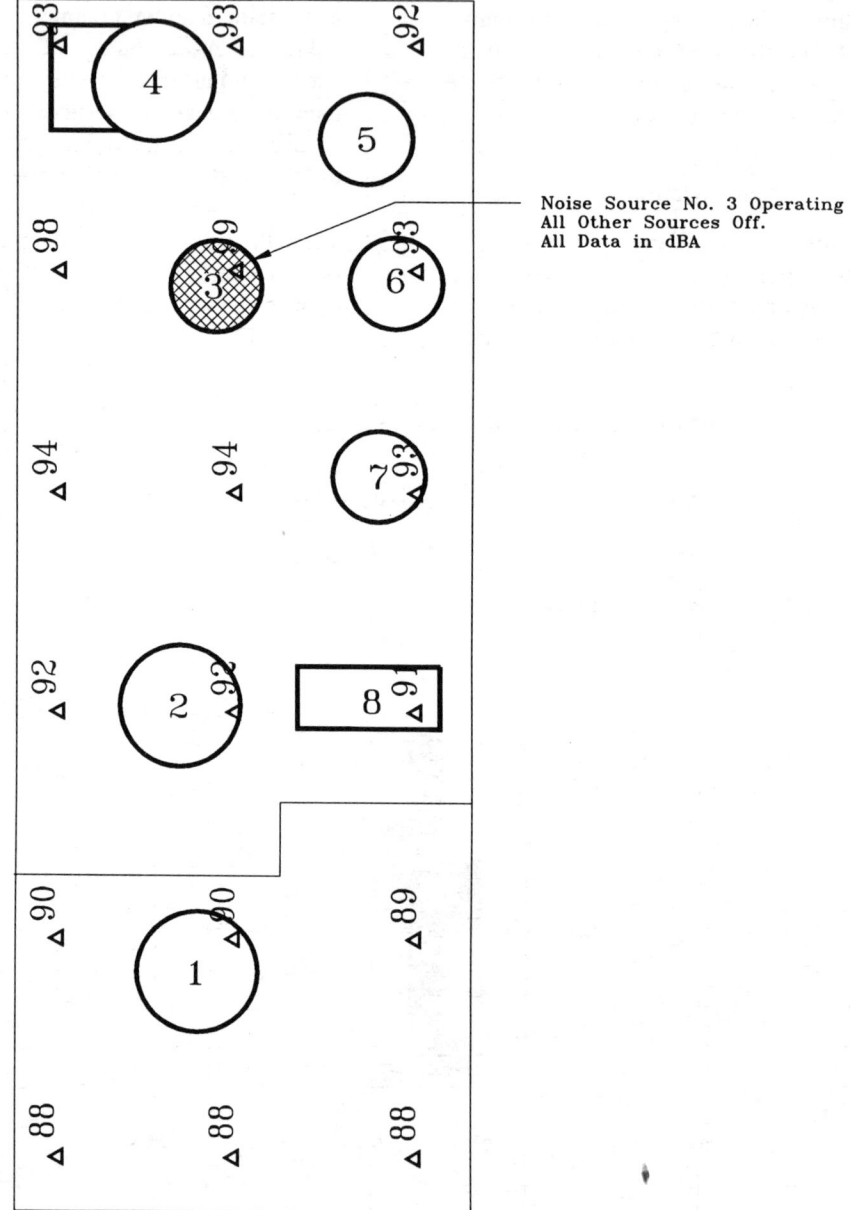

Noise Source No. 3 Operating
All Other Sources Off.
All Data in dBA

FIGURE 9.2   Grid method at plant column lines.

will be small or nonexistent. For areas having a high density of noise sources and areas highly populated by employees, a smaller zone size may be required. For example, contours recorded in steps of 2 dBA, as opposed to the previously described 5 dBA, may better define areas where use of hearing protection is required.

Grid Method

The grid technique is accomplished by sampling sound levels at preselected points within a plant or other area of interest, usually at column lines. This is shown in Fig. 9.2. The points should be located at equal intervals over the area. An accurate to-scale layout drawing of the survey area should be used to locate measurement points. The points are located by drawing a parallel line grid over a plot plan in both directions. The row and column spacing is determined by the desired detail of the sound level contours to be obtained. In the vicinity of a high density of noise sources or for indoor measurements, increments of five or ten feet should be chosen.

Once the data have been obtained at each accessible grid point, sound level contours are generated by passing contour lines through the field of points. The contour values are found by direct visual estimation of where they should be placed between grid point values.

This can be performed graphically by segmenting each line between two grid points according to the number of decibels difference in the two readings of interest. For example, with SLM readings of 93 dBA obtained at one grid point and 88 dBA at an adjacent point, the 90-dBA contour line can be interpolated between the measurement points away from the 88-dBA grid point.

There are also computer programs that can generate the contours from the data measured at grid points (Fig. 9.3). This method might allow an amount of noise reduction to be made in the computer program and results observed in a new contour plot.

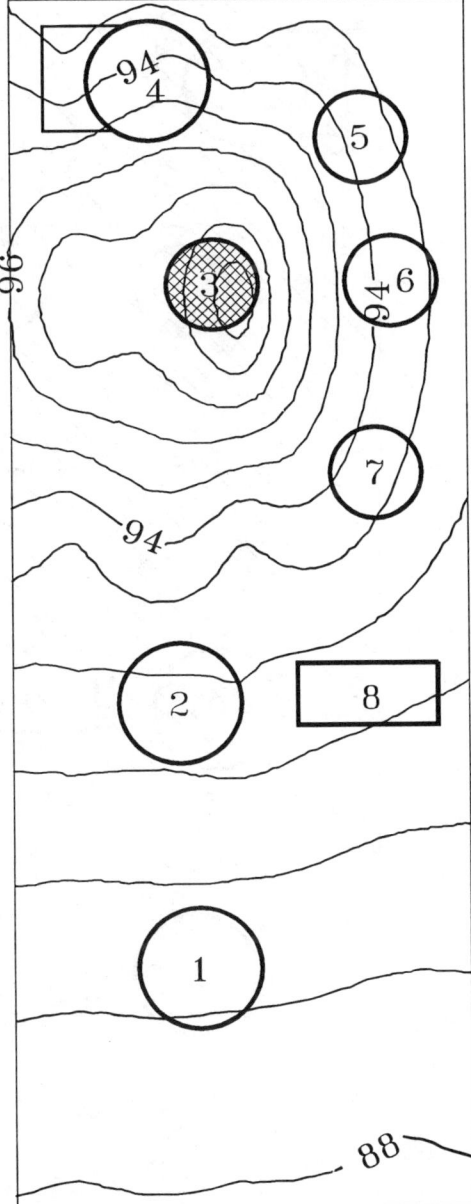

FIGURE 9.3   Computer-generated noise contours.

Finally, for a hearing conservation program it may be also advisable to use the grid method to show high and hazardous noise zones. Figure 9.4 shows an example of this method. As the engineering controls are installed the changes will be evident. These presentations can be used to show progress and results.

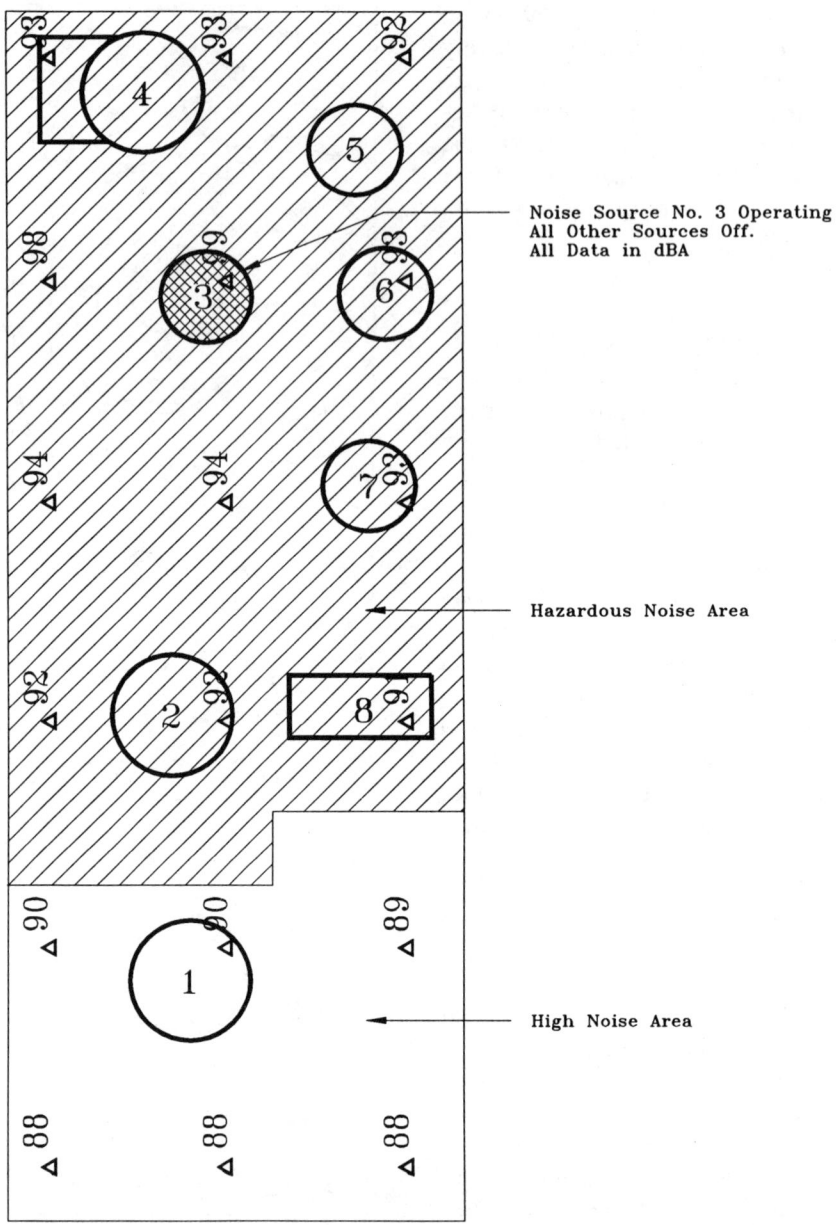

FIGURE 9.4   High and hazardous noise zones derived from column line measurements.

Engineering Noise Survey

A noise problem must first be defined as the initial step toward a solution. An engineering noise survey should be conducted to measure in detail the noise level of each source. The measurements must be of sufficient refinement to help separate the various noise contributors from one another so that each source can be appropriately treated. Many of these individual noise sources can be separated by differences in the frequency content of their characteristic noises. This leads to a requirement for a frequency analysis of the noise. An octave band

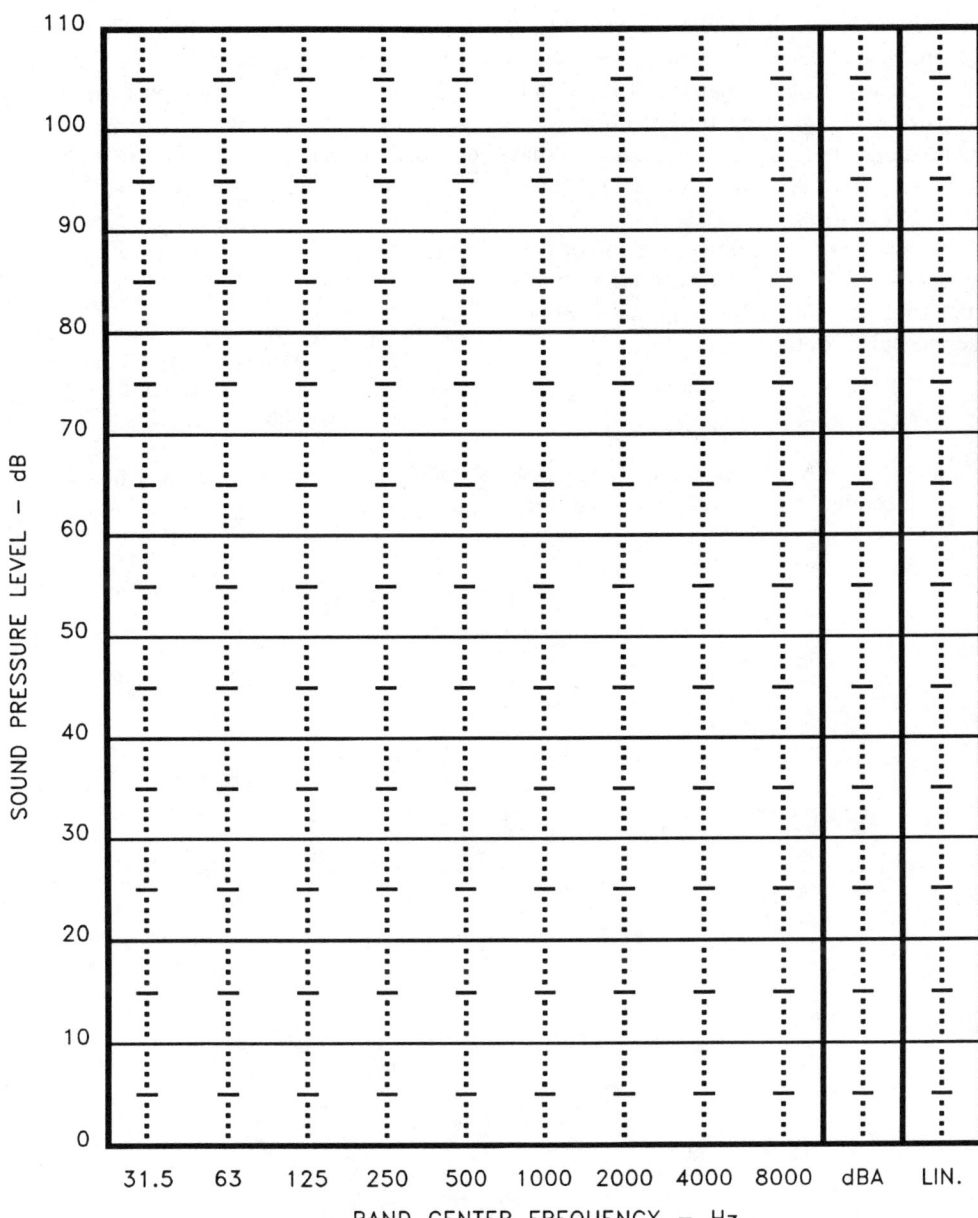

FIGURE 9.5   Octave band noise chart.

sound level meter can be used to make the frequency analysis.

Measurement of frequency data should be conducted in accordance with the instruction manual provided by the instrument manufacturer as well as the measurement techniques previously described. The frequency data can be recorded on a sheet similar to that found in Fig. 3.4. By plotting the frequency data on a chart, one can learn additional, often critical, information pertaining to the particular noise source being investigated. An octave band chart like that found in Fig. 9.5 can be used for this purpose.

In addition to octave band measurements, if the sound level meter has a one-third octave filter, each octave band can be divided into three parts revealing even further information about the noise source(s). In some cases, it may be possible to turn equipment on and off to determine the influence of a particular source. Thus the survey of overall octave band or one-third octave band data can be measured and plotted to assist in determining the best approach to noise control.

Acceptance Noise Survey

Attachment A in Appendix E outlines a procedure for conducting a noise survey after the equipment has been installed. This procedure can also be used for acceptance tests in an equipment manufacturers shop facilities. Adding this type of section to a specification adds emphasis to the seriousness which the Owner places on noise.

References

Bruel, and Kjaer, 1971. *Acoustic Noise Measurements*. Seborg, Denmark: K. Larson & Son.

Lord, Harold W., Gately, William S., and Evenson, Harold A. 1987. *Noise Control for Engineers*. Malabar, FL: Robert E. Krieger.

Peterson, Arnold P. G., ed. 1980. *Handbook of Noise Measurement*. Concord, MA: General Radio Company.

Part III

Case Histories

Chapter 10

Noise Control Measures for Specific Equipment: Case Studies and Noise Control Solutions

INTRODUCTION

In this chapter the guidelines for noise and vibration control outlined in this book will be applied to look at specific problems and case histories. The case histories cover a broad range and draw on 28 years of the author's experience. Some generic problems, such as barriers and enclosures, will be covered, as well as more specialized areas such as material handling and materials for clean rooms and food processing. The problem will be stated along with any data for further definition. The problem will be analyzed and solutions presented. In those cases where results are available they will be furnished, as well as materials used. The following areas will be covered by the case histories in this chapter:

Enclosures
Barriers
Combination barrier and enclosure
Acoustical lagging
Vibration damping
Silencers
Hydraulic pump systems
Source noise control systems
Dealing with noise flanking paths
Material-handling systems
Special consideration for the food-processing industry

Vibration isolation
Punch presses
Active noise control

Many more examples could be added to this list. It is hoped that this selection will show the concepts without getting into a great deal of detail. Most problems that are encountered require thoughtful consideration. The various constraints imposed by the actual situation may limit the noise control options in the final analysis.

ENCLOSURES

There are many types of noise control enclosures. Some of these include the close-fitting type to cover a single piece of equipment, the room type to house employees, and partial enclosures covering noisier items on a larger piece of equipment.

Close-Fitting Enclosures

Description—Cold Headers The function of a cold header (Fig. 10.1) in a metal-working operation was to produce a screw head and body by forming a head on a piece of wire using a cold forming process [Pelton and Storment, 1976]. This was the first step in making a screw

FIGURE 10.1   Cold header without enclosure.

and preceded the threading and head-slotting process. The six machines in this facility use $\frac{1}{8}$- and $\frac{3}{16}$-inch wire at a rate of 350 to 450 strokes per minute. This created a noise level of 103 dBA at the operator position. The machines were lined up on a rear wall, creating an oil mist in the process of making the screw body from the heat and cooling oil. The screw body

control. Several of the manufacturers were using close-fitting enclosures that did not interfere with the operation.

One of the added factors in this facility was the oil mist. It was in the air and covered equipment because the oil mist collector was ineffective. The following noise spectrum illustrates typical noise levels at the operator's position:

Freq (Hz)	63	125	250	500	1000	2000	4000	8000	dBA
L_p (dB)	83	92	94	95	100	98	97	96	103

fell into a basket where the oil vaporized. The room was very hard acoustically, with noise levels ranging from 95 to 100 dBA within 20 to 30 ft of the cold headers.

Analysis Detailed noise and vibration measurements were made to evaluate the noise sources within the machine. A strobe was also used to see the movement of the various components of the cold-heading process. In addition, various cold-header manufacturers were visited to determine their approach to noise

The criterion for this area was 85 dBA between machines in operation. Thus an individual enclosure must reduce the level to 82 dBA.

Design and Recommendations After a detailed review of all data, visits to cold-header manufacturers' plants, and discussions with production personnel, a close-fitting enclosure was determined to be the most feasible approach. A prototype enclosure (Fig. 10.2) was designed to provide the required access, motor cooling, and oil mist control. The results for the prototype were:

Freq (Hz)	63	125	250	500	1000	2000	4000	8000	dBA
L_p (dB)	77	80	75	72	73	76	77	76	82

The enclosure is constructed of light-gauge metal lined with acoustical foam. After several weeks of operation a design review meeting with production personnel was held in order to obtain their comments. The final design was developed for the remaining cold headers incorporating their comments. The final design incorporated an oil mist collector of a centrifugal design. The air was drawn through the enclosure to cool the drive motor. The final design is shown in Figs. 10.3 and 10.4.

The basic design was a square tube frame with 16-ga steel outer panels bolted to the frame. Gaskets were used to seal the panel to the frame. Thus any panels could be removed for easy access. The interior of the enclosure was covered with perforated metal to protect the acoustical foam. The acoustical foam used had a mylar film to keep oil out of the porus foam material. The enclosures have been in operation for 15 years. Noise measurements after 15 years of operation are shown in Fig. 10.5. The location is between two cold headers in operation.

During the years of operation the hardware such as hinges and latches have been upgraded. Since the seals have not been maintained, the noise levels have increased about 5 dBA in the last five years of operation. The author has taken readings of all the machines in this area for several years as a part of the annual noise survey in the facility. It is important to have heavy-duty hinges and latches, keep the seals maintained, and replace as necessary.

Room-Type Enclosures

Description—Personnel Room Operating personnel in a utility plant of a large manufac-

FIGURE 10.3   Final enclosure design with doors open.

FIGURE 10.4   Cold header in operation with enclosure.

FIGURE 10.2   Prototype enclosure.

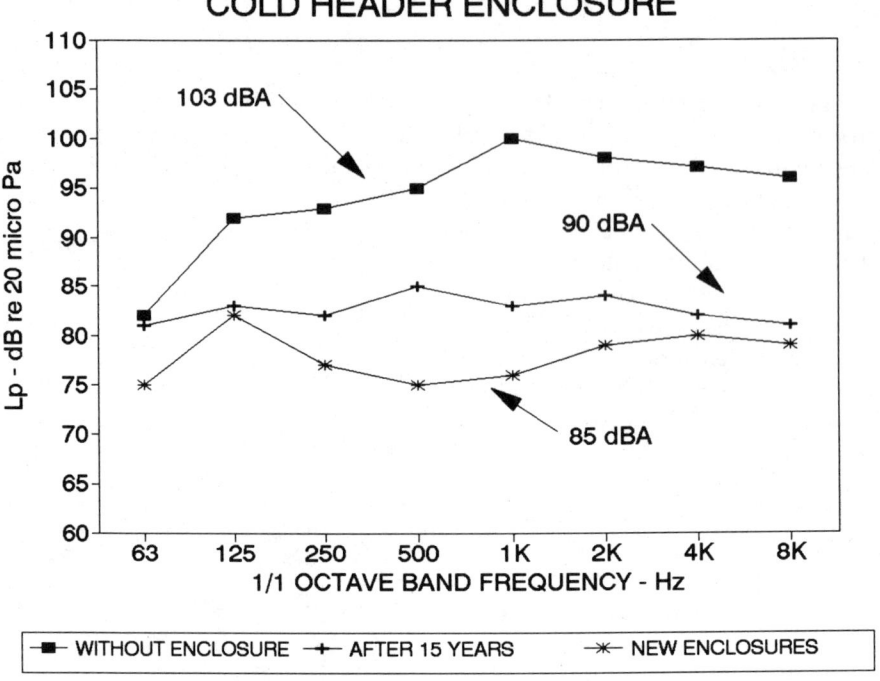

FIGURE 10.5   Noise control results.

turing plant were located in an open area. They were required to wear hearing protection throughout their shift. The noise levels in the plant ranged from 94 to 103 dBA.

The utility plant supplied chilled water, compressed air, and steam to the manufacturing plant. The utility plant contained four 600-hp centrifugal air compressors, three absorption chillers, one centrifugal chiller, and three low-pressure boilers.

Analysis   A noise survey in the plant was conducted to determine the noise spectra of the noise sources and in the area of the operator's desk. A plan of the equipment in the plant is shown in Fig. 10.6. The general noise levels are shown on this plan. In order to have a quiet place for the operators while they were not

making their rounds of the equipment, and to allow them to communicate with each other and on the telephone, a personnel room enclosure was determined to be the best approach.

The criteria for personnel and telephone communication for this area was determined to be in the 60- to 65-dBA range. Thus the enclosure had to reduce the plant noise by slightly over 30 dBA at the location selected. The noise level was 94 dBA before any treatment.

Design and Recommendations   Several noise control equipment manufacturers provide standard room-type enclosures for this application. A review of catalog material revealed that a standard 4-in. panel wall would provide the noise reduction required. The following noise spectra show the noise level in the area and the noise reduction of the room enclosure.

Freq (Hz)	63	125	250	500	1000	2000	4000	8000	dBA
L_p (dB)	89	90	88	85	90	86	88	84	94
N.R. (dB)	8	20	25	31	37	35	35	32	31
In room	81	70	63	54	53	51	53	52	63

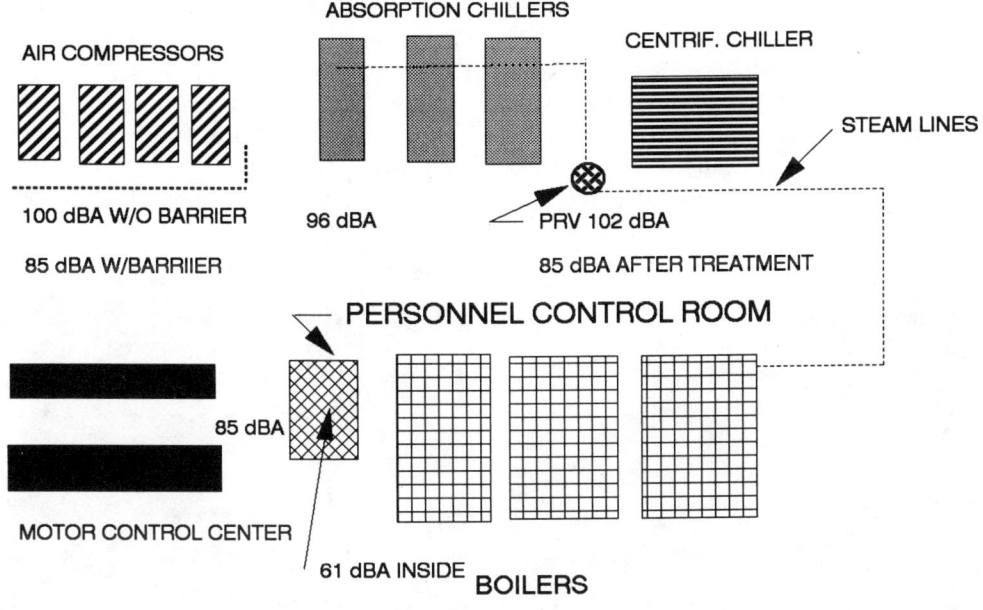

FIGURE 10.6   Utility plant with personnel control room.

The measured results in the room have varied from 61 to 65 dBA depending upon the outside level as well as the maintenance of the seals. Over time the seals will wear and need to be replaced.

The room located in the utility plant is shown in Fig. 10.7. These types of rooms are very easily located within the plant. In many plants the controls and gauges are located in the room as well. The operators are usually required to take some readings around the plant so they can see how the equipment is operating.

FIGURE 10.7   Personnel room.

BARRIERS

The basic purpose of a barrier is to block the line of sight from the noise source to the receiver. This can be accomplished with a variety of materials. Some of the more typical materials are concrete block, poured-in-place con-crete, plywood, metal acoustical panels, modular plastic panels, and loaded vinyl curtains. Examples of some of the materials are illustrated in Fig. 10.8. The materials selected will depend upon the noise source and the noise spectrum.

For example, the loaded vinyl curtains do

(c)

(d)

FIGURE 10.8   Typical materials for noise control barriers: (a) plywood; (b) high-density plastic modular panels (courtesy of Sound Fighter Systems, Shreveport, LA); (c) metal panel barrier; (d) quilted fiberglass protected material with 1 lb/ft^2 septum hung in panels as barrier (material courtesy of United Process Corp.).

not offer any low-frequency noise reduction with a weight of 1 lb/ft^2, whereas concrete block (45 1b/ft^2) is quite good in the low frequencies. Thus, an evaluation must be made of the spectrum rather than just measuring the dBA level.

The two case histories described below will describe an indoor condition in a utility plant of a large manufacturing facility, and an evaluation of an outdoor highway barrier.

Indoor Barrier

Description The function of a utility plant in a manufacturing facility was to provide steam, chilled water, and compressed air. The air compressors were identified as a major noise source related to the center aisle down the plant. The noise level was measured as 100 dBA. The four air compressors were 600-hp electric motor driven, high-speed, three-stage centrifugal units. The intake air was taken from the room through an oil bath filter [Pelton and Storment, 1975].

Analysis A series of noise measurements were taken around the equipment. The data indicated the noise sources around the compressor system were as follows: drive motor; air intake and radiation from the compressor and discharge piping. Discussions with the plant personnel indicated that their movement was from a control room behind the compressors to center aisle down into the plant. In addition, certain maintenance procedures had to be carried out on a regular basis. Thus, noise was to be controlled for one compressor while the others were operating.

Design and Recommendations In order to provide the noise control required and meet the maintenance constraints, a portable barrier was recommended. This was constructed of 4-in. acoustical panels with supports and casters to move them. The data presented in Table 10.1 show how the barrier calculations were made for a simple barrier calculation. Since the roof was about 30 ft and the aisle was close to the barrier, it was not necessary to consider the reflection from the roof to be added into the resulting noise levels.

The barrier is shown in Fig. 10.9. The panels were fitted around the air inlet, and that can be rolled aside to shield the adjacent compressor when it is off during maintenance.

The last recommendation was a barrier around the air inlet. This is shown in Fig. 10.10. The panels were constructed of 2-in. acoustical panels. They redirect the air as well as the sound from the compressor inlet. The results of this total approach are shown on the graph in Fig. 10.11.

Outdoor Barrier

Description The residents of a housing subdivision located next to a major freeway were concerned about noise from the traffic. The highway department had not extended the concrete barrier located to the south of them past their subdivision. The homeowners wanted to know the traffic noise level where the barrier should be located, and how much noise reduction would be afforded by a similar barrier.

TABLE 10.1   Barrier Noise Reduction

	A	B	D	Z	63	125	250	500	1000	2000	4000	8000
	8	12	15	5	17.94	9.04	4.52	2.26	1.13	0.57	0.28	0.14
Fresnel no. N					0.56	1.11	2.21	4.42	8.85	17.70	35.40	70.80
Barrier noise reduction					10	13	16	19	23	23	24	24

Note: A = source to top of barrier, B = top of barrier to receiver, D = source to receiver distance, $Z = A + B - D$, W = wavelength of sound, $N = 2Z/W$.

FIGURE 10.9   Metal roll-around barrier panels. (Reproduced from ASME Paper 75-WA/SAF-2.)

FIGURE 10.10   Air compressor inlet plenum barrier panels. (Reproduced from ASME Paper 75-WA/SAF-2.)

Analysis   Noise measurements were made next to the freeway to determine the traffic noise level. In addition, measurements were made behind the existing barrier at locations to simulate the homeowners' locations to determine the actual reduction the barrier could provide. Finally, measurements were made far from the freeway in the community to obtain some impression of the ambient level.

A freeway measurement location is shown in Fig. 10.12. The existing barrier is shown in Fig. 10.13. The relative location of the freeway and homeowners is shown in Fig. 10.14.

Data were compared to a study prepared by the highway department three years earlier showing that the noise levels had increased by approximately 5 dBA. This increase was due to increased traffic flow and was predicted by the highway department's study.

The noise level measured next to the freeway was 88 dBA. At the homeowners' property line the levels ranged from 68 to 74 dBA depending on the time of day and wind direction. Thus the distance effect was from 14 to 20 dBA. The noise reduction of the existing barrier of 7 ft 2 in. high, as well as the ambient level, is shown in Fig. 10.15.

Recommendation   Since the distance effect ranged from 14 to 20 dBA, it is possible to almost reach the ambient level when compared to the data in Fig. 10.15. The barrier height should be the same as the existing barrier and be placed next to the freeway. In general, barriers can reduce the level from 7 to 15 dBA depending on the noise spectrum and distances involved.

Combination Room and Barrier

Description   In a large manufacturing plant production required metal working and plastics manufacturing. In the latter process the scrap from the injection-molding process was reclaimed by grinding, extruding, and pelletizing. The new pellets were recycled back into the process. The more notable operation was the grinding of scrap [Pelton and Storment, 1976].

Analysis   The equipment layout is shown in Fig. 10.16. There were two conveyor-fed grinders and two small manually fed grinders. All of the equipment was located behind a con-

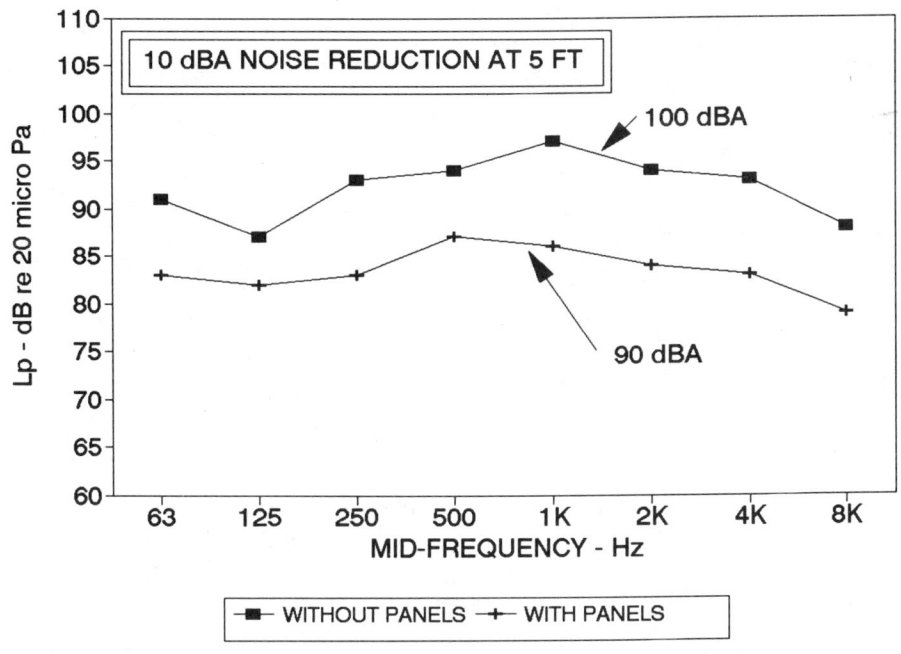

UTILITY PLANT AIR COMPRESSORS
BARRIER PANEL NOISE CONTROL

FIGURE 10.11   Results of noise control.   (Reproduced from ASME Paper 75-WA/SAF-2.)

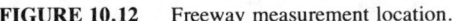

FIGURE 10.12   Freeway measurement location.

FIGURE 10.13   Existing concrete barrier.

FIGURE 10.14   Relative location of freeway and homeowners.

on the opposite side of the block wall they were, nonetheless, exposed to high levels of noise because of the large openings in the wall.

The conveyor-fed grinder had to have large openings for the operator to view the top of the chute. Thus, the direct noise of the grinder was received by the operator. Since the room housing the grinders was hard, the noise built up at least 5 to 10 dBA due to reflections.

The inlet chute of the small grinders extended through the hole framed in the wall. The operator was exposed directly to the noise of the grinder. The inlet was not sealed in the wall and the room noise leaked past as well.

Typical operator noise levels were as follows:

Freq (Hz)	63	125	250	500	1000	2000	4000	8000	dBA
L_p (dB)	82	92	99	103	103	102	100	90	108

crete block wall with openings for material to pass through. While the operators were located

Typical levels in the grinder room ranged from 108 to 112 dBA.

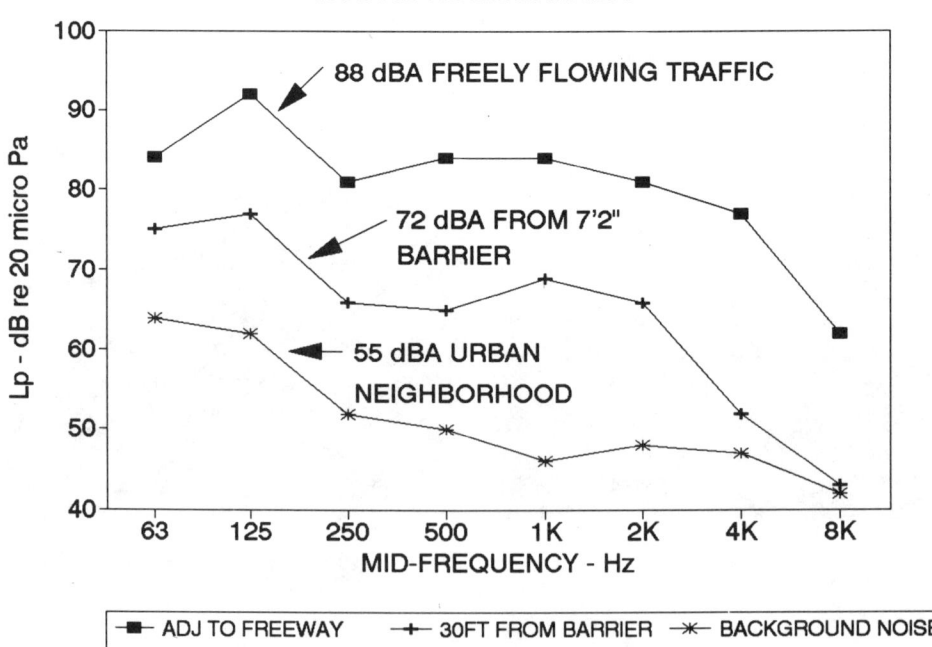

FIGURE 10.15   Noise spectrum.

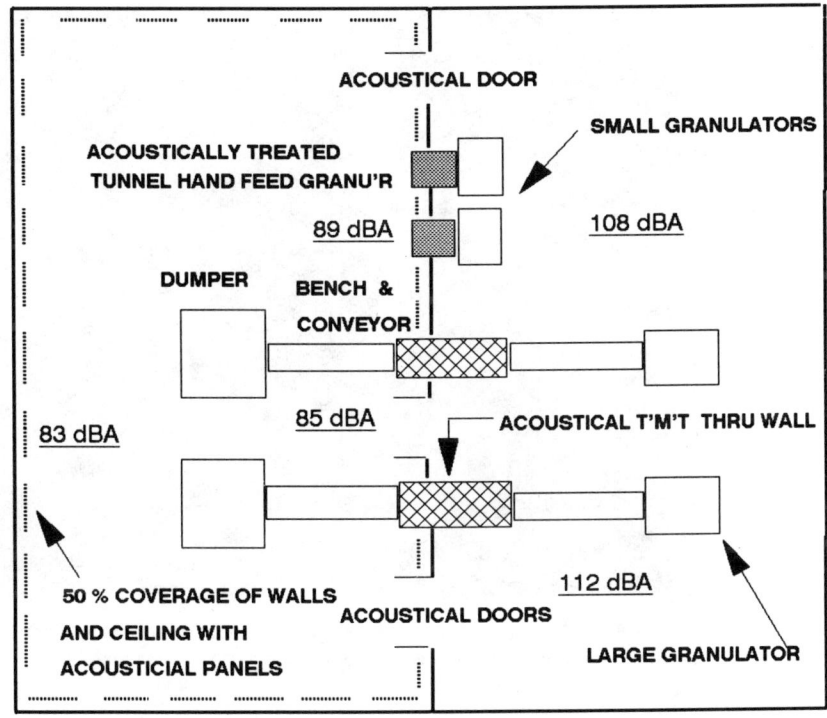

FIGURE 10.16 Plastic grinding area. (Reprinted by permission from *Sound and Vibration* magazine.)

When the plant was constructed this wall was built and acoustical doors were installed. However, the seals and closers were never installed and the doors usually remained open.

Recommendations After several discussions with the operators and observations of the operation it was apparent that the operator did not need to see the top of the in-feed chute to the conveyor. Therefore the solution was to acoustically seal the openings in the wall to bring it up to its full potential. The following steps were carried out:

1. Acoustical baffles were installed at the belt conveyor feed opening, not only to partially close the opening but also to acoustically close the opening. The baffles would act as an absorptive tunnel. This is shown in Figs. 10.17 and 10.18.
2. The acoustical door seals were refitted and closers added so that they were op-

erating at their full potential and would not stand open.
3. The operator was placed on an elevated platform to move the hearing zone up and keep it out of the direct path of the noise. This also helped the operator, from an ergonomic standpoint, placing the material on the conveyor belt.
4. For the small grinders a tunnel was built and sealed into the wall. This is illustrated in Fig. 10.19. This tunnel was constructed of heavy steel, was lined internally with acoustical foam and perforated metal, and had a series of three flaps for the material to be pushed through. This not only reduced the opening, but also allowed the last flap to be closed at the throat of the grinder to contain as much noise as possible. The operator could reach down into the box and pick up a handful of material, push it into the tunnel, then use a pusher to get the

FIGURE 10.17 Acoustical baffles at conveyor feed openings at operator's position. (Reprinted by permission from *Sound and Vibration* magazine.)

material through. This action also made him step back, placing the hearing zone further away from the source.

5. Acoustical absorbing panels were added to 50 percent of the ceiling and walls of the operator area to reduce the reverberant buildup of noise. These can be seen in Figs. 10.17 and 10.19.

The resulting levels were 89 dBA at the small grinder and 85 dBA at the table of the large grinder in-feed. If the operator of the small grinder could have been at the same distance as the operator of the large grinder, the level would have been 85 dBA.

For this illustration simple solutions were available to bring a room enclosure up to its

FIGURE 10.18 Acoustical baffles at conveyor feed openings inside grinder room. (Reprinted by permission from *Sound and Vibration* magazine.)

FIGURE 10.19 Small granulator acoustically treated in-feed tunnel. (Reprinted by permission from *Sound and Vibration* magazine.)

full potential. The observations of the work process and discussion with the operating personnel are very important to provide a practical solution. This type of problem does not require rigorous calculations, because in some cases the problem does not lend itself to this type of analysis. The potential wall transmission loss can be determined as well as the effect openings will have on the result.

As a rule of thumb the acoustical panels added to the room will reduce the levels 4 to 6 dBA if the ceiling height is not over 14 to 16 ft and there is no other absorbing material in the space. The only other unknown is the effect of the tunnel opening, since the doors and the small grinders will be sealed. If the opening can be closed up 50 percent, the level can be reduced by 3 dBA. If the opening can be closed up to 10 percent of the original size, the level can be reduced by 10 dBA. The chart in Fig. 5.8 will be of assistance in evaluating this. The other approach for calculation is to use barrier calculation for the tunnel opening and then add this to the noise reduction from the opening.

With an absorptive tunnel the levels can be reduced another 3 to 5 dBA. The longer the tunnel the more reduction. It is apparent from the photo in Fig. 10.19 that more tunnel could have been added if necessary. However, with the level at 85 dBA this was sufficient. It is important to understand that many problems such as these are approached by trial and error, since the materials are relatively inexpensive. Consideration of the practical nature of the solution is very important since the operating and maintenance personnel must work around this "solution" all the time.

ACOUSTICAL LAGGING

Acoustical materials can provide a limited amount of noise control when placed in a hard reverberant room. These materials are usually fiberglass, rock wool, and acoustical foam. They can also be used in other ways to reduce noise. For example, they are placed within the cavity of gypsum board partitions to improve the transmission loss. They can also be used as part of an acoustical lagging system as shown in Fig. 10.20. This system is used to cover such items as pipes, duct work, large process vessels, and chiller units.

The general acoustical lagging system is composed of an impervious barrier separated from the surface by the fibrous material forming an airspace. This acts as an enclosure for the item of concern. The general amount of noise reduction ranges from 10 to 20 dBA. This decrease will vary depending upon the frequency of the source, the thickness of the acoustical material, and the surface weight of the outer cover. A standard acoustical enclosure can provide a noise reduction in the range of 20 to 30 dBA as a comparison.

The following are two case histories involving acoustical lagging.

880-Ton Air-Conditioning Chiller

Description   A large chiller was located on the top-floor mechanical equipment room. The chiller was in an alcove between the curtain wall and a concrete block wall. This is shown in the photograph in Fig. 10.21. On the other side of the block wall was a law library for a major petrochemical manufacturer. The chiller and associated piping were well isolated; therefore the airborne noise was the only concern.

Analysis   Noise measurements were taken at the machine and in the adjacent office space. These data are shown in Fig. 10.22. The data show that the chiller was very high frequency in nature with a level of 98 dBA. The chiller's high-speed compressor created the noise, which was also transmitted to the condenser and evaporator casing of the chiller. Noise was radiated from all of these surfaces.

Noise was traveling from the equipment room to the adjacent space through many cracks and openings in the building. It was determined that in order to obtain a satisfactory noise level in the law library, more than sealing the openings would be required in order to make the acoustical treatment more effective.

Recommendations   Acoustical lagging was selected to treat the chiller. This would also re-

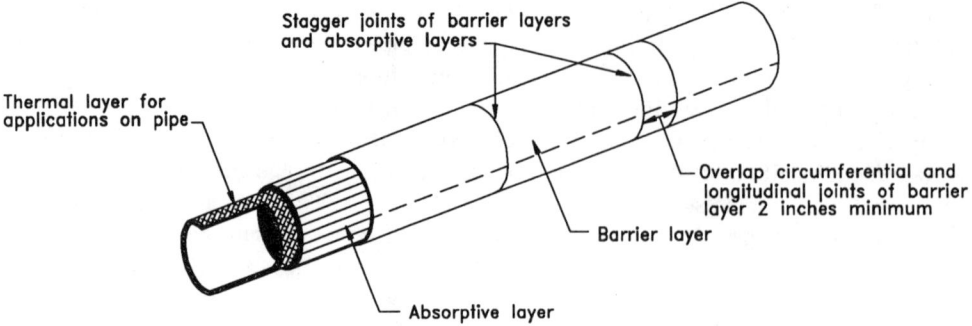

STRAIGHT RUN PIPE LAGGING

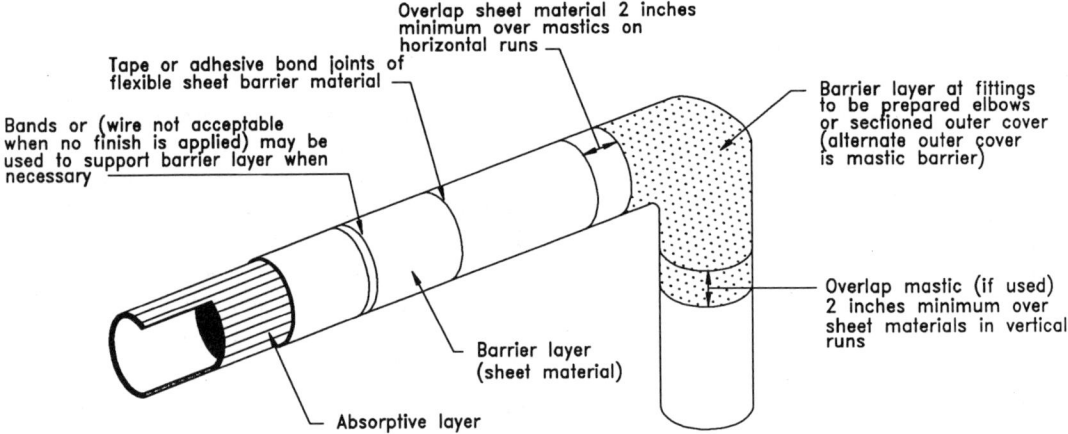

TYPICAL FITTINGS LAGGING
(For Tees and Valves Flexible Acoustical Blanket can)

FIGURE 10.20   Typical acoustical lagging.

duce noise in the mechanical equipment room, providing a safer work environment, and assist in reducing the noise in the law library. The lagging consisted of 2 in. of 3 lb/ft^3 fiberglass semirigid board banded in place on the various surfaces of the chiller. Over that was placed a lead-aluminum laminate installed much as other insulator's aluminum, but with all joints sealed. This is shown in Fig. 10.23. The installation is shown in Fig. 10.24.

The results of the treatment in the space are shown in Fig. 10.25. The overall level was reduced from 98 dBA to 83 dBA. This reduction eliminated the need for hearing protection while working in the area. The noise in the adjacent law library was reduced by applying other architectural acoustical treatment, such as adding gypsum board to the walls and sealing all the openings. Acoustical absorbing material was also installed on the walls next to the chiller.

FIGURE 10.21   880-Ton chiller.

line pressure to that required for the gas-fired boilers. The gas yard noise level was 109 dBA.

Analysis Noise measurements taken around the gas yard indicated a very-high-frequency source for the PRVs. These measurements are shown in Fig. 10.26, a typical spectrum for this type of valve application. Furthermore, it was necessary to have access to the PRVs for maintenance.

Recommendation   This type of noise source is a good application for acoustical lagging. The lagging material consisted of a preformed fiberglass pipe insulation covered with a lead-aluminum laminate with the joint sealed. Since the valves had to be accessible it was not practical to lag them in the conventional manner. Therefore, a flexible acoustical blanket was constructed and installed. This blanket allowed complete accessibility to the valve. The flexible blanket was constructed of a quilted fiberglass material covered with a 1 lb/ft^2 loaded

Pressure-Reducing Valve Station

Description A series of natural gas pressure-reducing valves (PRVs) were located in a gas yard of a power plant. They reduced the pipe-

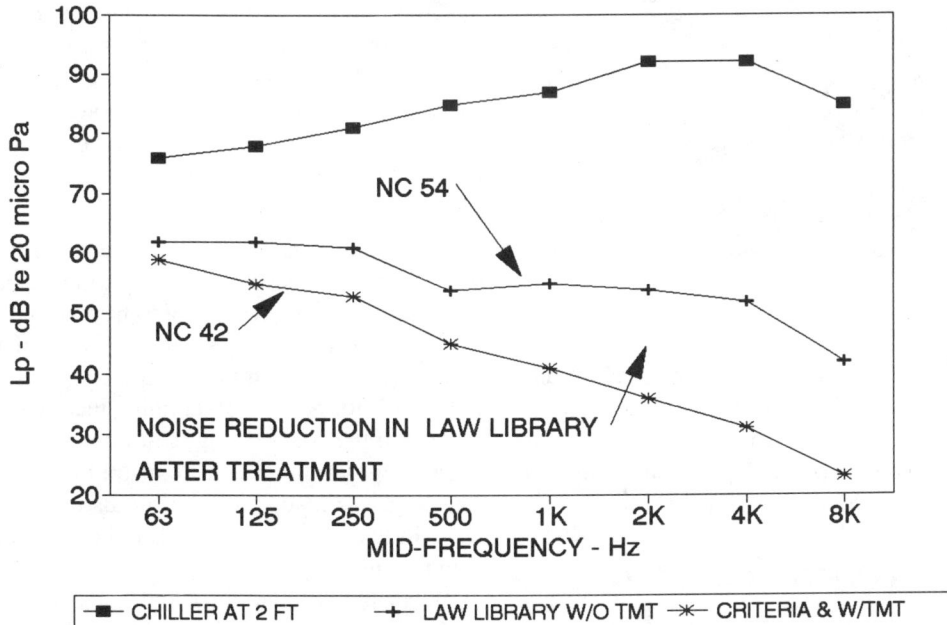

FIGURE 10.22   Chiller noise measurements.

FIGURE 10.23   Lead-aluminum laminate used in acoustical lagging.

FIGURE 10.24   Lagging of an 880-ton chiller.

vinyl material and a rubberized glass cloth material. This installation is shown in the photograph in Fig. 10.27. The results after treatment are shown in Fig. 10.26. In this case an 18-dBA noise reduction was achieved.

VIBRATION DAMPING

Description   A large disk brake system produced a level of 130 dBA with the dominant peak frequency at 980 Hz. A number of em-

ployees were located near the piece of equipment containing the disk brake. The noise occurred as the brakes were applied, which could be for extended periods of time. The noise spectrum is illustrated in Fig. 10.28 [Pelton, 1989].

Analysis   The brake assembly is illustrated in Fig. 10.29. The disk was mounted on an adapter ring that was attached to the rest of the equipment. A study of the equipment drawings indicated that the adapter ring was a likely candidate as a major noise source.

The major source of noise was caused by a phenomenon known as slip-stick. The brake sticks then slips at a rapid rate, creating vibration excitation that can radiate as noise. A review of the literature revealed that vibration damping could provide some assistance in noise reduction.

The first step in confirming the suspected noise source was to conduct a series of noise and vibration measurements made on the adapter ring. This ring was separated from the assembly. A force hammer and accelerometer coupled to a two-channel spectrum analyzer were used to determine the normal vibration modes of the adapter ring. The data showed that there were significant responses at 940, 1250, and 1825 Hz; the 940-Hz mode was the most significant radiator. Other components were tested in the same manner but did not exhibit this type of response.

Recommendations   Since this was a resonant response of the adapter ring, a vibration-damping treatment was determined to be a partial solution to the problem. A cross section of the constrained layer damping treatment is shown in Fig. 10.30. A layer of damping material was applied to the adapter ring held in place with a metal ring serving as the constraining layer. The ring was placed around the damping material in two halves and bolted to compress the damping material.

When this was applied and the equipment started, the noise level was reduced by 4 dBA. While this was not very encouraging, additional measurement revealed that a great deal

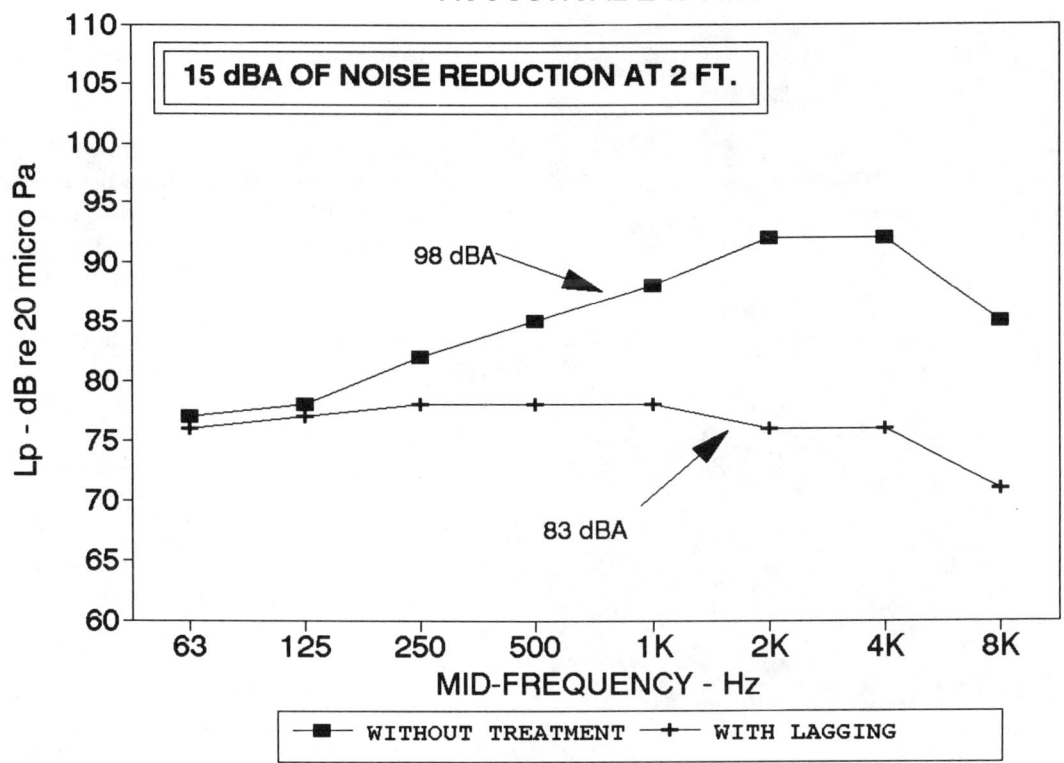

FIGURE 10.25   Noise control results of chiller acoustical lagging.

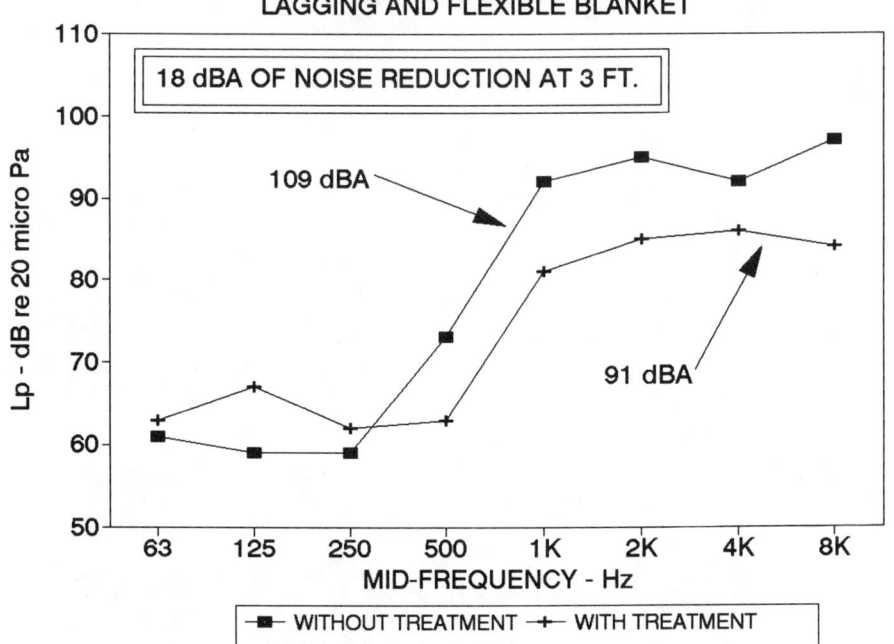

FIGURE 10.26   Pressure-reducing valve noise spectra with and without acoustical lagging.

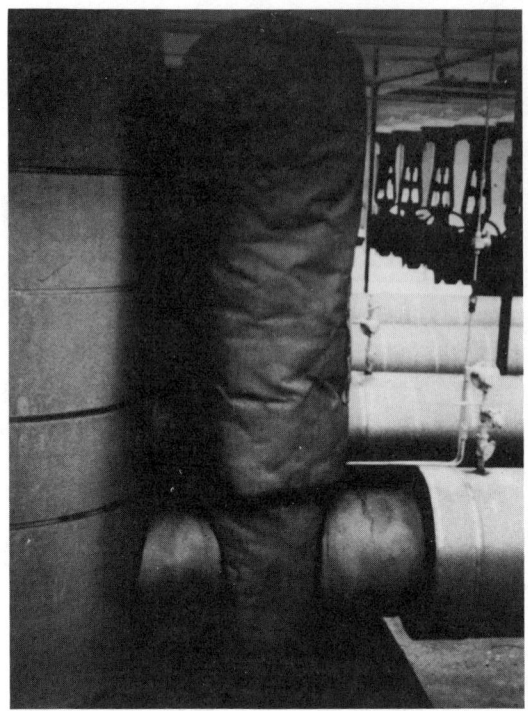

of energy was coming from the brake pads and being transmitted throughout the structural system. Some of the extra damping material, on hand, was placed between the brake pads and the hydraulic pistons to absorb the flow of energy through this path. This reduced the noise 12 to 15 dBA with a level of 110 dBA. In addition, acoustical absorbing treatment was applied to adjacent reflective surfaces on the machine and various cracks around the housing were closed. This reduced the level further to 108 dBA. The last step was to build a full enclosure over other components with an expected level of 100 dBA. Compared to the ambient noise level of 94 dBA, caused by other sources in the area, a reduction from 130 to 100 dBA was quite substantial. This was much more acceptable to the employees. The final results of the prototype treatments are shown in Fig. 10.28.

SILENCERS

FIGURE 10.27   Pressure-reducing valve acoustical lagging.

Silencers are used to attenuate noise from air jets, steam jets, engines, compressors, fans,

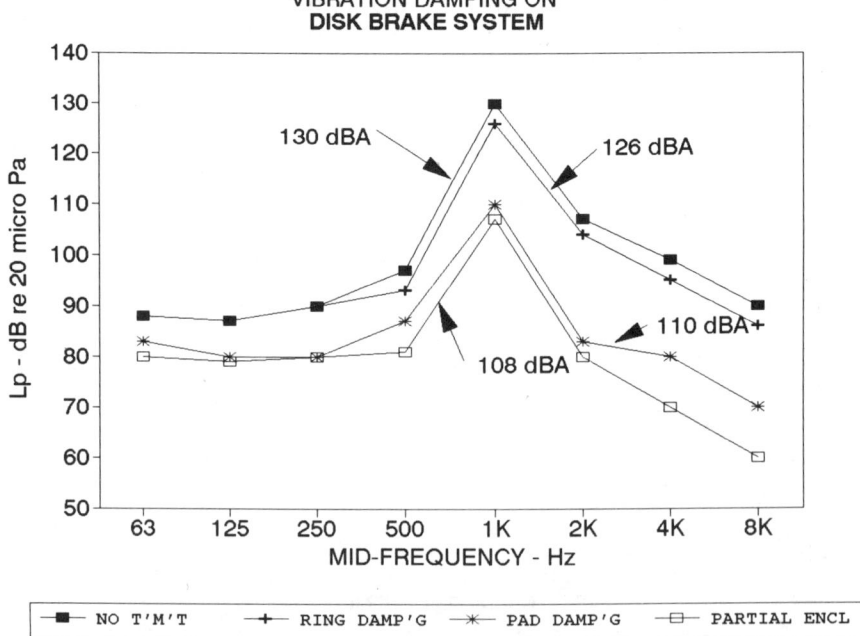

FIGURE 10.28   Noise spectrum of a disk brake system. (Reprinted by permission from *Sound and Vibration* magazine.)

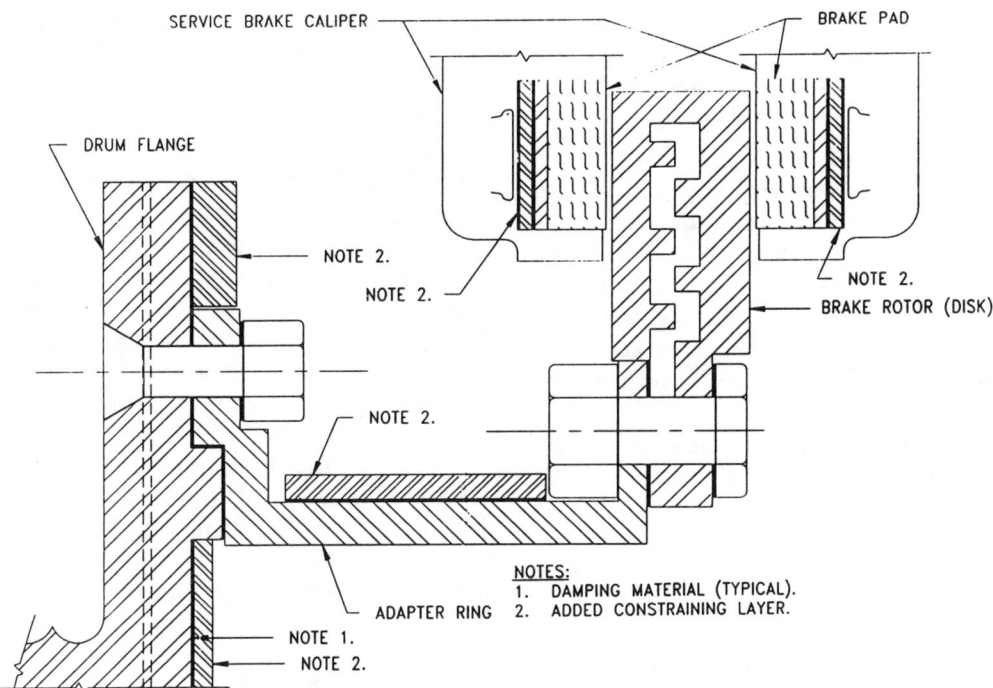

FIGURE 10.29   Disk brake assembly. (Reprinted by permission from *Sound and Vibration* magazine.)

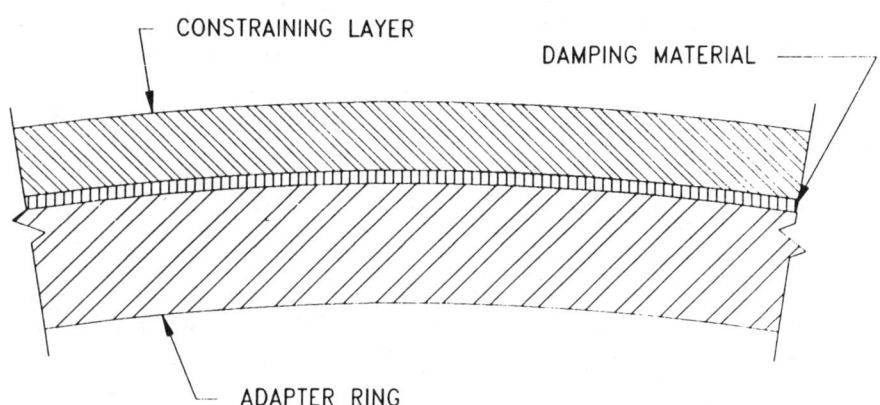

FIGURE 10.30   Constrained layer damping. (Reprinted by permission from *Sound and Vibration* magazine.)

blowers, gas turbines, electric motor cooling fans, and so forth. These types of sources can easily be fitted with silencers that range in size from quite small for air jets to very large gas turbines. The following case histories provide examples of silencers for engine noise, steam jet, and exhaust fans.

Engine Noise

Description   A hospital was expanding its utility plant, which required adding three additional emergency generators. The utility plant was located adjacent to a residential area. Noise had to comply with the city's noise ordinance at the property line of the hospital. Data from the engine manufacturer were obtained and an ambient noise survey was conducted.

Analysis   The city noise ordinance limited the residential property line levels to 49 dBA nighttime and 56 dBA daytime. It was decided to operate the engines once a month (other than for emergencies) during the daytime only. Emergency operations could not be controlled. However, every attempt would be made to meet the nighttime levels.

The nighttime ambient levels at the property line ranged from 57 to 69 dBA. Thus, the design noise level was to be at least 47 dBA, so this would not add to the ambient level. Since this hospital was located in a large city the ambient levels were very typical.

The equipment layout is illustrated in Fig. 10.31. This shows the plan and section through the space. Both the exhaust noise and the engine radiated noise were required to be treated. Since the engine was cooled by the cooling tower it did not have a radiator.

The engine radiated and exhaust noise spectra are shown below:

Recommendations   An acoustical enclosure was installed over the engines with acoustical louvers to attenuate the combustion air inlet opening. The engine exhaust noise was controlled with a silencer similar to that shown in Fig. 10.32. The installed silencer is shown in the photo in Fig. 10.33. It is a multiple-chamber device with internal baffles and side tubes to direct and attenuate the exhaust noise. This is usually called a hospital or critical grade silencer. In addition, a parapet wall is used to partially shield the system from the street. An end view of the installation is shown in Fig. 10.34.

The result after installation during an exercise operation was 54 dBA with an ambient level of 53 dBA. Subtracting the levels results in a system noise level of 47 dBA. This was below the nighttime criterion of 49 dBA.

STEAM VENT

Description   A steam pressure control and vent valve was part of a steam system in a resource recovery plant that burned garbage to make steam. The steam was supplied to an adjacent food-processing plant and was controlled by the

Exhaust Noise

Freq (Hz)	63	125	250	500	1000	2000	4000	8000	dBA
L_p (dB)	106	117	113	112	109	109	109	109	116

Radiated Noise

Freq (Hz)	63	125	250	500	1000	2000	4000	8000	dBA
L_p (dB)	94	91	93	101	101	100	94	87	106

Brief calculations are shown below for the exhaust silencer:

Freq (Hz)	63	125	250	500	1000	2000	4000	8000	dBA
Exh. noise	106	117	113	112	109	109	109	109	116
Silencer	32	42	42	39	38	35	35	35	
Distance	26	26	26	26	26	26	26	26	
3 engines	5	5	5	5	5	5	5	5	
Results at street	43	34	40	42	40	43	43	41	49

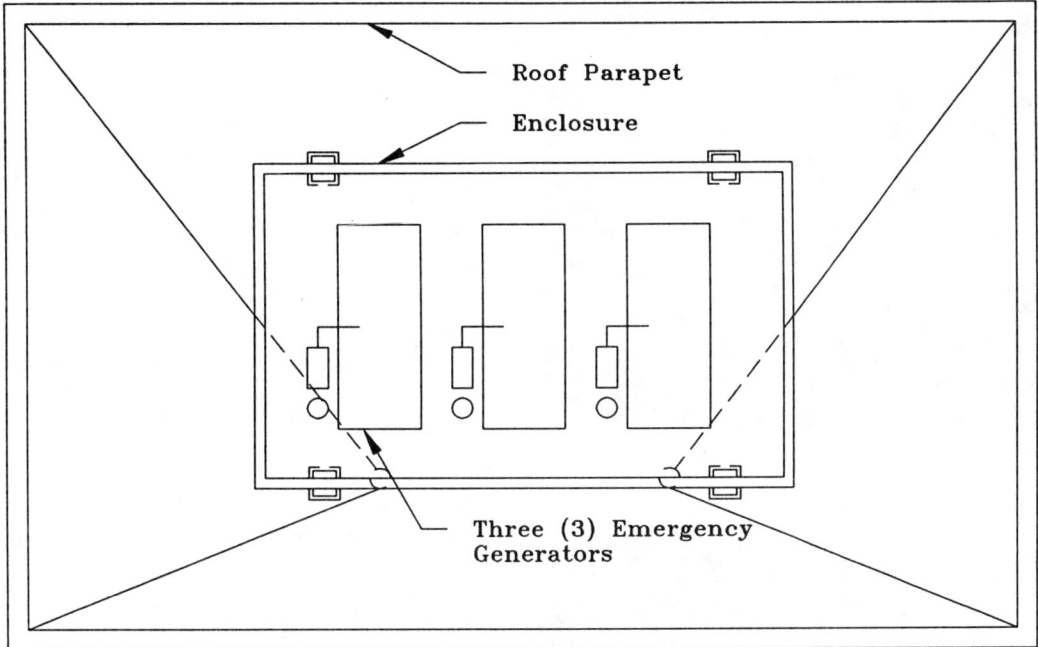

FIGURE 10.31   Plan view of emergency generators.

steam valve. At times steam had to be vented in order to maintain pressure and flow depending on the demand of the food-processing plant. During some portions of a 24-hour period all of the steam produced could be used by the plant. The remainder of the time steam had to be vented. The venting noise was creating complaints from the surrounding neighborhood even though there was a silencer installed downstream from the valve. Finally, the noise level inside the occupied areas of the plant was too high due to valve noise radiation.

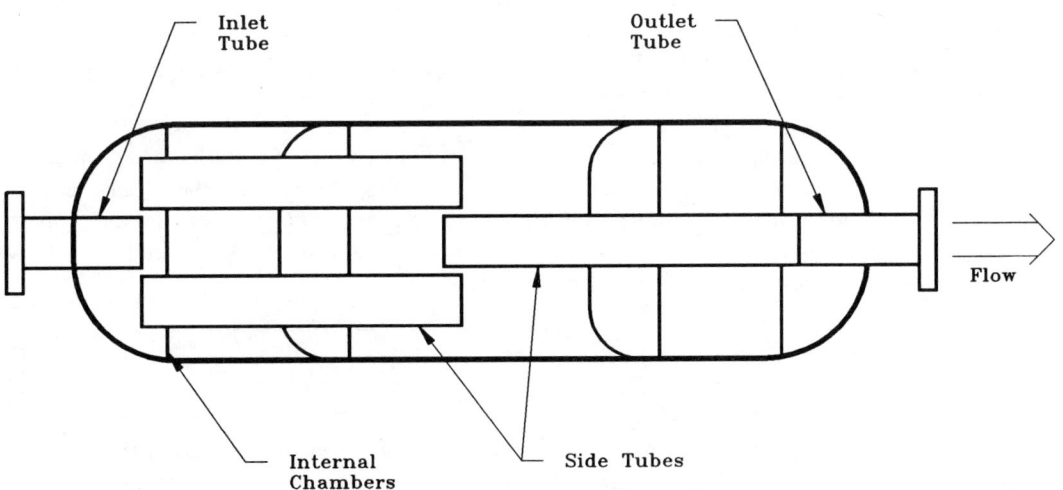

FIGURE 10.32   Typical engine exhaust silencer.

FIGURE 10.33   Installation of engine exhaust silencer.

FIGURE 10.34   End view of emergency generator installation.

Analysis   A noise survey was conducted in the vicinity of the plant to determine levels at various locations. A noise spectrum is shown in Fig. 10.35 for a location behind the plant at the south fence line and about 0.1 mile to the northwest. The steam venting was also audible at 0.4 mile to the northeast at the complainant's residence.

Since there was a silencer already installed, why was it not reducing the noise? A velocity calculation was made for the flow rate and indicated that it far exceeded normal velocities found in these types of silencers. The maxi-

mum velocity for a silencer of this type should be limited to 200 fps or 12,000 fpm. The silencer was sized too small for the application. In addition, it was determined that it was only a straight-through absorptive type. For a vent silencer application a combination reactive and absorptive type would be more appropriate. A silencer of this type is illustrated in Fig. 10.36. The initial high-velocity flow is forced through a perforated diffuser, dispersing the valve jet flow. The reactive interaction of the diffuser and inlet section converts the low frequencies to high frequencies that the silencer can more easily absorb.

Recommendations   A new properly sized silencer was recommended to be installed at the end of the existing piping. In addition, the valve, piping, and silencer were recommended to be acoustically lagged. This is shown in Figs. 10.37 and 10.38. The noise spectrum after installation is shown in Fig. 10.35 for a location about 0.1 mile northwest of the plant. The noise levels in dBA, at three locations, before and after installation are shown below:

Location	Before	After
S. fence	92	70
0.1 mi NW	62	55
0.4 mi NE	57	51

MANUFACTURING PLANT EXHAUST FANS

Description   When a manufacturing plant was being planned next to a residential area, the zoning requirements provided that the property line noise levels should not exceed the following:

Freq (Hz)	63	125	250	500	1000	2000	4000	8000
PL levels	70	64	58	55	52	48	42	37

After the plant was in operation the residents complained about the noise from the plant. The plant industrial hygienist measured the noise levels at the property line at various locations.

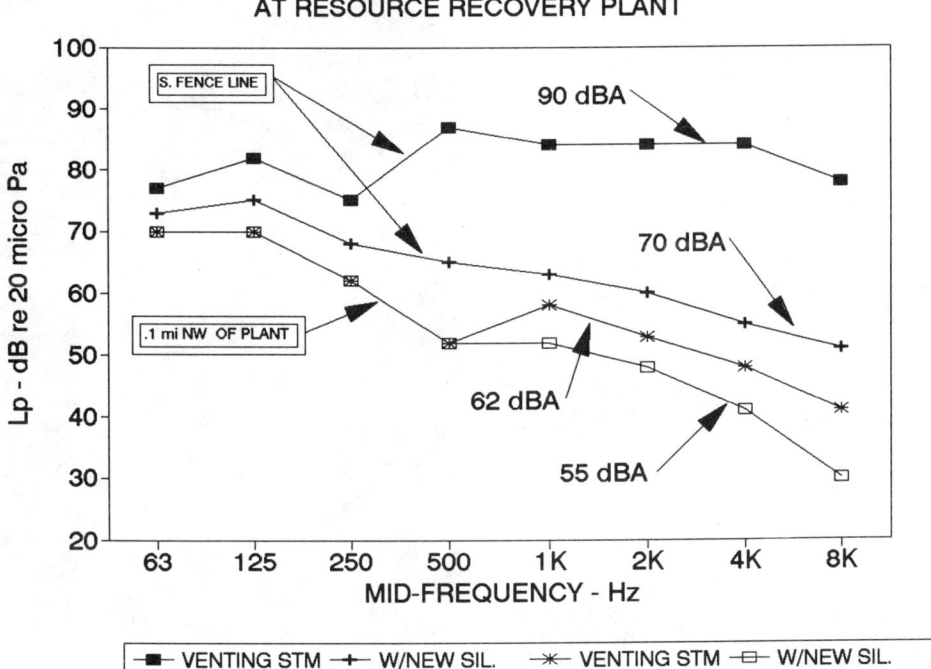

STEAM VENT SILENCER
AT RESOURCE RECOVERY PLANT

FIGURE 10.35   Typical levels at two locations in community with and without noise control treatment.

These are shown in Fig. 10.39 compared to the noise level requirements.

Analysis   A portion of the plant and residential area is shown in Fig. 10.40. On the roof of the plant are a series of scrubber exhaust fans that remove the cleaned air from the process areas. There are two stacks per fan depending on the scrubber being used. Listening to the noise at the stack and in the neighborhood made it clear that the exhaust fans were the offending noise source. There were clear tonal components that could be heard. Another complicating factor was that the gas from the scrubber was corrosive. All of the fan stacks and duct work were made of fiberglass to resist corrosion. Thus the noise control device had to be of a similar material.

A detailed noise study was conducted. The study involved one-twelfth octave band noise measurements at the top of the stack of each fan system. The results were then compared to the noise spectra in the neighborhood to determine the predominant tones from each fan. This comparison allowed a determination about how much noise reduction was required to meet the property line limits. The noise reduction required for the exhaust fan stacks was as follows:

Frequency	63	125	250
Noise reduction	10	12	15

Recommendations   An exhaust fan stack silencer was recommended. This silencer was an absorptive type to meet the flow and pressure drop requirements. Due to the low- and mid-frequency noise reduction requirements a splitter panel type of silencer was recommended. Since both stacks were physically very close together it would be necessary to have two silencers that could fit the existing opening or one

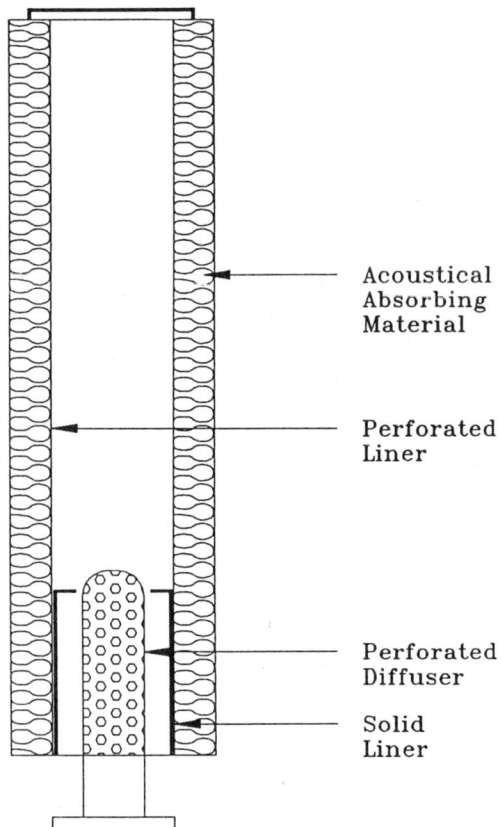

FIGURE 10.36   Cross section of steam vent silencer.

FIGURE 10.37   Acoustical lagging of PRV and piping.

larger silencer with a blade damper to switch between scrubbers.

Inquiries were made to silencer manufacturers based on the design requirements. Two manufacturers responded with different designs. One was a single rectangular splitter panel silencer sitting on a large box with a damper to switch between scrubbers. The other was a round silencer with internal round splitter panel baffles inserted in the outer shell. All were constructed of structural fiberglass with fiberglass lining to absorb the noise. The latter design is shown in Fig. 10.41.

A prototype was ordered and installed (Fig. 10.42). The noise measurements made to check the guaranteed performance revealed that it was satisfactory for the application. The flow and

FIGURE 10.38   Installation of new vent silencer.

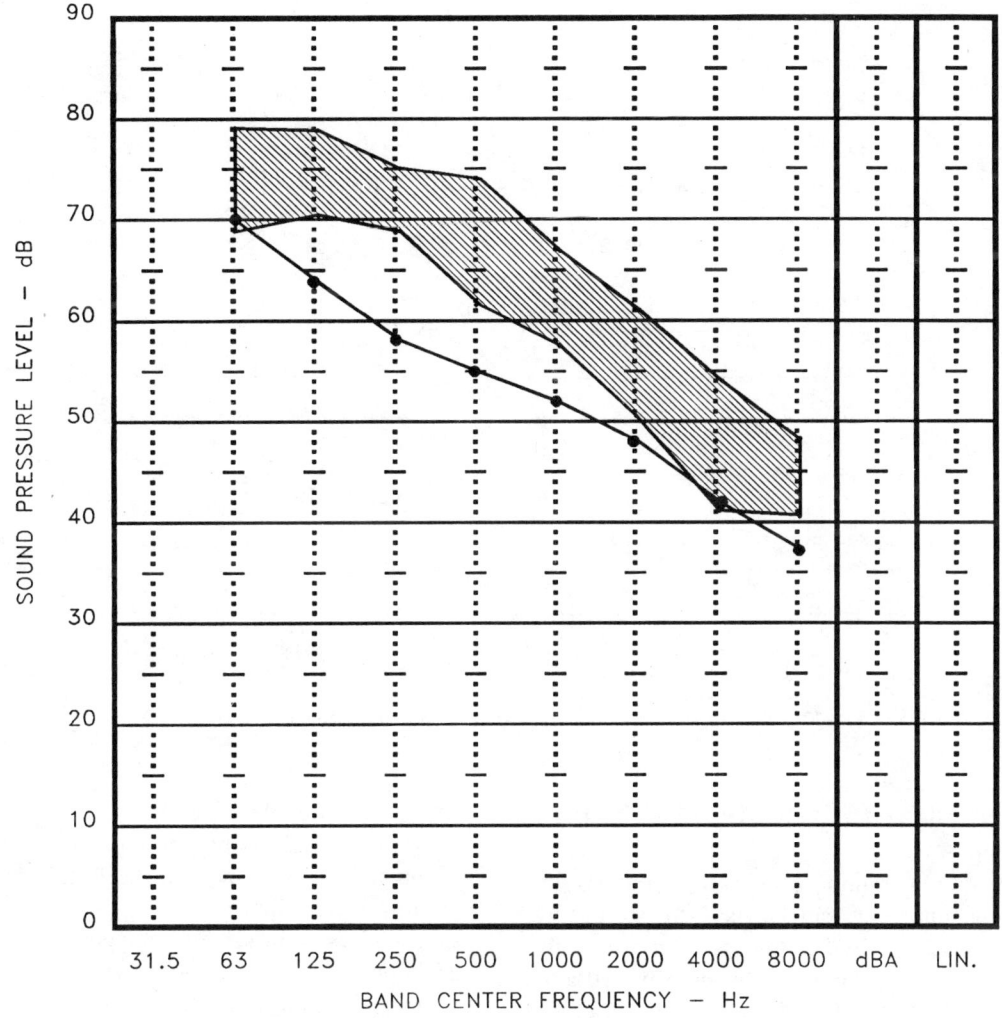

Property Line Data

▨ Range of Data at Property Line

● Planned Development Zoning Requirements

FIGURE 10.39   Property line planned development district zoning noise level requirements.

pressure drop was also checked during these measurements and found to be satisfactory. It was recommended that the remaining silencers be purchased and installed.

The results of the final measurements are shown in Fig. 10.43 compared to the required property line noise limits. The time required to complete this project from the initial identification of the problem to the final installation and measurements was 18 months. For a project of this nature for a large corporation this amount of time is not unusual.

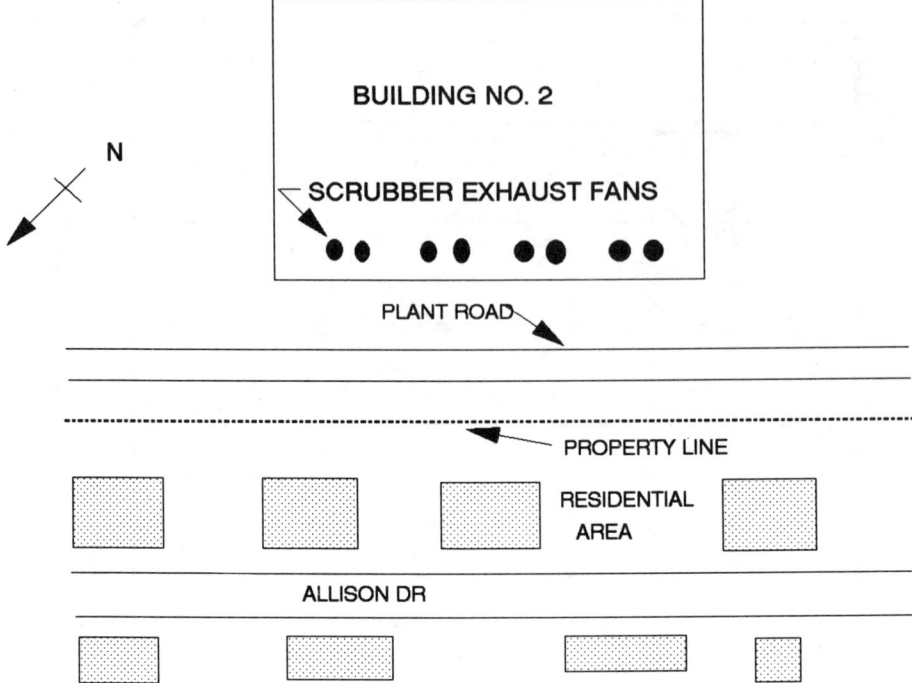

FIGURE 10.40   Plan view of plant and adjacent community area.

HYDRAULIC PUMP SYSTEMS

A hydraulic pump is used to compress hydraulic fluid that can be used to operate hydraulic motors and cylinders to move equipment. There are several reasons why hydraulic pumps, shown in Fig. 10.44, make noise. The following are typical causes and solutions:

Cause—Partly closed intake line or restricted intake hydraulic lines.
Solution—Pump must receive intake fluid freely or cavitation takes place.
Cause—Air leaks at pump intake pipe joints.
Solution—Tighten as required. Pour fluid on joints while listening for changes in the sound of operation.

Cause—Air bubbles in fluid.
Solution—Check to be certain return lines are below fluid level and well separated from intake line.
Cause—Reservoir air vent plugged.
Solution—Inspect and unplug as required.
Cause—Coupling misalignment.
Solution—Check for damaged shaft bearing or other parts. If necessary, replace and realign the coupled shafts.

All hydraulic pumps that have higher than normal noise levels should be routinely checked for items outlined above. If no satisfactory remedy is achieved, then the hydraulic pump might be totally enclosed. The following noise spectra are from data measured from hydraulic pumps at 3 ft with and without an enclosure:

Freq (Hz)	63	125	250	500	1000	2000	4000	8000	dBA
L_p (dB w/o encl.)	74	74	80	77	77	87	78	76	90
L_p (dB w/encl.)	72	74	79	80	75	72	70	68	80
Noise reduction	2	0	1	7	2	15	8	8	10

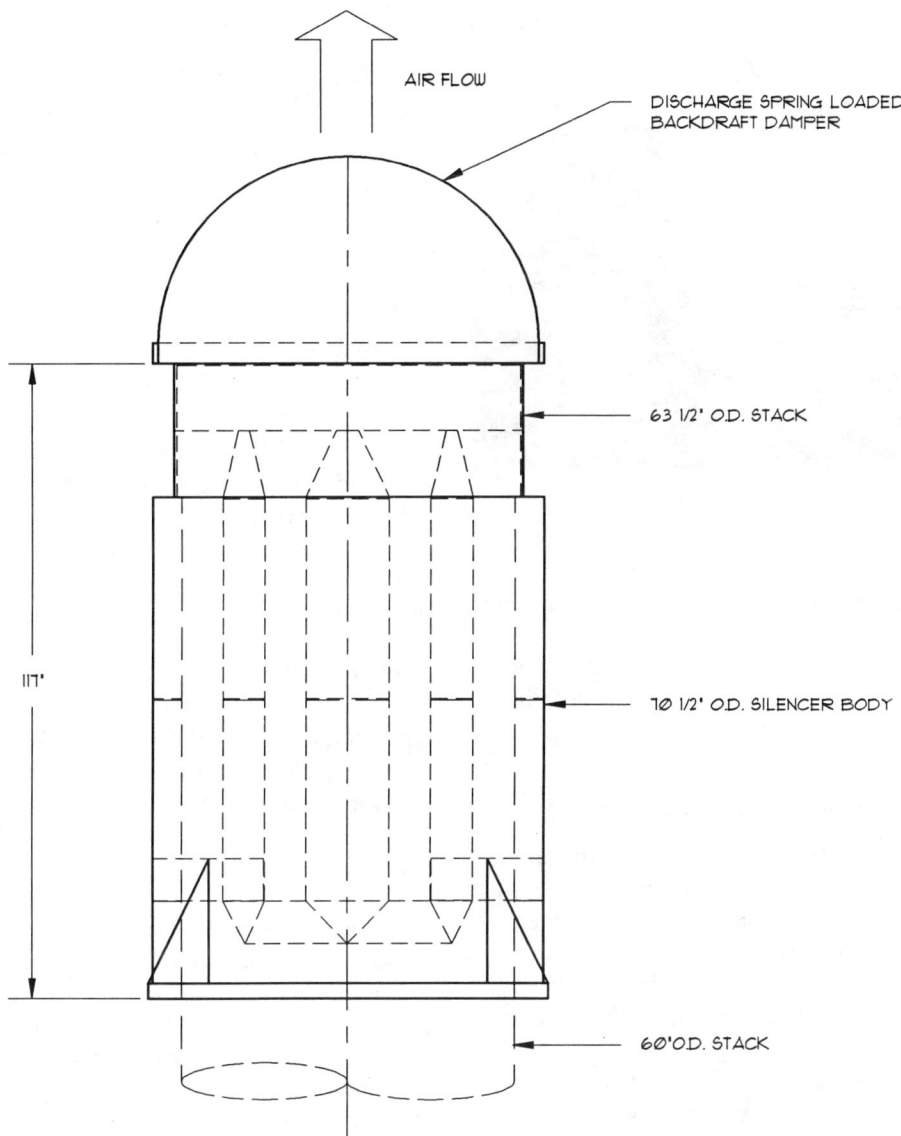

AIR FLOW

DISCHARGE SPRING LOADED
BACKDRAFT DAMPER

63 1/2' O.D. STACK

70 1/2' O.D. SILENCER BODY

117'

60' O.D. STACK

FIGURE 10.41   Cross section of fiberglass exhaust silencer.

The enclosure that is placed over the hydraulic pump provides a 10-dBA noise reduction. The enclosure is constructed of 16-ga steel lined with 1-in. acoustical foam covered with perforated metal.

SOURCE NOISE CONTROL MEASURES

Many times the influence of the noise source can be changed to affect the results. For ex-

ample, if a fan is operating at a high rpm it may be possible to reduce the speed, resulting in lower noise levels. The L_w of the fan is a function of 56 log (blade tip speed, fpm/1000) [API, 1973]. If the desired noise reduction is known, the final tip speed can be calculated to achieve the noise reduction. That is to say, the blade pitch is changed to provide the same amount of airflow from the fan.

A radiator fan, propeller type, on an engine might operate at an rpm resulting in a tip speed

FIGURE 10.42   Prototype silencer installation.

of 15,000 to 18,000 fpm. If the fan rpm was lowered to a tip speed of 9,000 fpm, the noise level could be reduced about 17 dBA. The relationship above was used to calculate the reduction.

Fan Duct Work Configuration

Description   In applications of air-handling units many times the discharge duct work can cause low-frequency noise problems when it is improperly designed and installed. A fan room located next to an executive office was too high. The air-handling unit was located in a room and the return air was brought in through openings above the ceiling of the office area. The supply duct went out through another wall.

Analysis   Octave band noise measurements compared to the required noise criteria (NC) are plotted in Fig. 10.45. A review of the duct work both as installed and from the design documents indicated a major difference. These ducts are shown in Figs. 10.46 and 10.47, respectively. The installed duct work had an angle that was not square. This resulted in the low-frequency noise, generated by turbulent airflow, in the occupied space.

Recommendation   In order to reduce the noise at the source, a rounded elbow with turning

vanes was recommended. This elbow, shown in Fig. 10.48, reduced the system pressure drop, allowing more airflow. The fan was slowed down by 15 percent to obtain airflow. The resulting noise level is plotted in Fig. 10.45 compared to the NC curve for that type of occupied space. In the designing of any type of duct system the aerodynamics must be considered. If the air is forced to turn abrupt corners or flow through restricted spaces the result will be turbulence and low-frequency rumble.

Engine Radiator Fan

Description   An engine radiator has a high noise level. If the engine is enclosed, the radiator fan must be considered as part of the overall noise control design. The radiator is shown in Fig. 10.49.

Analysis   A noise measurement taken at the fan is shown in Fig. 10.50. Observations of the fan configuration and the relationship to the shroud indicated a great deal of turbulence. Smoothing the airflow could have an effect on the noise levels. Other fans have been observed with better aerodynamic characteristics.

Recommendations   A redesign of the fan shroud was carried out and installed. The resulting noise measurements are shown in Fig. 10.51. Thoughtful consideration is required in the design of any air-moving system. If the air is required to make abrupt turns turbulence and noise will result.

DEALING WITH NOISE FLANKING PATHS

A flanking path is one that allows noise to leak past or around something. For example, if an air duct passes through a partition and the duct is not sealed there is a flanking path that allows the noise to leak past. Or if an air duct or pipe is rigidly sealed into a partition the noise will be transferred to the partition, reradiate, and flank around the pipe or duct that may be containing the noise.

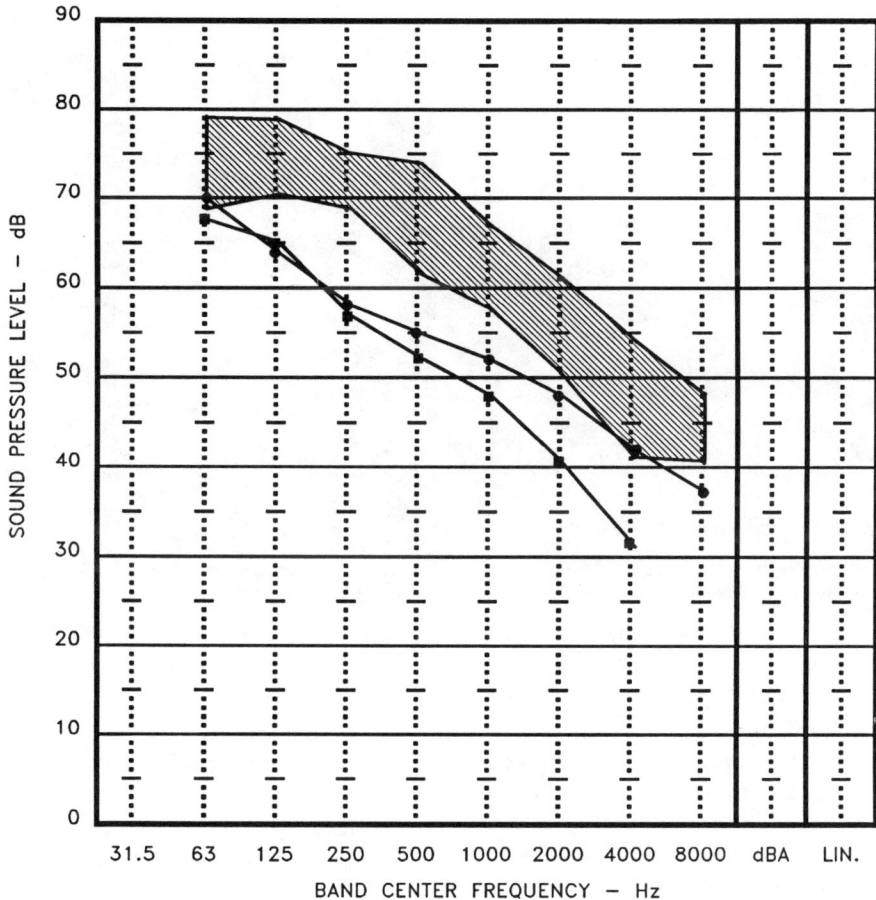

Property Line Data

▨ Range of Data at Property Line

● Planned Development Zoning Requirements

■ Property Line Fence with Exhaust Fan Silencer Installed

FIGURE 10.43   Noise level results after installation of silencers.

Description In an office building an 880-ton water chiller was located on an upper floor. Below the space was an occupied office area. The space between the glass curtain wall and the floor slab allowed the chiller noise to flank around the slab. This is illustrated in Fig. 10.52. The noise level in the occupied space is shown in Fig. 10.53 compared to the noise criteria (NC) curve for this type of space.

Recommendation The recommendation is illustrated in Fig. 10.54. It was an easy matter to close up the space between the glass curtain wall and the slab with sheetrock and caulking. In addition, the space above the slab up to the mullion was closed off as well. The results are shown in Fig. 10.53. When noise is being contained in any type of enclosure, it is necessary to seal all flanking paths.

FIGURE 10.44   Hydraulic pumps: (a) hydraulic pump system driving a can-filling machine and (b) hydraulic pump located on a plastic injection-molding machine.

MATERIAL-HANDLING-SYSTEMS

In manufacturing facilities and process plants, raw materials, materials in process, and finished goods must be moved or transported from one location to another within the plant. Many systems are used for this purpose, depending on the nature of the product or process. Some examples of these typical material-handling systems are:

Belt and roller conveyors
Vibrating conveyor
Pneumatic conveyer
Bucket elevator
Dumping into hoppers
Trucks and forklifts

If the materials are metal, rock, plastic, or some other hard material a great deal of noise can be generated. General noise control methods for some of these types of conveyance systems are:

Apply acoustical lagging on pneumatic piping
Line internal surfaces of conveyors, hoppers, and chutes by applying vibration-damping material to chute surfaces
Reduce fall height of materials dropping into hoppers

In the last example above, if a steel plate is struck by an object, the plate vibrates and makes noise. The sound level is determined by the weight of the object and the fall height. Since the weight is constant a change in the fall height can reduce noise. If the fall height can be reduced from 16 ft to 2 in. the level can be reduced 20 dBA. Fall heights of 16 ft are not that unusual when considering material-handling systems for steel parts, coal, lignite, rock, and so on.

Two case histories are presented in this section to illustrate the noise control approach. These are a pneumatic conveying system for plastic pellets, and a handling system for a carbon filter manufacturing process plant.

Pneumatic Conveying System

Description   An intermediate product in the manufacture of plastics is a pellet that can be

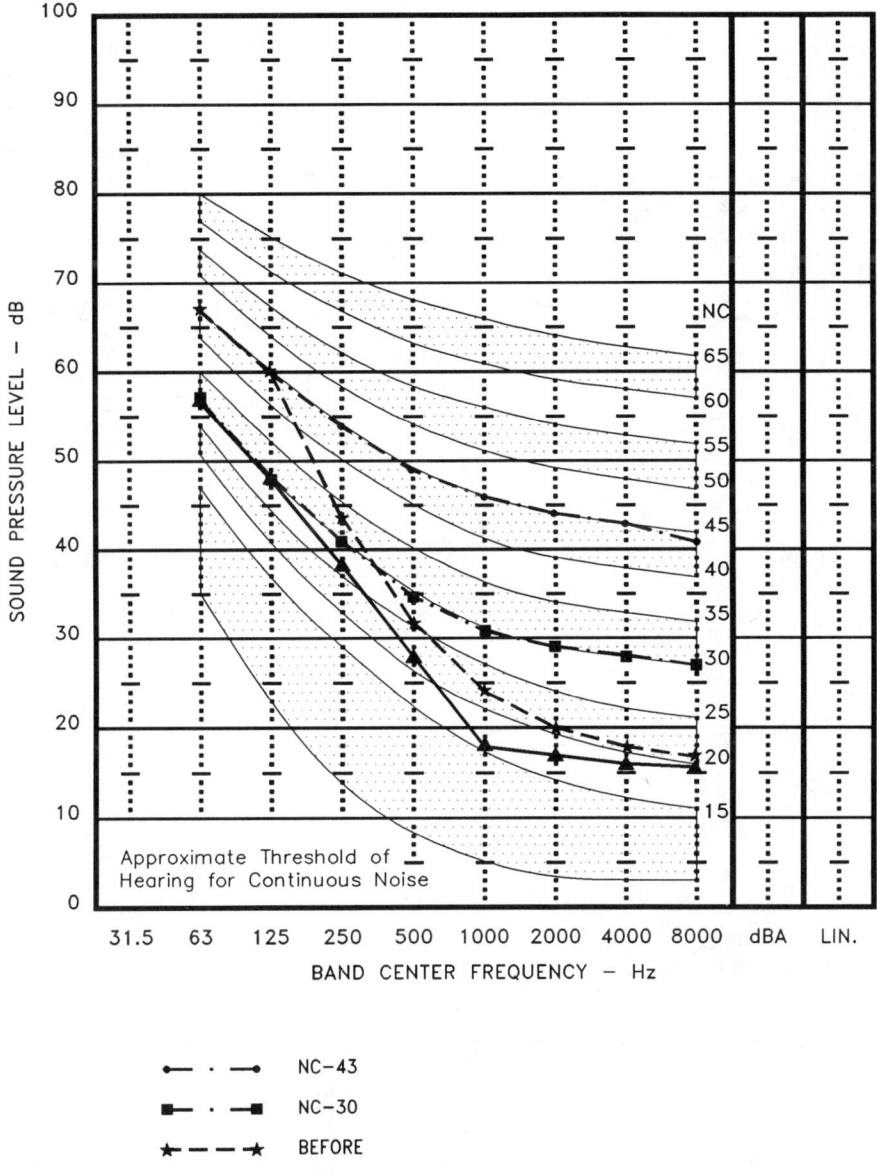

FIGURE 10.45   Air handler noise spectrum.

used in injection-molding machines to form the final product. The plastic pellet is the final product for the process plant that may use natural gas or other petroleum feedstock to obtain the various powders that are mixed, heated, and extruded into pellets. The pellets must be conveyed from the extruder to a storage silo, then to packaging equipment. Therein lies the problem, since the small hard pellets are moved through steel piping by pneumatic conveying systems.

Analysis   Thousands of pellets randomly impact the interior of the pipe, creating noise that is generally high frequency in nature and is like a "swishing" sound. A typical spectrum is

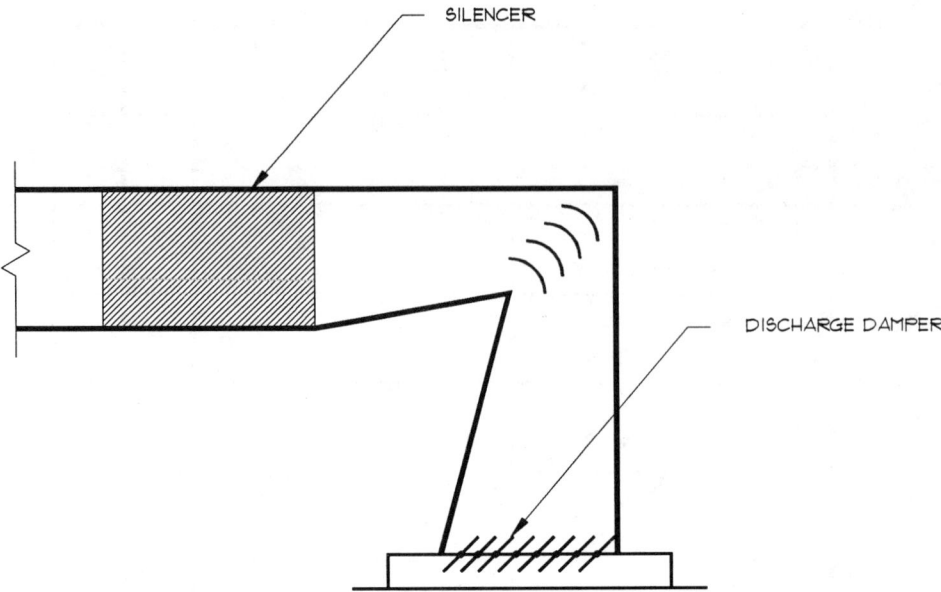

FIGURE 10.46   Air handler discharge duct elbow as installed.

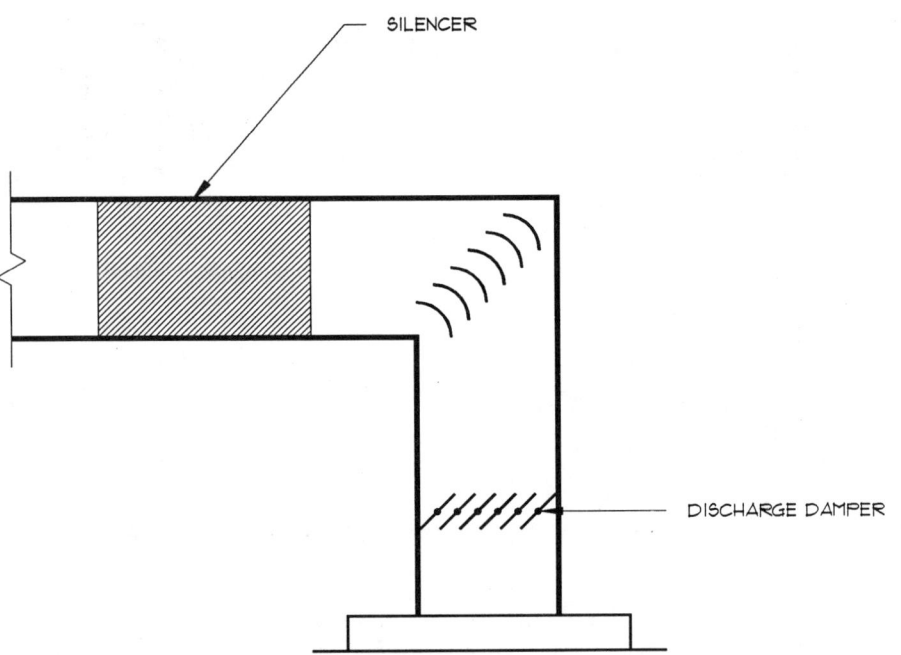

FIGURE 10.47   Air handler discharge duct elbow as designed.

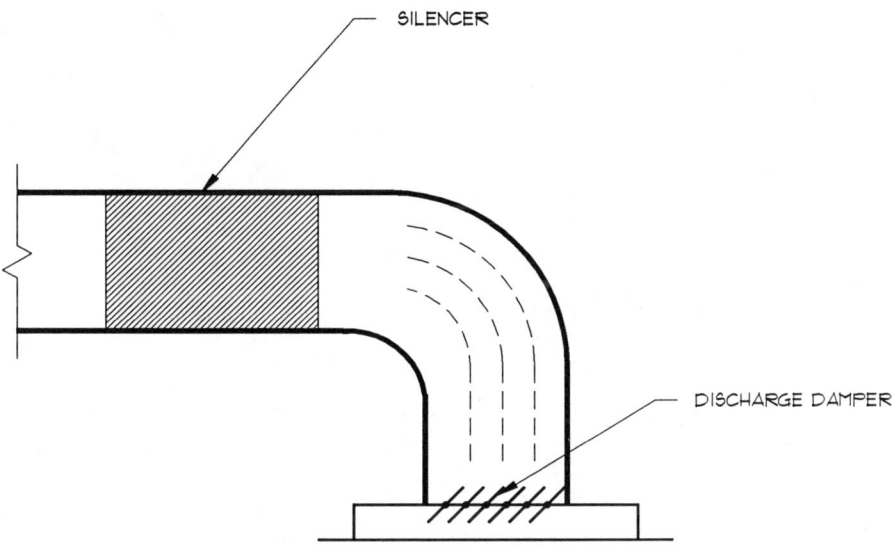

FIGURE 10.48   Revised air handler discharge duct elbow.

shown in Fig. 10.55. Pneumatic conveying lines can be located above the work area in a pipe rack. This allows noise to spread throughout the work area very easily.

Recommendations   The pipe can be covered with an acoustical pipe lagging similar to that illustrated in Fig. 10.20. This lagging can provide a 10- to 20-dBA noise reduction depending on the spectrum shape, outer pipe covering, and size of pipe. The outer covering of the lagging treatment can be a loaded vinyl material, lead-aluminum laminate, normal aluminum in-

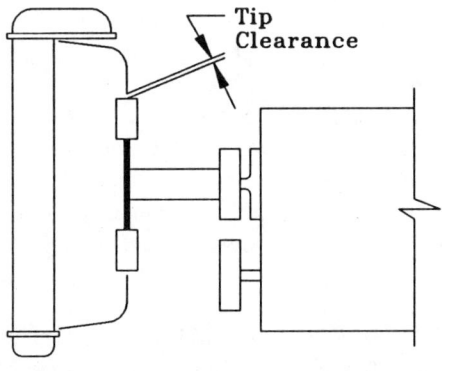

FIGURE 10.49   Poor radiator fan shroud design.

sulating covering, stainless steel, or galvanized steel. The outer jacket must be sealed to obtain a good result. The result of this treatment is also shown in Fig. 10.55.

Coal-Handling System Noise

Description   A process plant uses a low-heating-value coal as a raw material in the manufacture of a filter medium. This coal is found near the surface and is strip-mined. The raw material is delivered to the plant by truck, dumped into a hopper, then transported by belt conveyors to a storage silo. It then goes from the silo to the crusher by means of more belt conveyors and a series of transfer chutes. The noise is generated as the coal impacts on the sides of the apron conveyor at the bottom of the storage silo and transfer chutes at the top of the crusher.

Analysis   A series of noise measurements were made at the bottom of the silo and around the crusher and transfer chutes. These data are shown in Fig. 10.56 with the overall dBA levels, which range from 93 to 104 dBA. The ambient noise with the coal-handling system off

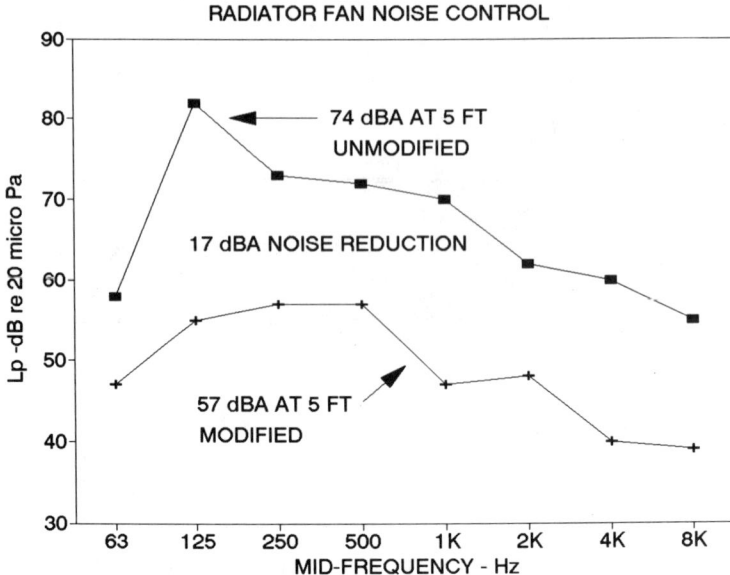

FIGURE 10.50   Noise spectrum.

was 87 dBA, a level produced by other noise sources in the plant. From observations of the operation it was apparent that the noise was being generated from the coal impacting on the metal chutes.

An investigation was undertaken into materials that could be used to line these metal chutes and withstand the impact of the material directly. A report of the Department of Interior on coal-crushing operation and noise control, and various neoprene and urethane material suppliers' literature, illustrated that a noise re-

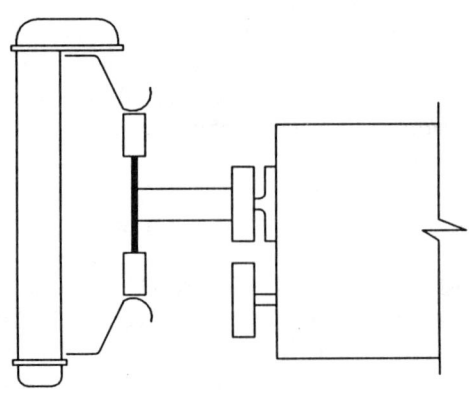

FIGURE 10.51   Revised radiator fan shroud design.

duction of 2 to 30 dBA could be achieved. A test using metal shot impacting on a steel plate at a controlled rate and angle was used as a baseline. Various materials were then placed on the steel plate and the noise reduction was determined. Some examples of the more common materials and results are shown in Table 10.2 [Rubin et al., 1982].

Several manufacturers of neoprene and urethane were contacted to obtain samples and literature for evaluation. Much of their literature illustrated the installation technique for these types of materials. Generally they will wear better than steel if properly installed. Figure 10.57 shows the effect of an impact angle on these resilient materials.

A typical chute arrangement is shown in Fig. 10.58. Usually the material falls from a large height and impacts on the sides of the chute. Thus the use of tough, resilient materials can provide a significant noise reduction.

Recommendations   A tough, resilient material was installed on the surfaces of the chutes. This urethane material, which comes in sheet form, was bolted directly to the chutes. One chute in-

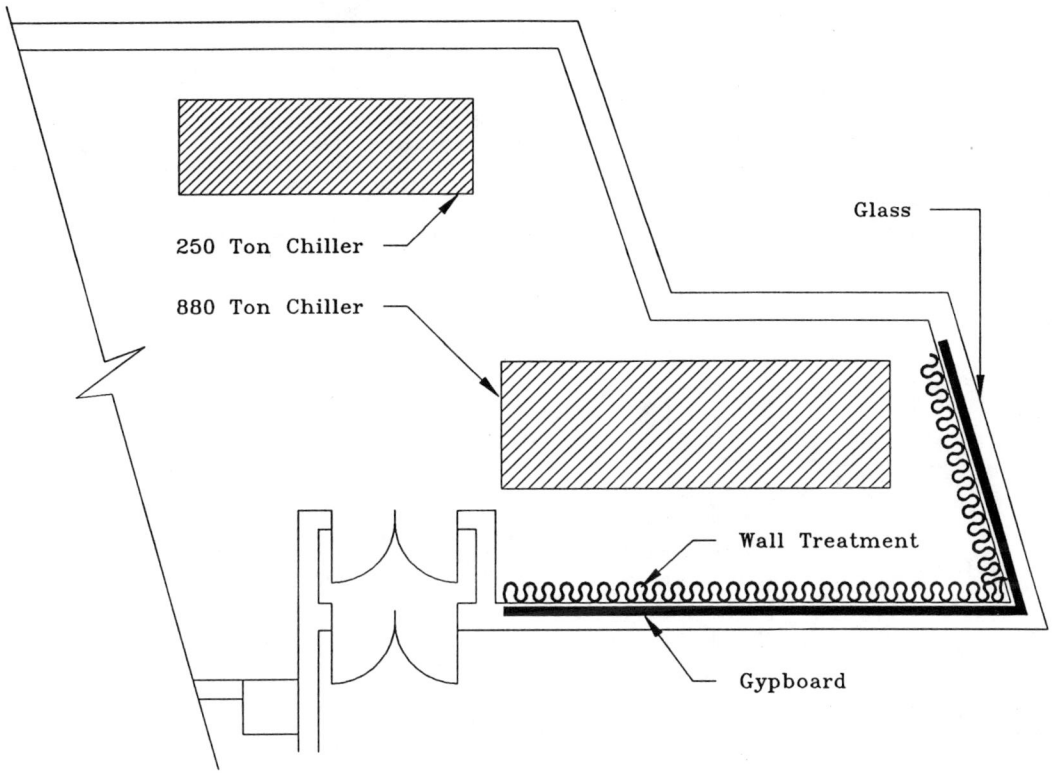

FIGURE 10.52 Opening between curtain wall and floor slab.

CLOSING FLANKING PATHS
MECHANICAL EQUIPMENT ROOM

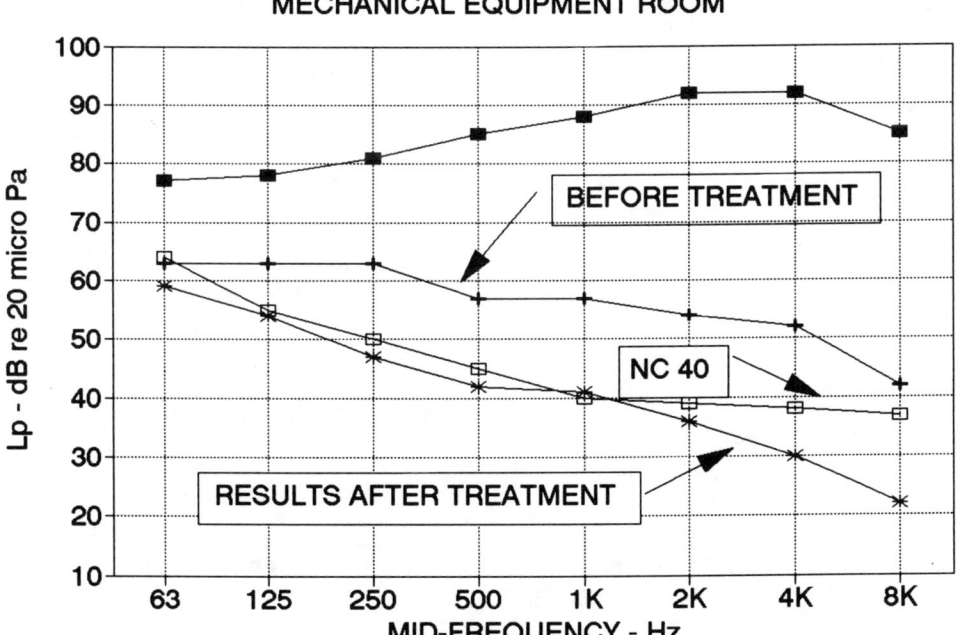

FIGURE 10.53 Noise spectrum in occupied space.

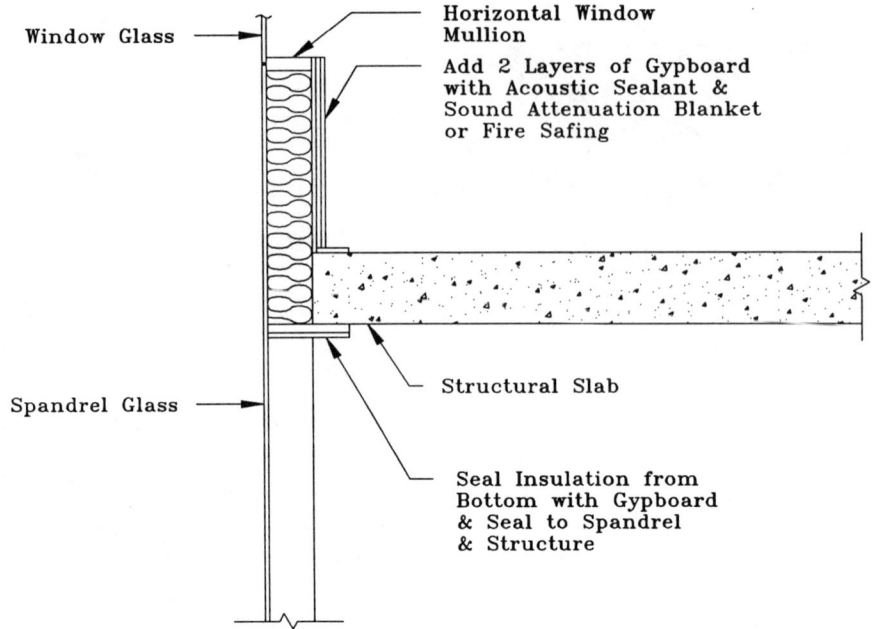

FIGURE 10.54   Seal at floor opening.

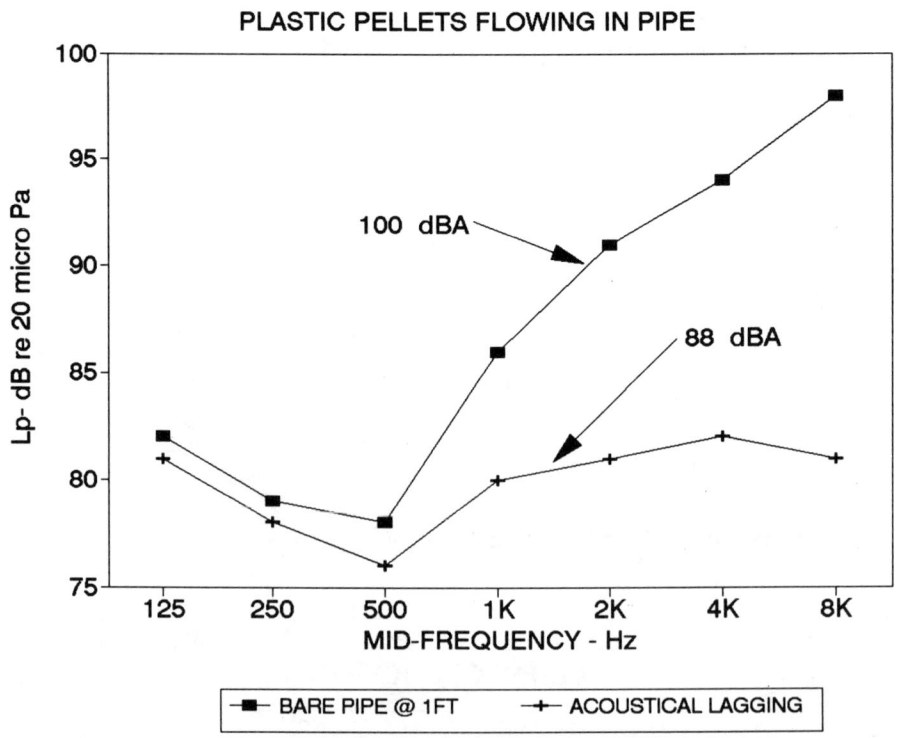

FIGURE 10.55   Noise spectrum of plastic pellets flowing in pipe.

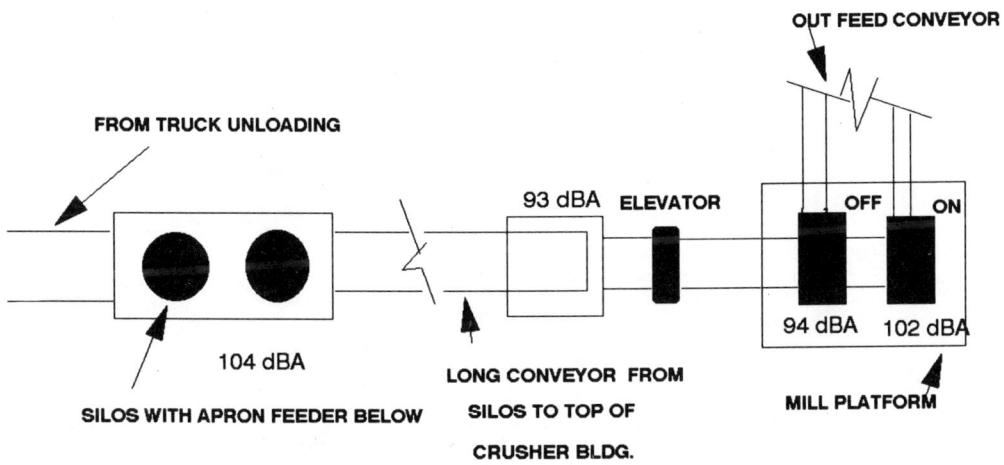

FIGURE 10.56   Coal conveyor system layout with dBA levels.

stallation is shown in Fig. 10.59. The results of the noise control are as follows:

Apron feeder—without lining	103 dBA
with chute lining	88 dBA
Noise reduction	15 dBA
Top of conveyor—without lining	105 dBA
with chute lining	96 dBA
Noise reduction	9 dBA

The top of the conveyor did not have as much noise reduction because of an opening at the top. This opening was to be closed. A noise spectrum of the apron feeder conveyor noise and reduced spectrum is shown in Fig. 10.60. Note the mid- and high-frequency reduction. These materials can provide a significant amount of noise reduction if properly installed.

SPECIAL CONSIDERATIONS FOR THE FOOD-PROCESSING INDUSTRY

There are some unique problems associated with noise control in the food-processing in-dustry. Most noise control solutions require some type of soft absorptive material incorporated into the design. The food industry has a strict requirement for sanitation. This means hard, nonporous surfaces. Thus the use of porus acoustical absorbing material is not usually compatible where food products are in contact or even in the vicinity of this type of noise control material.

Standard noise control methods and materials cannot be used to solve the problems. Thus a more creative approach must be considered. There will be some problems that cannot be solved, and hearing protection must be used. Maintaining the sanitary conditions must always take first priority.

A few manufacturers have developed cleanable absorptive materials. They may use a standard product and cover it with a protective film that is suitable for cleaning. One manufacturer uses a standard 2- or 4-in. acoustical panel with a solid back and seals the perforated face with a plastic film suitable for steam cleaning. This is shown in Fig. 10.61. Other manufacturers will use a 1-in. fiberglass board covered with a

TABLE 10.2   Impact Test Noise Date for Rubber and Plastics

Material	dBA	Noise Reduction (ref: 1018 stl)
1018 steel (16 ga)	91	—
Gum rubber ($\frac{1}{8}$″, 60 shore A)	62	30
Neoprene rubber ($\frac{1}{8}$″, 60 shore A)	66	25
Neoprene rubber ($\frac{1}{4}$″, 60 shore A)	72	19
Polyurethane ($\frac{1}{8}$″)	76	15
PVC rigid ($\frac{1}{8}$″)	84	7
Polycarbonate ($\frac{1}{4}$″)	89	2

Source: U.S. Department of Interior Report, *Techniques for the Design of Coal Preparation Plants*, 1982.

white plastic film. This material is amenable to heavy cleaning, but due to its size and shape can be used in lay-in ceiling grids. Examples of two of these products are shown in Fig. 10.62. Further information about these manufacturers can be found in Appendix F.

Acoustical foams are also used in some cases. For example, the foam shown in Fig.

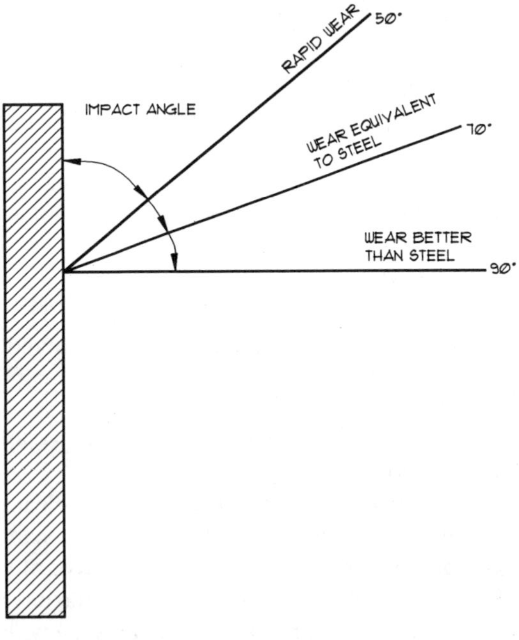

FIGURE 10.57   Installation of abrasion-resistant neoprene.

10.63 has a mylar film and can be easily cleaned. If it is used inside an enclosure all edges and joints must be sealed. The joints and edges can be sealed with a silicone caulking material.

In some cases the enclosure must be hard on the inside. This requires the enclosure to be stronger acoustically; that is, the walls and openings must be designed to account for the increased noise level inside.

There are noise sources for which standard methods and materials can be used. For example, if a product is being conveyed within a pneumatic system the exterior can be acoustically lagged, as described in the material-handling section. There are some process areas that have sanitary conditions, but the product is kept internal to equipment. Vibrating surfaces can have damping materials applied that are readily cleanable. The utility plants with the food-processing plant will use standard methods as well.

Thus it is important to look at each problem as an opportunity to be creative in developing noise control methods that can meet both the sanitary and noise criteria. The example provided involved a corn grinding operation.

Description   A hammermill was used to grind corn as the first step in the manufacturing process. The hammermill was on a steel platform about 12 ft above the floor. A prototype design was developed for the pilot plant prior to full-scale production.

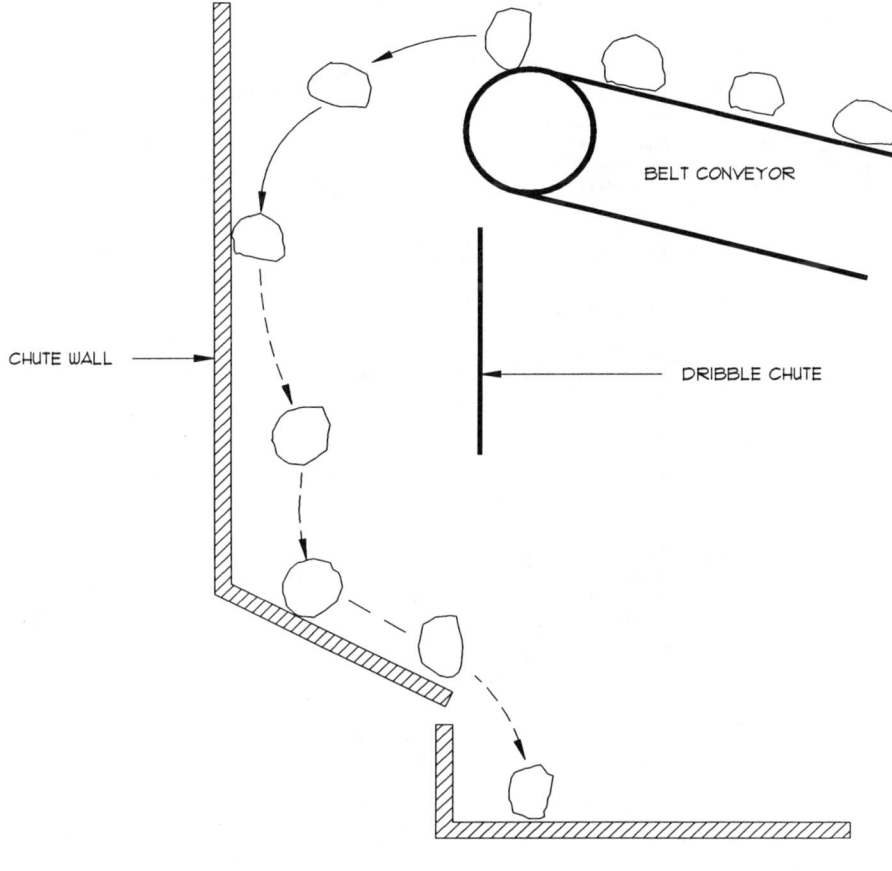

FIGURE 10.58   Typical chute.

Analysis   A series of noise measurements was made to determine the level and spectrum shape for the hammermill. The mill was driven by a 3600-rpm motor with a belt drive on a 1/1 ratio. A belt guard was attached to the frame. The data in Fig. 10.64 illustrate the typical noise spectrum. The mill had a characteristic sound based on the number of hammers passing the anvil. A hammer passing frequency can be calculated by: rpm/60 × no. of hammers on the shaft.

The noise was coming from the inlet, discharge, and casing radiation. In the installation the inlet and discharge would be connected to duct work with flexible connections. Noise could pass through the flexible connection, but this would break the direct vibration path. The duct work radiation would need to be considered since it would be constructed of light-gauge stainless steel. The mill was usually opened one or two times a day for a thorough cleaning with high-pressure water hoses and detergent.

Recommendations   The most efficient solution was to provide an enclosure over the mill with inlet and discharge silencers for cooling air provided by a fan. The enclosure was constructed of steel tubing with acoustical panels bolted and sealed in place. The panels were solid on both sides and made of 16-ga stainless steel. To increase the sound transmission loss

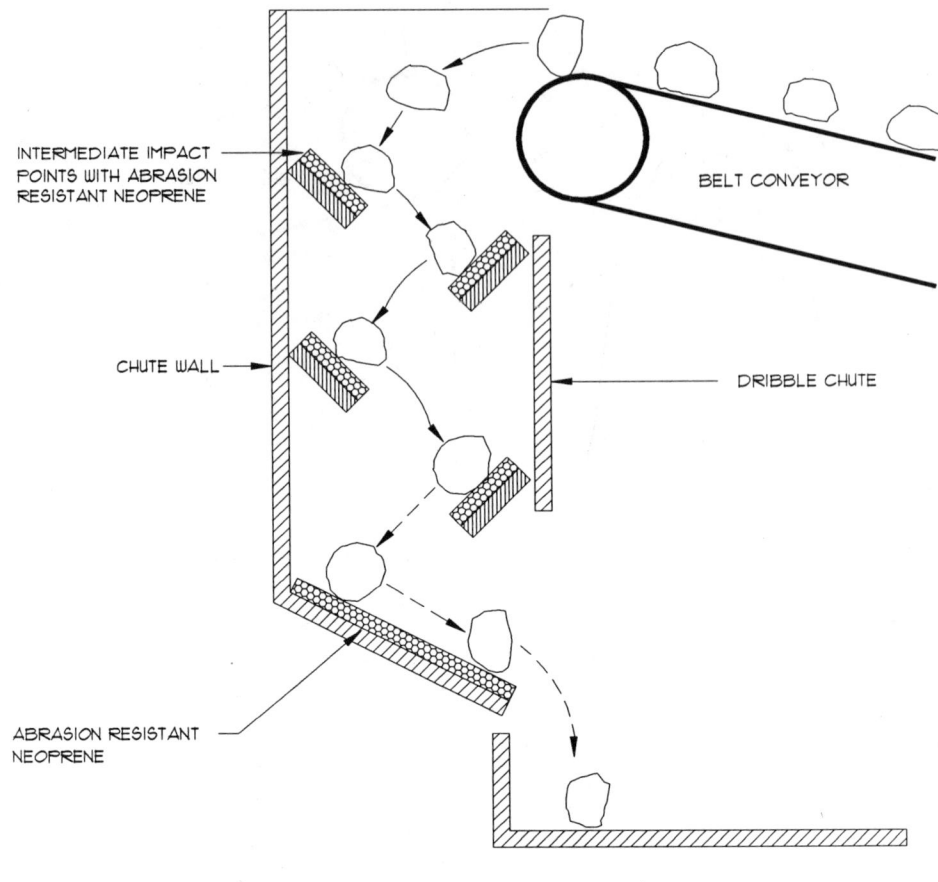

INTERMEDIATE IMPACT
POINTS WITH ABRASION
RESISTANT NEOPRENE

BELT CONVEYOR

CHUTE WALL →

← DRIBBLE CHUTE

ABRASION RESISTANT
NEOPRENE

SCREEN AREA

FIGURE 10.59   Chute with fall breaks and abrasion-resistant neoprene.

a piece of $\frac{1}{2}$-in. sheetrock was to be placed inside the panel in addition to the fiberglass fill. The noise inside a small hard enclosure will increase 15 to 20 dBA. An overview of the enclosure is shown in Fig. 10.65. The prototype enclosure reduced the noise level to less than 85 dBA.

VIBRATION ISOLATION

A variety of equipment in industrial plants, hotels, hospitals, and several other types of structures requires isolation to prevent vibration from interfering with equipment operations, causing an unpleasant working environment, or

creating unwanted noise. Many types of isolators are available to reduce unwanted vibration. The following is a list of some of these:

Cork pads
Neoprene pads
Neoprene mounts and hangers
Spring mounts and hangers
Pneumatic mounts
Active pneumatic systems
Floating floor systems

All of these have specific uses. Manufacturers found in Appendix F can provide their catalog materials that define use, function, and costs. As with all noise and vibration control mate-

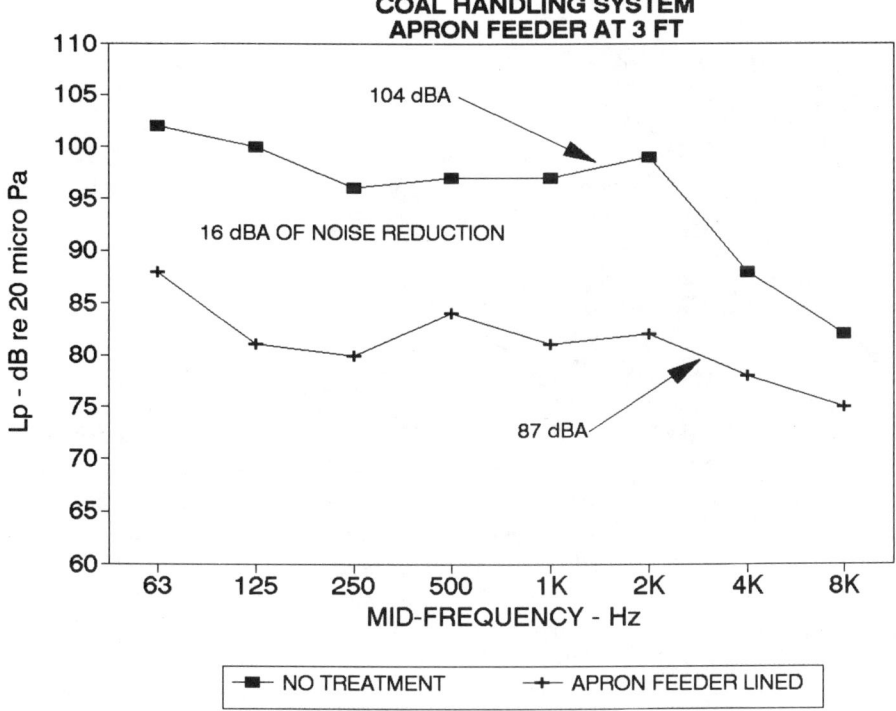

FIGURE 10.60   Results of adding abrasion-resistant neoprene.

rials there are misapplications or incomplete applications of the products.

For example, a large cooling tower located on the twenty-fourth floor of a high-rise building was well isolated with heavy-duty restrained (holding the springs in place when the water in the tower is emptied) spring mounts.

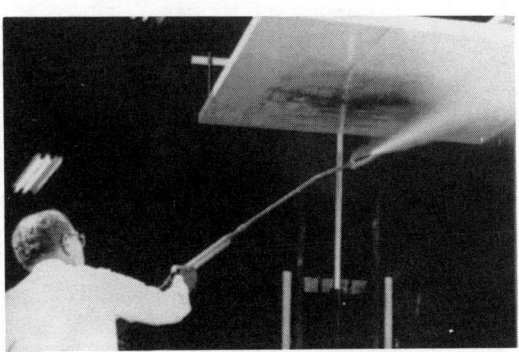

FIGURE 10.61   Cleanable acoustical panel. (Reprinted by permission from Industrial Acoustics Co.)

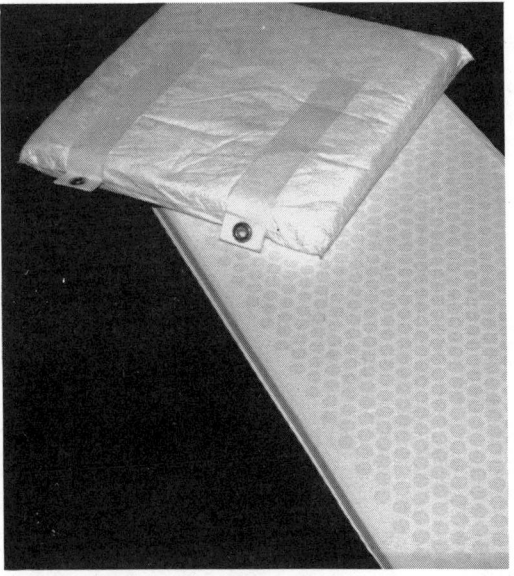

FIGURE 10.62   Cleanable acoustical panel. (Materials courtesy of United-McGill Corp. and Kinetics Noise Control Corp.)

FIGURE 10.63   Acoustical foam absorbing material. (Reprinted by permission from Kinetics Noise Control Corp.)

The large condenser water lines and pumps to the tower were also isolated from the equipment room floor with spring mounts. However, the piping was also rigidly attached, at two points, to the bottom of the steel structure supporting the tower, rendering all the isolation installed ineffective. Vibration was flanking into the structure and could be easily heard in the occupied space below.

The following case history will describe an application of three types of isolation materials. The whole project spanned several years because the use of the space in question changed over time.

Description   A high-rise office building had electrical distribution unit substations located on various floors. The word-processing room of a law office was located next to the unit substation room. Over time this room was changed in a mini-courtroom where lawyers and witnesses could practice. In addition to the unit substation room the building mechanical rooms were located above and below this room. This arrangement is illustrated in Fig. 10.66. The building is a lightweight steel frame structure, which added to the problem.

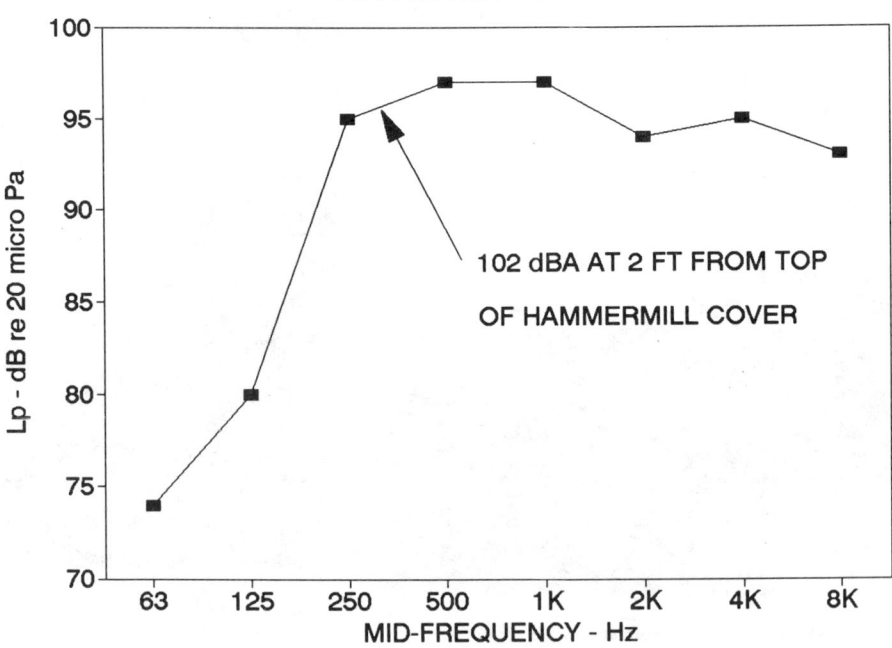

FIGURE 10.64   Hammermill noise spectrum.

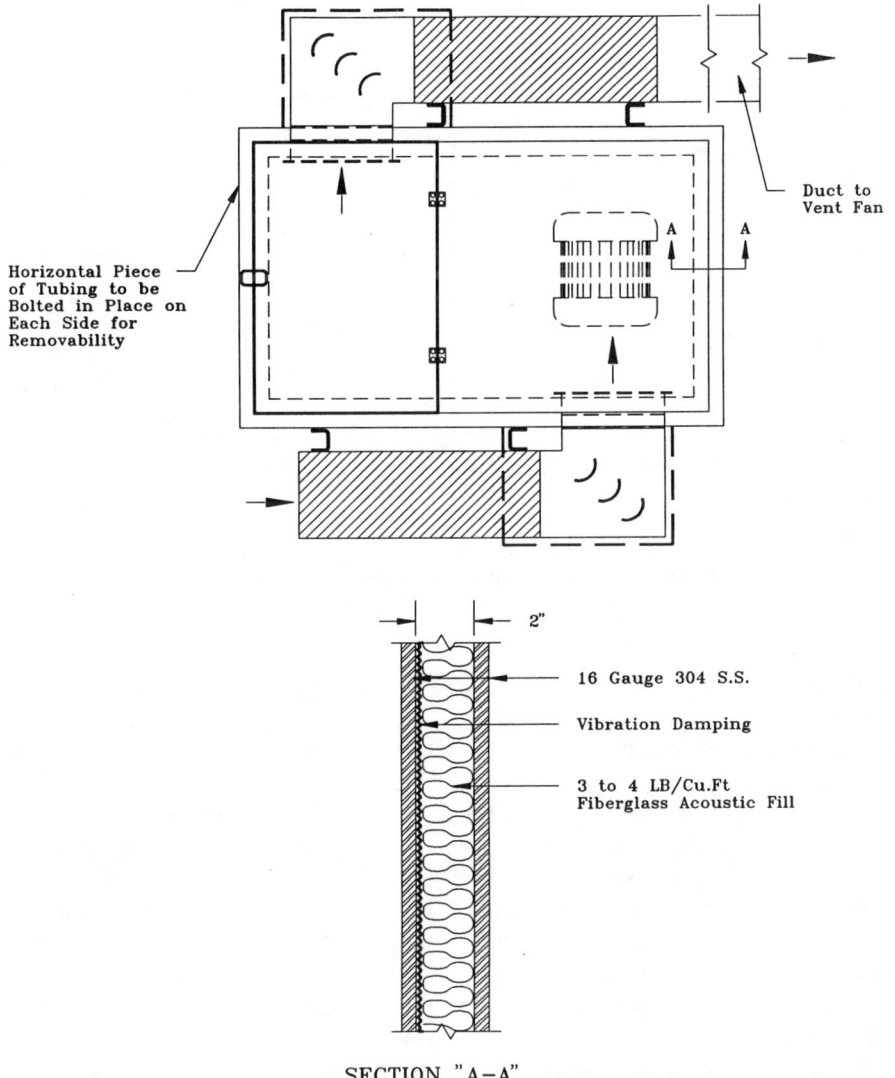

SECTION "A−A"

FIGURE 10.65   Plan view of enclosure and panel cross section.

The criteria for the original use of the space was an NC 45. When the use changed this criterion was not sufficient. In addition the original complaint was about the "buzz" that could be easily heard.

Analysis A noise and vibration survey revealed that the noise complaint was justified and that indeed the transformer "buzz' could easily be heard. This spectrum is shown in Fig. 10.67. Vibration was easily felt in the walls and

floor. Measurement revealed that the vibration was from the transformer and mechanical equipment above and below. The analysis was done using a two-channel spectrum analyzer that could pinpoint the offending frequency. This assisted in determining the noise and vibration paths.

Recommendations The first use of the space was for word processing, and a nominal amount of noise reduction was required. The unit sub-

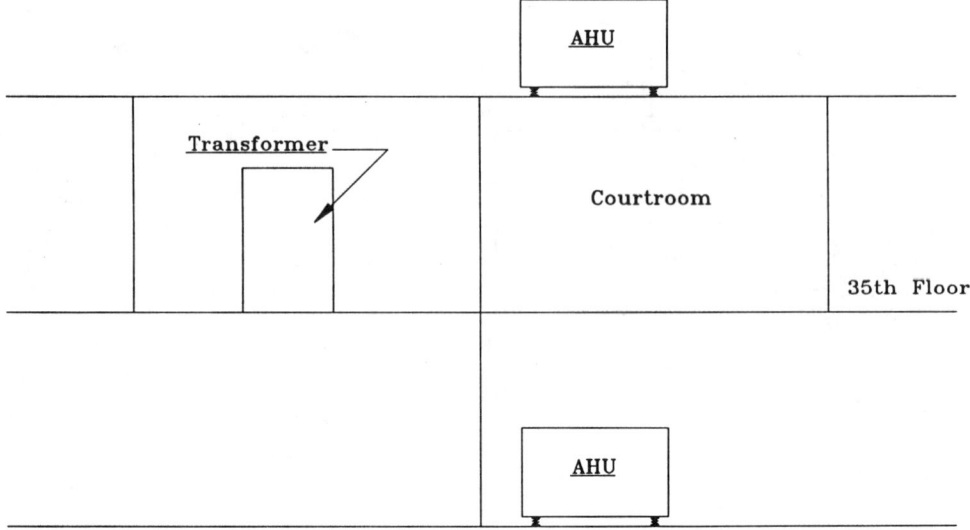

FIGURE 10.66   View of unit substation equipment room and adjacent spaces.

station design allowed the transformer core to be isolated. A pneumatic-type isolator was selected to isolate the core from the metal cabinet. The only hard connections remaining were through the electrical bus bar system. This reduced the noise level in the adjacent room as shown in Fig. 10.67 and was satisfactory for the intended use, even though one could slightly hear the transformer due to its pure-tone nature. There was enough activity in the room, and the vibration was reduced enough not to be a problem.

When the use of the space was changed to a mini-courtroom and all the activity in the space was reduced, the "buzz' was quite evident. Additional measurements were made to pinpoint more precisely the remaining paths. These measurements revealed that there was residual vibration from the bus bars, airborne excitation of the light-gauge metal unit substation case, electrical control panels on the common wall, electrical cable hangers, and the lack of pipe supports from the air.

The following items were carried out:

Isolation of the unit substation with a combination neoprene and steel layered pad

Isolation of all electrical control panel boxes with a neoprene mount

Isolation of all electrical cables with neoprene hangers

Isolation of the chilled water piping on the floor below with spring hangers

These measures reduced the noise level in the mini-courtroom to an NC 35, eliminating the "buzz." Because of the lightweight structure this was about the limit unless major structural changes were made, which was not possible on the thirty-fifth floor of the building. The results were satisfactory, however. The stepwise results are illustrated in Fig. 10.67. Usually this type of approach is required to see the results for each step. The vibration isolators used for the substation are shown in Fig. 10.68. This type of problem can occur in many facilities.

PUNCH PRESSES

The complexity of the problem of reducing punch press noise becomes apparent as soon as the total problem is studied. The difficulty of providing a simple solution to the problem is complicated by the wide variety of presses in

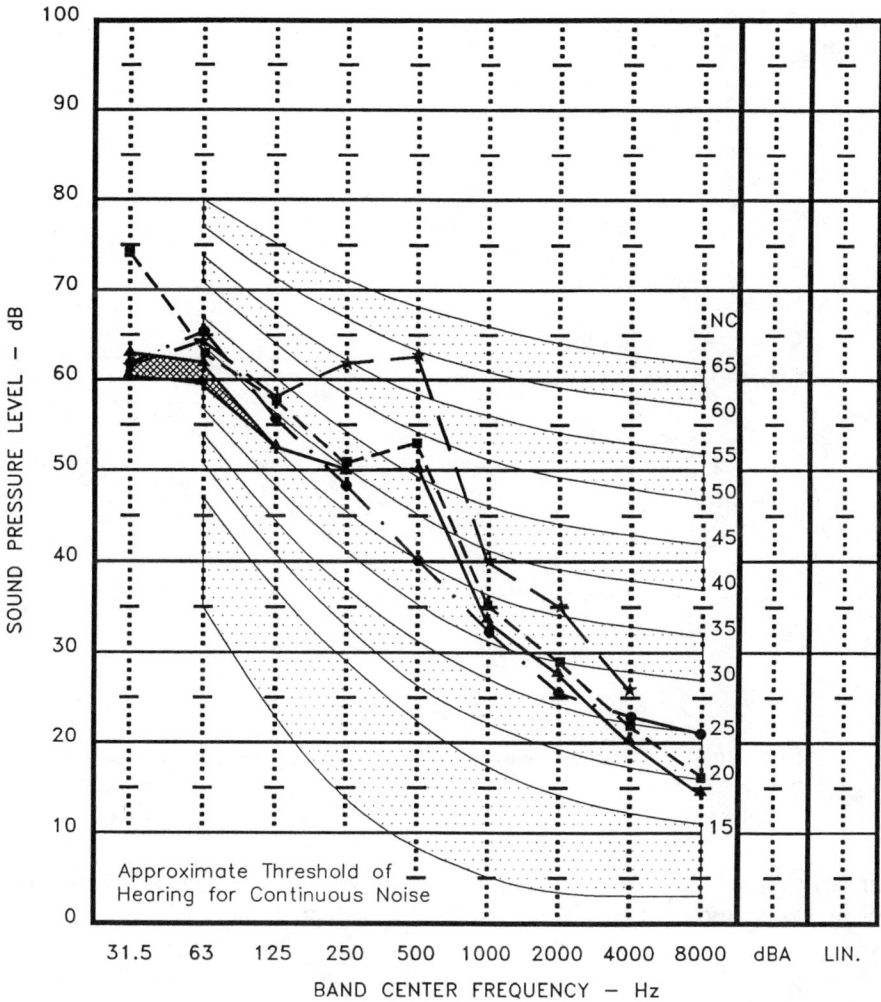

SOUND PRESSURE LEVEL — dB

BAND CENTER FREQUENCY — Hz

Legend

★   Word Processing Room

■   Air Mounts Added

▲   Piping and Duct Work AHU on
    34 Control Panels and Exhaust Fan
    in Transformer Room Isolated

●   Transformer Isolated

FIGURE 10.67   Results of noise control in occupied space of adjacent unit substation.

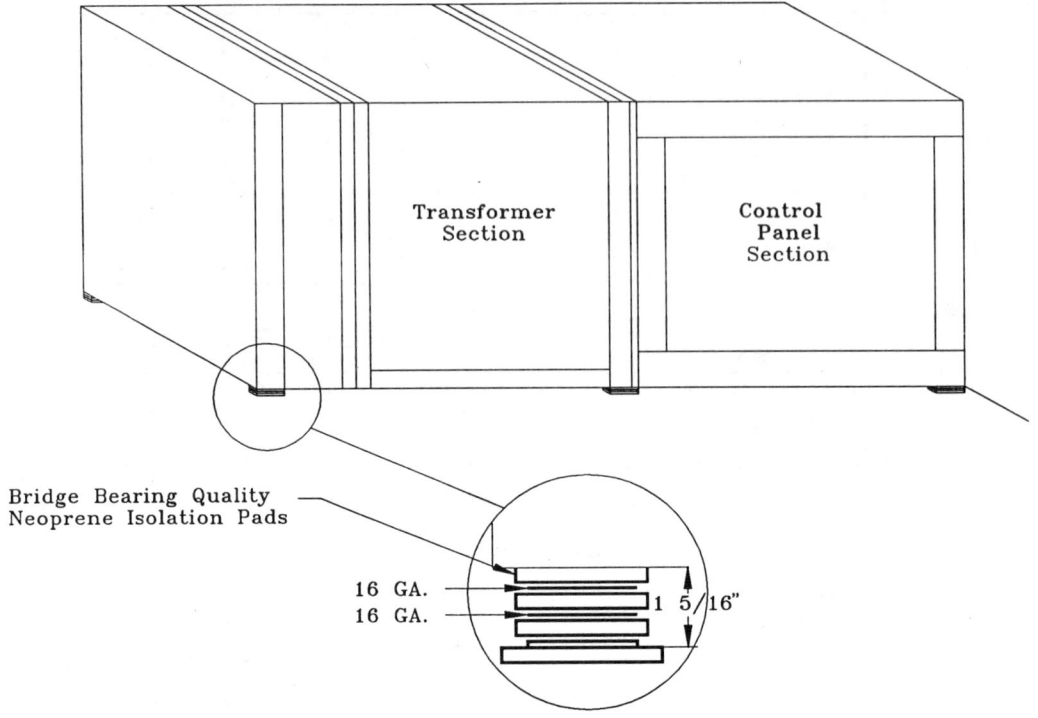

Bridge Bearing Quality
Neoprene Isolation Pads

16 GA.
16 GA.   1 5/16"

FIGURE 10.68   Transformer substation vibration isolation material.   (Courtesy of Mason-Dallas.)

use as well as the different types of tooling used for various press operations. It is further complicated by space configurations and allocations for punch presses, the number of presses involved, the multiple noise sources within a press, and problems associated with retrofitting noise abatement equipment.

In view of the complexity of this problem, how should it be approached, or what is the best method of controlling punch press noise? The first step in controlling punch press (or any other) noise is to identify the specific noise source(s). There are several primary noise sources associated with punch presses. They may be identified as follows:

1. Impacts associated with die operations
2. Turbulence due to air operations
3. Metal-to-metal impacts of parts leaving the press
4. Automatic feed mechanisms
5. Scrap-cutting attachments to the press

6. Vibration of sheet metal parts fastened to the press
7. Clutch and brake mechanism on the drive shaft
8. Transmission of vibration from press to control or other units (structure-borne transmission)
9. Vibration of the surfaces of the press itself (frame radiated noise)

There are several recognized methods of noise reduction that may be considered. Some engineering methods of controlling punch press noise are:

1. Full room-type enclosures
2. Air noise reduction
3. Constrained layer damping of press fame
4. Modification of tools
5. Vibration isolation
6. Partial enclosures

The specific method of punch press noise reduction should be selected using a systems approach. That is, all limiting factors, constraints, and local problems must be taken into consideration before the control method is selected. In most instances, more than one of the methods listed will be used to effectively control punch press noise.

Partial Enclosures for Punch Presses

Description Partial enclosure is most effective on automatic punch presses. On the presses studied, the primary noise sources were determined to be die area noise (including air), sheet metal parts fastened to the press, and metal-to-metal impacts of parts leaving the press. The frame-radiated noise was determined not to be a significant problem except in some blanking operations. Generally, frame-radiated noise results more from blanking operations than from forming, embossing, and so on. This type of noise appears to develop when the tonnage rating of the specific operation reaches 30 to 50 percent of the rated capacity of the press [Pelton and Storment, 1979].

Since the frame of the press was determined not to be a major noise source, it does not have to be covered or treated. In fact, using the partial-enclosure concept, the press frame serves as the basic structure for the enclosure. With the openings within the press frame enclosed, the noise generated in the die area is effectively contained within the frame enclosure.

Analysis The first step in testing this approach was to construct a mock-up noise enclosure on a smaller 25-ton automatic punch press. Figure 10.69 illustrates the mock-up. A four-ply corrugated fiberboard was used for panels. These panels were lined with one-inch acoustical foam and placed over the front, rear, and side openings of the punch press frame. The subjective response of the operator was that it reduced the noise coming from the tool area as well as the noise generated by the air ejectors. Figure 10.70 illustrates the amount of noise reduction achieved by the mock-up enclosure. Although the effectiveness was more apparent for the re-

FIGURE 10.69   25-Ton punch press noise control mock-up. (Reprinted from ASME Paper 79-WA/DCS-1.)

duction of air noise, there was still a significant reduction of the middle and low frequencies afforded by the mock-up treatment.

The transition from the mock-up design to a prototype noise enclosure was achieved with the use of functionally designed panels. The standard panel was constructed of 16-ga outside solid sheet metal that had internal damping placed on it. The acoustical material was a one-inch glass fiber material covered with mylar or other such protective film with the edges sealed. This was covered with 22-ga perforated sheet metal to provide the protection needed for the absorption material (Fig. 10.71). The panels were placed over the die area opening, the lower opening between the press legs on the front of the press, and over similar openings on the rear of the press. The panel over the die area was designed to swing open as a door and served a dual purpose as a die area guard once it was electrically interlocked. The hardware selected was the heavy-duty type. Lift-off

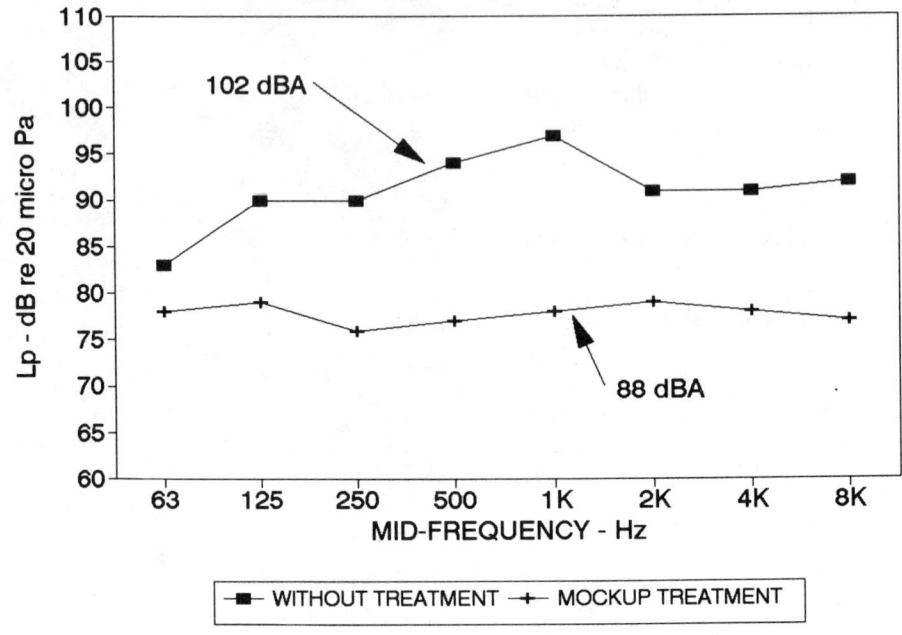

FIGURE 10.70   Spectrum of 25-ton punch press mock-up noise control.   (Reprinted from ASME Paper 79-WA/DCS-1.)

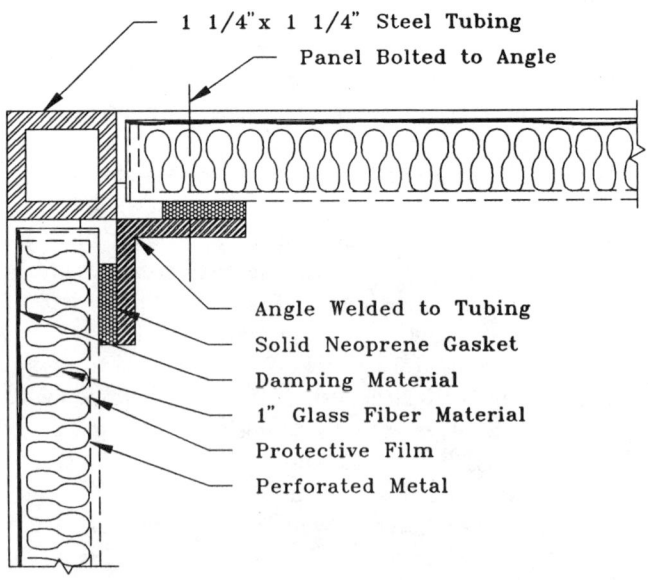

FIGURE 10.71   Typical acoustical panel construction. (Reprinted from ASME Paper 79-WA/DCS-1.)

hinges were provided for setup accessibility. Gaskets were the bulb type that can be compressed when the door is closed. Vision panels were 1/4-in. Lexan and were mounted in neoprene glazing gaskets. The side openings were covered with solid sheet metal panels dampened with a viscoelastic material. Slots were provided in the side panel to allow for coil stock going into the die and scrap coming out. The overall design concept was to afford complete accessibility to the press during press operation and setup operations. The first prototype of the partial enclosure is illustrated in Fig. 10.72.

Evaluation and Results of Partial Enclosures An evaluation period of several months produced some observed weaknesses in the system. These are listed as follows:

1. The side panels did not provide for vertical adjustment of the tool bed and feed mechanism.
2. The lower front door and rear cover did not allow enough floor space for parts and scrap pans. Consequently, operators began leaving doors open to obtain additional space for pans.
3. The in-feed roller/stock straightener mechanism and the scrap cutter located on the out-feed of the press were determined to be new dominant noise sources.
4. The seals began to wear after a period of time, causing the noise level to increase.
5. Blanking tools caused some frame-radiated noise when they were in use.

Since this prototype was applied to one of seven similar presses, a second-generation prototype design was begun. This prototype is shown in Fig. 10.73. In order to circumvent problems created by the side plates as well as to contain the noise generated by in-feed/out-feed mechanisms, the perimeter of the enclosure was expanded to completely enclose these mechanisms. Similarly, the rear portion of the press enclosure was expanded to provide for additional space for parts and scrap pans. This expanded enclosure used $1\frac{1}{4}$-in.-square tubing as the primary structure to which standard 1-in. panels were attached. Access doors and vision

FIGURE 10.72 Prototype of punch press noise control partial enclosure. (Reprinted from ASME Paper 79-WA/DCS-1.)

FIGURE 10.73 Second-generation prototype punch press noise control partial enclosure. (Reprinted from ASME Paper 79-WA/DCS-1.)

FIGURE 10.74   View of punch presses after final installation.   (Reprinted from ASME Paper 79-WA/
DCS-1.)

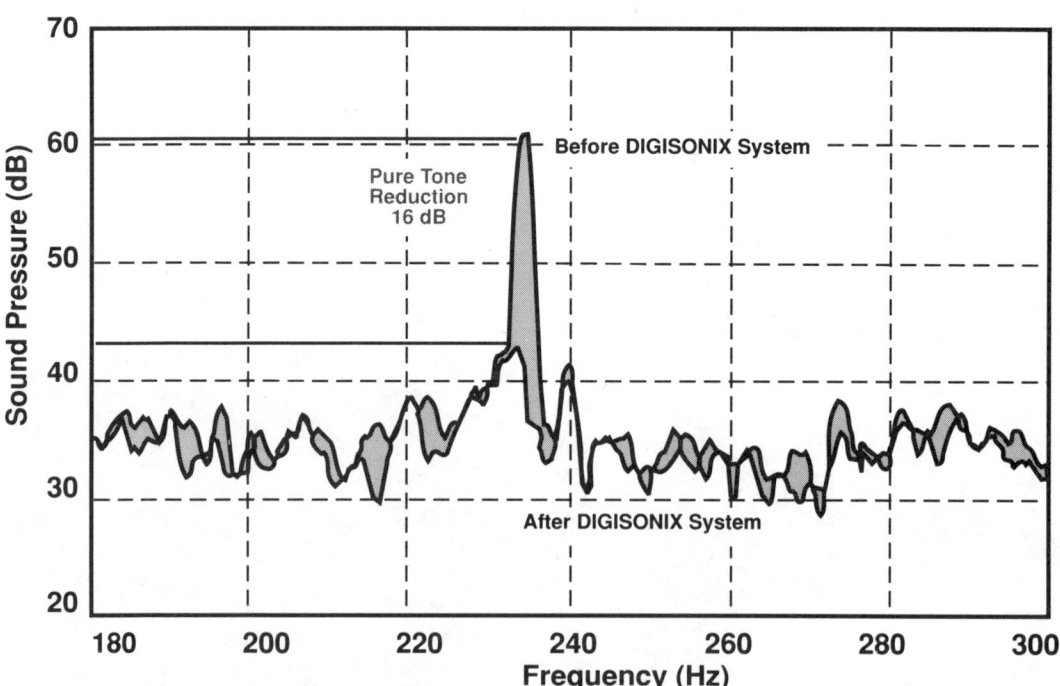

FIGURE 10.75   Narrow-band spectrum of wet scrubber exhaust approximately one-quarter from plant.   (Reprinted by permission from DIGISONIX, Division of Nelson Industries.)

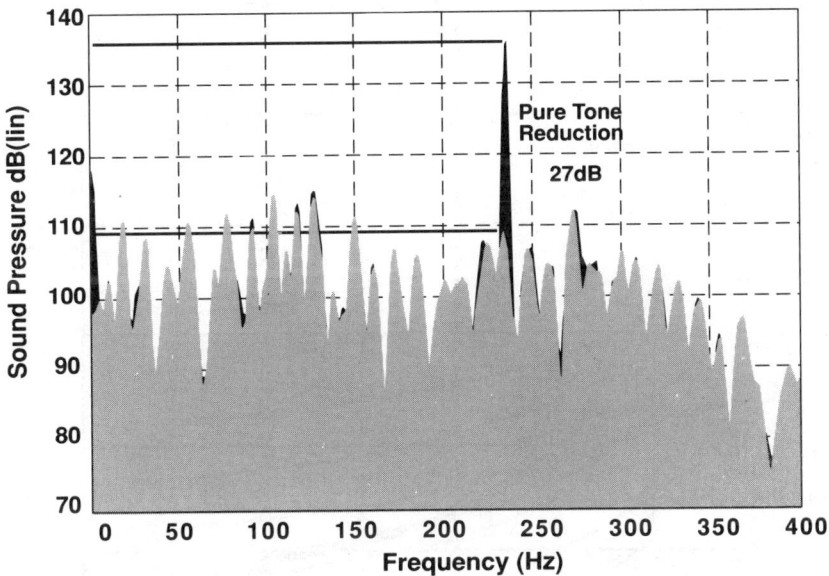

FIGURE 10.76   Narrow-band spectrum of wet scrubber exhaust in duct from plant.   (Reprinted by permission from DIGISONIX, Division of Nelson Industries.)

panels were placed at appropriate locations. The resultant enclosure still did not extend beyond the original perimeter of the punch press.

After the second-generation prototype enclosure was installed and evaluated, the remainder of the presses were treated in this manner. Figure 10.74 shows the line of presses with the enclosures installed. The subjective operator response to the enclosures is that a conversation can be carried on in this area without shouting—something that was not previously possible.

ACTIVE NOISE CONTROL

The general description of active noise control or noise cancellation devices are outlined at the end of Chapter 4. Two case histories of active noise control will describe how this new technology can be used in place of passive (absorptive) silencers. One major advantage is that the silencer offers no more pressure drop than a straight piece of duct or pipe. Thus, the amount of horsepower is less, resulting in lower electrical consumption and costs. Over time the silencer can pay for itself and make money for the user. The information in the following case histories was provided courtesy of and reprinted by permission of DIGISONIX Division of Nelson Silencer Co.

Wet Scrubber Induced Draft Fan Many scrubber systems used in air pollution control

FIGURE 10.77   Installation of active noise silencer in wet scrubber exhaust.   (Reprinted by permission from DIGISONIX, Division of Nelson Industries.)

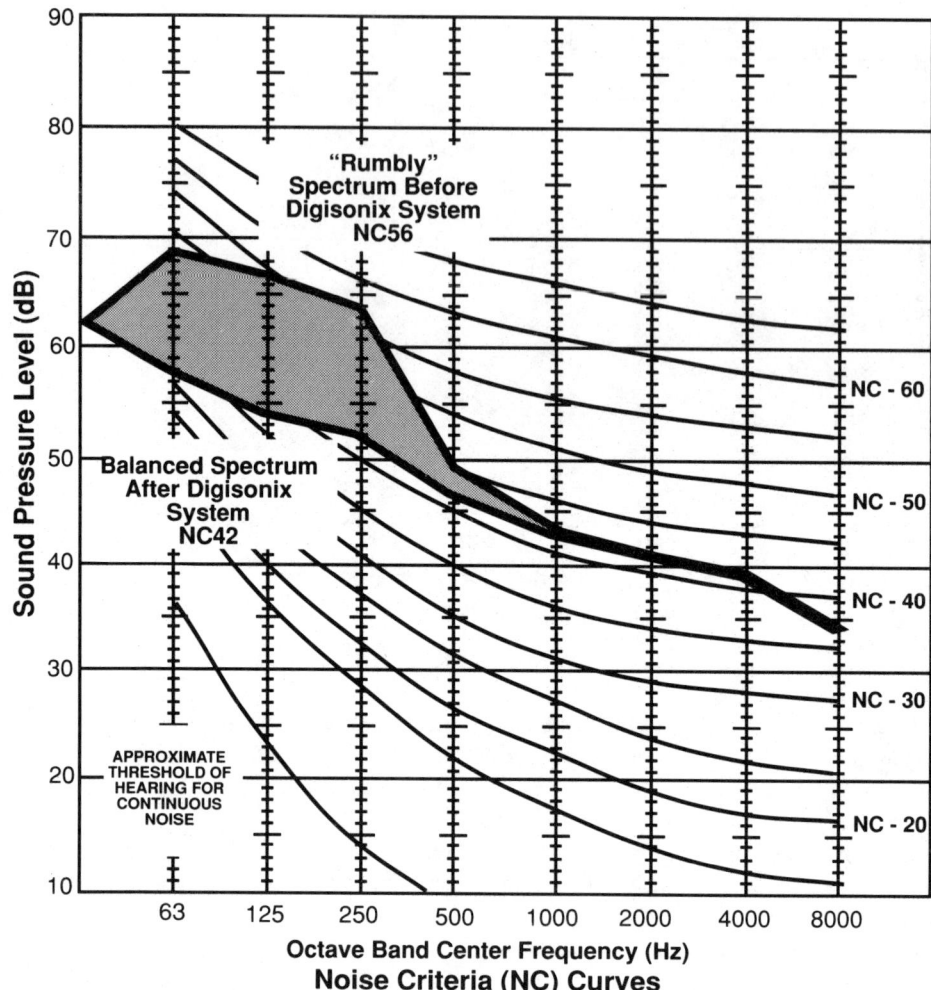

FIGURE 10.78   Octave band spectrum showing noise criteria (NC) curves before and after installing an active noise silencer in a supply fan duct system.   (Reprinted by permission from DIGISONIX, Division of Nelson Industries.)

systems use radial blade induced draft fans. These fans, which are usually located near the end of the duct work in the system, have tonal components. After the installation of the wet scrubber system at the plant, the neighbors complained about the noise. The major complaint was the tone of the fan. The narrow-band noise spectrum shown in Fig. 10.75 was measured at about one-quarter mile from the plant. The tone was 61 dB at 235 Hz, or about 51 dBA. Due to the pure tone and its height above the background noise, it was quite noticeable.

To confirm the fan as the noise source, measurements were taken in the fan exhaust stack, as shown in Fig. 10.76. The tone was 135 dB at 235 Hz. An active silencer was installed in the wet scrubber exhaust as shown in Fig. 10.77. The tone was reduced by 27 dB. The result in the community was a level in the 40- to 45-dBA range, which was quite acceptable since the plant blended in with the background noise.

Air-Conditioning Supply Fan System   A miscellaneous storage and laboratory space in the basement of a building was to be converted into

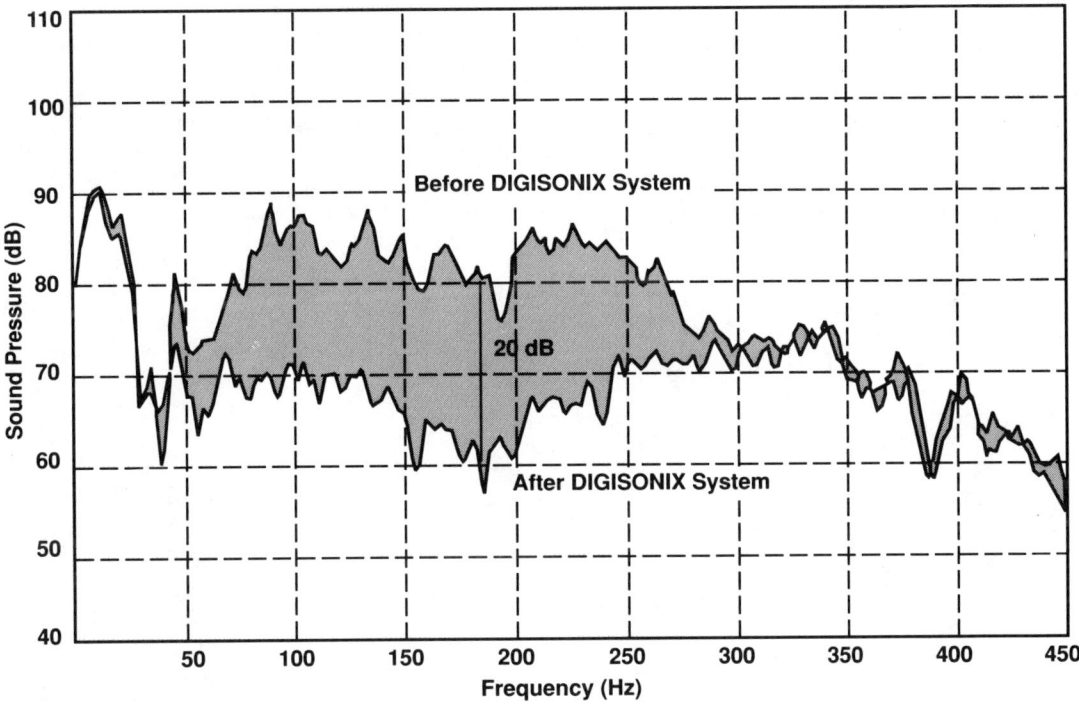

FIGURE 10.79   Narrow-band spectrum installing an active noise silencer in a supply fan duct system. (Reprinted by permission from DIGISONIX, Division of Nelson Industries.)

office space. Noise from the adjacent air handler was breaking out of the flat oval duct creating a low-frequency rumble in the proposed office space. The fan was an airfoil type producing 6000 cfm. The air passed through the flat oval duct with 1-in. acoustical insulation.

Sound measurements made in the space were plotted on an NC chart, revealing an NC 56 as shown in Fig. 10.78. Office space of this type should be designed for a criterion of NC 40. All of the noise energy is in the 63-, 125-, and 250-Hz octave bands. A narrow-band spectrum in Fig. 10.79 also revealed the low-frequency nature of the noise.

Due to space limitations and pressure drop considerations a passive silencer was judged not suitable. The active silencer shown in Fig. 10.80 was installed in the duct system and reduced the noise by 14 NC points to NC 42 (Fig. 10.78). In the narrow-band spectrum a 20-dB reduction was achieved as shown in Fig. 10.79.

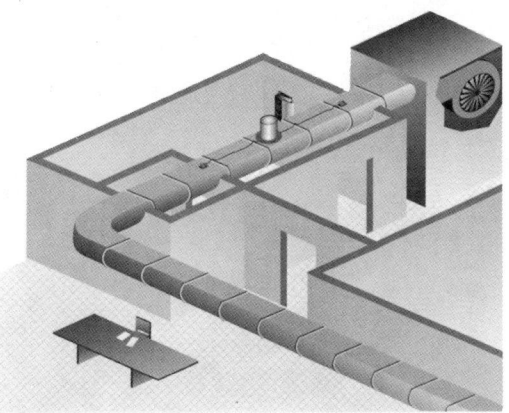

FIGURE 10.80   Installation of active noise silencer in duct system.   (Reprinted by permission from DIGISONIX, Division of Nelson Industries.)

References

API. 1973. *Guidelines on Noise*. American Petroleum Institute, Medical Research Report EA7301.

Pelton, H. K. 1989. Solving Disk Brake Squeal on Oil Drilling Rigs. *Sound and Vibration*, October, pp. 14–18.

Pelton, H. K., and Storment, J. W. 1975. *Meeting OSHA Noise Criteria in a Central Utility Plant of a Large Manufacturing Facility*. ASME 75-WA/SAF-2.

Pelton, H. K., and Storment, J. W. 1976. Practical Noise Control in a Large Manufacturing Plant. *Sound and Vibration*, May, pp. 22–32.

Pelton, H. K., and Storment, J. W. 1979. *Partial Enclosures—An Effective Method of Controlling Punch Press Noise*. ASME 79-WA/DCS-2.

Rubin, M., et al. 1982. *Techniques for the Design of Coal Preparation Plants*. U.S. Department of Interior, Bureau of Mines.

Appendix A

Glossary of Acoustical Terms

Acoustic An adjective used in conjunction with a basic property of sound, such as "acoustic energy."

Acoustic center The location of an assumed point source that has the same total sound power level as the total collection of all sources of a plant in regard to sound emitted to remote received positions.

Acoustical treatment The use of acoustical absorption, acoustical isolation, or any changes or additions to a structure to correct acoustical faults or improve the acoustical environment.

Ambient sound The existing background environmental sounds, usually interpreted as excluding the noise from a power plant under consideration (or other "new intruding noise source"). Ambient sound may be expressed as sound pressure level in octave frequency bands or as A-weighted sound levels. Time-varying ambient levels may be designated by L_1, L_{10}, L_{eq}, etc.

Articulation index A numerical value (0 to 1.0) of speech intelligibility—derived from an analysis of background sound, expected speech effort, and the acoustical qualities of the area and its components. An A.I. of 0.1 is low, indicating that little, if any, of a conversation will be intelligible to listeners. An area with an A.I. of 0.6, on the other hand, will make for poor speech privacy.

A-weighted sound level The value of a sound pressure signal after it has been passed through the A-weighted filter response described in American National Standards Institute S1.4-1971 Specification for Sound Level Meters. A-weighting is often used because its value correlates well with subjective interpretations of loudness, annoyance, etc. A-weighted sound levels are expressed in decibels and are designated by the symbol dBA.

Community area All the habitable area outside a plant property.

Decibel (dB) The basic unit of sound level. The decibel denotes a ratio of the intensity of one sound and the lower intensity of a reference sound. On the decibel scale, small differences become highly critical. Only 5 dB separate the level of a normal conversation from the din created by nine typewriters.

Decile sound level, L_n The sound level (usually A-weighted) that is equaled or exceeded for a specific percentage of the time period of interest; L_1, L_{10}, L_{50}, etc., are the sound levels exceeded for 1%, 10%, 50%, etc., of the time period, respectively.

Equivalent sound level, L_{eq} The equivalent steady sound level that, if continuous during the time period of interest (t_1 to t_2), would contain the same total sound energy as the actual time-varying sound.

Frequency   The number of complete cycles per second of a vibration (or other periodic motion). Usually stated in Hertz (Hz). The frequency of the human voice can range from 100 to 10,000 Hz, though the frequencies of intelligible speech lie between 400 and 2,000 Hz.

Insertion loss (I.L.)   The difference, in decibels, between two sound pressure levels which are measured at the same point in space before and after a muffler or barrier is inserted between the measurement point and the noise source.

Level, sound/noise   A measure of sound pressure level as determined by electrical equipment meeting ANSI requirements. Unless specifically stated otherwise, levels refer to root mean square of sinusoidally varying level.

Level meter, sound   An electrical instrument for determining sound pressure level.

NC (noise criteria curve or level)   A curve which describes sound levels that are acceptable over a range of frequencies for a specific building function. The ear is less sensitive to low-frequency sound, so the permissible sound levels at low frequencies can be relatively high without causing problems. On the curve NC 40, for example, a 66-dB level is permissible at 63 Hz. At 2000 Hz, however the acceptable level is only 40 dB because the ear is more sensitive to higher frequencies. The NC 40 curve, not incidentally, describes an acceptable background sound level from all of the sound sources in a normal office.

Noise reduction coefficient (NRC)   An average of the sound-absorptive properties of a material at frequencies of 250, 500, 1000, and 2000 Hz.

Octave   The frequency interval between two frequencies whose ratio is 2:1.

Octave band   The range of frequencies in an octave, usually banded by two of the frequencies in the sequence . . . 62.5, 125, 250, 500. . . .

Omnidirectional source   A sound source that radiates its sound power approximately uniformly in all directions.

Point source   A sound source so small that it may be treated as though all its power radiates from a point, hence has no directivity variations and obeys ideally the inverse square law (6 dB sound pressure level rate of decrease with each doubling of distance) for all distances; sometimes defined as a source having a dimension that is much smaller than one-sixth of a wavelength.

Principal sound source   A source whose outdoor-radiated sound power level is among the highest 8 to 10 power levels of all the plant sources.

Sound power level (PWL) $L_w = 10 \log_{10} W/W_{ref}$, where W is the sound power radiated by a source, expressed in watts, and W_{ref} is the reference power of 10^{-12} watts; L_w is expressed in decibel (dB) units. This quantity cannot be measured directly (i.e., with a single measurement).

Sound pressure level (SPL) $L_p = 20 \log_{10} P/P_{ref}$, where p is the rms pressure in a sound wave, expressed in Pascal (Pa), and P_{ref} is the reference pressure of 2×10^{-5} Pa (Pa = N/m^2); L_p is expressed in decibel (dB) units. Single microphones and most other sound-measuring equipment are capable of measuring sound pressure level directly (i.e., with a single measurement).

Sound transmission class (STC)   A general method of categorization of partitions by transmission loss performance; a good single-number descriptor for noise such as speech, radio, and TV but not for mechanical equipment, HVAC, etc.

Speech interference level (SIL)   A way of rating the speech-masking affects of noise based on measurements of the noise in each of the octave bands centered at 500, 1000, and 2000 Hz.

Transmission loss (T.L.)   A measure of the sound-insulating properties of a wall, floor, ceiling, window, or door that is characteristic of the partition itself and not of the room that it bounds. It cannot be measured directly.

Windscreen   A shield placed around a microphone to prevent turbulent eddies (from wind with enough velocity to cause turbulent flow over the microphone) from impinging on the diaphragm, which would cause pressure variations similar to those produced by a high noise level.

Appendix B

Occupational Noise Exposure
Compliance Assistance
Guideline

U.S. Department of Labor
Occupational Safety and Health Administration

OCCUPATIONAL NOISE EXPOSURE COMPLIANCE ASSISTANCE GUIDELINE

GUIDELINE ONLY: THIS OCCUPATIONAL NOISE EXPOSURE COMPLIANCE GUIDELINE IS INTENDED AS A
GUIDE TO ASSIST EMPLOYERS IN DEVELOPING AN INITIAL OCCUPATIONAL NOISE EXPOSURE PROGRAM
OR FOR EVALUATING AN EXISTING OCCUPATIONAL NOISE EXPOSURE PROGRAM. COMPLETE TEXT OF
OCCUPATIONAL NOISE EXPOSURE STANDARD MUST BE CONSULTED FOR COMPLIANCE WITH THE RULE.

OBTAIN A COPY OF OCCUPATIONAL NOISE EXPOSURE STANDARD, 29 CFR 1910.95. READ/REVIEW.

> The employer shall protect all workers from occupational noise exposure that exceeds
> an 8-hour time weighted average (TWA) of 90 decibels (dBA).

To protect workers the employer shall: (a) monitor noise exposure, (b) institute
control measures, and (c) implement a hearing conservation program (HCP) when
occupational noise exposure exceeds an 8-hour TWA of 85 dBA.

MONITOR NOISE EXPOSURE

- Monitor noise to determine level of exposure to employees.
- Calibrate all sound measuring equipment before and after each use according to the
  manufacturer's instructions.
- Measure noise exposure levels with a dosimeter and/or a sound level measuring
  instrument with an A-weighting network.

CONTROLS

The employer shall institute engineering and/or administrative controls whenever
possible. If these controls fail to reduce employee noise exposures to an 8-hour TWA
of 90 dBA or less, then the employer shall provide and enforce the use of hearing
protectors that attenuate employee exposure to at least an 8-hour TWA of 90 dBA.

ENGINEERING CONTROLS
- Use technology to reduce noise levels.
- Keep machinery in good maintenance repair to reduce noise.
- Erect total or partial barriers to confine noise.

ADMINISTRATIVE CONTROLS
- Limit employees scheduled work time in a noisy area.
- Limit noisy operations and activities per shift.

PERSONAL PROTECTIVE EQUIPMENT
- Provide at no cost to the employee a selection of hearing protection appropriate for
  noise levels in the environment.
- Provide training on the selection, fitting, use, and care of hearing protectors.
- Ensure that protectors are worn.

IMPLEMENT A HEARING CONSERVATION PROGRAM

To protect workers whose noise exposure equals or exceeds an 8-hour TWA of 85 dBA
the employer shall implement a continuing, effective hearing conservation program
(HCP).

MONITORING NOISE EXPOSURE
- Use only measuring instruments that meet the American National Standard
  Institute (ANSI) specifications.
- Use a sampling strategy that will pick up all continuous, intermittent, and
  impulsive sound levels from 80-130 dBA, and include all of these sound levels in
  the total noise measurement.
- Permit employees or their representatives to observe monitoring.
- Notify employees of noise exposure at or above 8-hour TWA of 85 dBA.

INFORMATION SOURCES

Title 29, *Code of Federal Regulations* (CFR) Part 1910.95 (Occupational Noise Exposure).
OSHA-2056 *All About OSHA*
OSHA-2098 *OSHA Inspections*
OSHA-3074 *Hearing Conservation*
OSHA-3077 *Personal Protective Equipment*
OSHA-3021 *OSHA: Employee Workplace Rights*
OSHA-3000 *Employer Rights & Responsibilities Following an OSHA Inspection*
OSHA-3110 *Access to Medical and Exposure Records*

A single free copy of the above materials can be obtained from the OSHA Publications
Office, Room N3101, 200 Constitution Ave. N.W., Washington, DC, 20210, (202) 523-9667;
or call your local OSHA Area Office (listed under the U.S. Department of Labor in the
telephone book).

OCCUPATIONAL SAFETY AND HEALTH   1910.94(d)(12)(iv)

■ NOTE: Asterick Denotes stayed Material   STANDARDS AND INTERPRETATIONS

no leakage of solvent when they are closed.

(13) Scope.

(i) This paragraph (d) applies to all operations involving the immersion of materials in liquids, or in the vapors of such liquids, for the purpose of cleaning or altering their surfaces, or adding or imparting a finish thereto, or changing the character of the materials, and their subsequent removal from the liquids or vapors, draining, and drying. Such operations include washing, electroplating, anodizing, pickling, quenching, dyeing, dipping, tanning, dressing, bleaching, degreasing, alkaline cleaning, stripping, rinsing, digesting, and other similar operations, but do not include molten materials handling operations, or surface coating operations.

(ii) "Molten materials handling operations" means all operations, other than welding, burning, and soldering operations, involving the use, melting, smelting, or pouring of metals, alloys, salts, or other similar substances in the molten state. Such operations also include heat treating baths, descaling baths, die casting stereotyping, galvanizing, tinning, and similar operations.

(iii) "Surface coating operations" means all operations involving the application of protective, decorative, adhesive, or strengthening coating or impregnation to one or more surfaces, or into the interstices of any object or material, by means of spraying, spreading, flowing, brushing, roll coating, pouring, cementing, or similar means; and any subsequent draining or drying operations, excluding open-tank operations.

1910.95—OCCUPATIONAL NOISE EXPOSURE

(a) Protection against the effects of noise exposure shall be provided when the sound levels exceed those shown in Table G–16 when measured on the A scale of a standard sound level meter at slow response. When noise levels are determined by octave band analysis, the equivalent A-weighted sound level may be determined as follows:

TABLE G–16—PERMISSIBLE NOISE EXPOSURES [1]

Duration per day, hours	Sound level dBA slow response
8	90
6	92
4	95
3	97
2	100
1½	102
1	105
½	110
¼ or less	115

[1] When the daily noise exposure is composed of two or more periods of noise exposure of different levels, their combined effect should be considered, rather than the individual effect of each. If the sum of the following fractions: $C_1/T_1 + C_2/T_2 ... C_n/T_n$ exceeds unity, then, the mixed exposure should be considered to exceed the limit value. C_n indicates the total time of exposure at a specified noise level, and T_n indicates the total time of exposure permitted at that level.

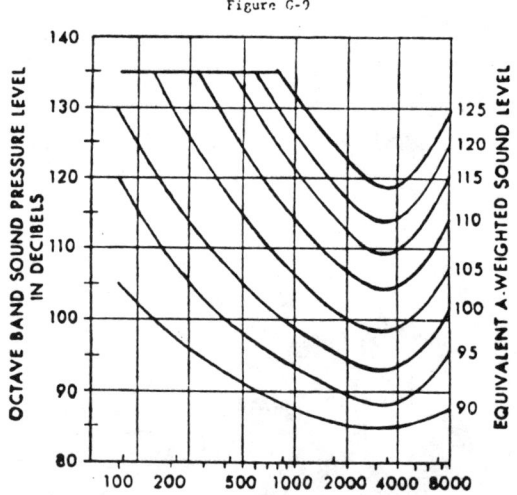

Figure G-9

BAND CENTER FREQUENCY IN CYCLES PER SECOND

Equivalent sound level contours. Octave band sound pressure levels may be converted to the equivalent A-weighted sound level by plotting them on this graph and noting the A-weighted sound level corresponding to the point of highest penetration into the sound level contours. This equivalent A-weighted sound level, which may differ from the actual A-weighted sound level of the noise, is used to determine exposure limits from Table G–16.

STANDARDS AND INTERPRETATIONS

(b)

(1) When employees are subjected to sound exceeding those listed in Table G-16, feasible administrative or engineering controls shall be utilized. If such controls fail to reduce sound levels within the levels of Table G-16, personal protective equipment shall be provided and used to reduce sound levels within the levels of the table.

Exposure to impulsive or impact noise should not exceed 140 dB peak sound pressure level.

(2) If the variations in noise level involve maxima at intervals of 1 second or less, it is to be considered continuous.

(c) Hearing conservation program.

(1) The employer shall administer a continuing, effective hearing conservation program, as described in paragraphs (c) through (o) of this section, whenever employee noise exposures equal or exceed an 8-hour time-weighted average sound level (TWA) of 85 decibels measured on the A scale (slow response) or, equivalently, a dose of fifty percent. For purposes of the hearing conservation program, employee noise exposures shall be computed in accordance with Appendix A and Table G-16a, and without regard to any attenuation provided by the use of personal protective equipment.

(2) For purposes of paragraphs (c) through (n) of this section, an 8-hour time-weighted average of 85 decibels or a dose of fifty percent shall also be referred to as the action level.

(d) Monitoring.

(1) When information indicates that any employee's exposure may equal or exceed an 8-hour time-weighted average of 85 decibels, the employer shall develop and implement a monitoring program.

(i) The sampling strategy shall be designed to identify employees for inclusion in the hearing conservation program and to enable the proper selection of hearing protectors.

(ii) Where circumstances such as high worker mobility, significant variations in sound level, or a significant component of impulse noise make area monitoring generally inappropriate, the employer shall use representative personal sampling to comply with the monitoring requirements of this paragraph unless the employer can show that area sampling produces equivalent results.

(2)

(i) All continuous, intermittent and impulsive sound levels from 80 decibels to 130 decibels shall be integrated into the noise measurements.

(ii) Instruments used to measure employee noise exposure shall be calibrated to ensure measurement accuracy.

(3) Monitoring shall be repeated whenever a change in production, process, equipment or controls increases noise exposures to the extent that:

(i) Additional employees may be exposed at or above the action level; or

(ii) The attenuation provided by hearing protectors being used by employees may be rendered inadequate to meet the requirements of paragraph (j) of this section.

(e) Employee notification. The employer shall notify each employee exposed at or above an 8-hour time-weighted average of 85 decibels of the results of the monitoring.

(f) Observation of monitoring. The employer shall provide affected employees or their representatives with an opportunity to observe any noise measurements conducted pursuant to this section.

(g) Audiometric testing program.

(1) The employer shall establish and maintain an audiometric testing program as provided in this paragraph by making audiometric testing

available to all employees whose exposures equal or exceed an 8-hour time-weighted average of 85 decibels.

(2) The program shall be provided at no cost to employees.

(3) Audiometric tests shall be performed by a licensed or certified audiologist, otolaryngologist, or other physician, or by a technician who is certified by the Council of Accreditation in Occupational Hearing Conservation, or who has satisfactorily demonstrated competence in administering audiometric examinations, obtaining valid audiograms, and properly using, maintaining and checking calibration and proper functioning of the audiometers being used. A technician who operates microprocessor audiometers does not need to be certified. A technician who performs audiometric tests must be responsible to an audiologist, otolaryngologist or physician.

(4) All audiograms obtained pursuant to this section shall meet the requirements of Appendix C: *Audiometric Measuring Instruments.*

(5) Baseline audiogram.

(i) Within 6 months of an employee's first exposure at or above the action level, the employer shall establish a valid baseline audiogram against which subsequent audiograms can be compared.

(ii) Mobile test van exception. Where mobile test vans are used to meet the audiometric testing obligations, the employer shall obtain a valid baseline audiogram within 1 year of an employee's first exposure at or above the action level. Where baseline audiograms are obtained more than 6 months after the employee's first exposure at or above the action level, employees shall wear hearing protectors for any period exceeding six months after first exposure until the baseline audiogram is obtained.

(iii) Testing to establish a baseline audiogram shall be preceded by at least 14 hours without exposure to workplace noise. Hearing protectors may be used as a substitute for the requirement that baseline audiograms

be preceded by 14 hours without exposure to workplace noise.

(iv) The employer shall notify employees of the need to avoid high levels of non-occupational noise exposure during the 14-hour period immediately preceding the audiometric examination.

(6) Annual audiogram: At least annually after obtaining the baseline audiogram, the employer shall obtain a new audiogram for each employee exposed at or above an 8-hour time-weighted average of 85 decibels.

(7) Evaluation of audiogram.

(i) Each employee's annual audiogram shall be compared to that employee's baseline audiogram to determine if the audiogram is valid and if a standard threshold shift as defined in paragraph (g)(10) of this section has occurred. This comparison may be done by a technician.

(ii) If the annual audiogram shows that an employee has suffered a standard threshold shift, the employer may obtain a retest within 30 days and consider the results of the retest as the annual audiogram.

(iii) The audiologist, otolaryngologist, or physician shall review problem audiograms and shall determine whether there is a need for further evaluation. The employer shall provide to the person performing this evaluation the following information:

(a) A copy of the requirements for hearing conservation as set forth in paragraphs (c) through (n) of this section;

(b) The baseline audiogram and most recent audiogram of the employee to be evaluated;

(c) Measurements of background sound pressure levels in the audiometric test room as required in Appendix D: Audiometric Test Rooms.

(d) Records of audiometer calibrations required by paragraph (h)(5) of this section.

STANDARDS AND INTERPRETATIONS

(8) Follow-up procedures.

(i) If a comparison of the annual audiogram to the baseline audiogram indicates a standard threshold shift as defined in paragraph (g)(10) of this section has occurred, the employee shall be informed of this fact in writing, within 21 days of the determination.

(ii) Unless a physician determines that the standard threshold shift is not work related or aggravated by occupational noise exposure, the employer shall ensure that the following steps are taken when a standard threshold shift occurs:

(a) Employees not using hearing protectors shall be fitted with hearing protectors, trained in their use and care, and required to use them.

(b) Employees already using hearing protectors shall be refitted and retained in the use of hearing protectors and provided with hearing protectors offering greater attenuation if necessary.

(c) The employee shall be referred for a clinical audiological evaluation or an otological examination, as appropriate, if additional testing is necessary or if the employer suspects that a medical pathology of the ear is caused or aggravated by the wearing of hearing protectors.

(d) The employee is informed of the need for an otological examination if a medical pathology of the ear that is unrelated to the use of hearing protectors is suspected.

(iii) If subsequent audiometric testing of an employee whose exposure to noise is less than an 8-hour TWA of 90 decibels indicates that a standard threshold shift is not persistent, the employer:

(a) Shall inform the employee of the new audiometric interpretation; and

(b) May discontinue the required use of hearing protectors for that employee.

(9) Revised baseline. An annual audiogram may be substituted for the baseline audiogram

when, in the judgment of the audiologist, otolaryngologist or physician who is evaluating the audiogram:

(i) The standard threshold shift revealed by the audiogram is persistent; or

(ii) The hearing threshold shown in the annual audiogram indicates significant improvement over the baseline audiogram.

(10) Standard threshold shift.

(i) As used in this section, a standard threshold shift is a change in hearing threshold relative to the baseline audiogram of an average of 10 dB or more at 2000, 3000, and 4000 Hz in either ear.

(ii) In determining whether a standard threshold shift has occurred, allowance may be made for the contribution of aging (presbycusis) to the change in hearing level by correcting the annual audiogram according to the procedure described in Appendix F: *Calculation and Application of Age Correction to Audiograms.*

(h) Audiometric test requirements

(1) Audiometric tests shall be pure tone, air conduction, hearing threshold examinations, with test frequencies including as a minimum 500, 1000, 2000, 3000, 4000, and 6000 Hz. Tests at each frequency shall be taken separately for each ear.

(2) Audiometric tests shall be conducted with audiometers (including microprocessor audiometers) that meet the specifications of, and are maintained and used in accordance with, American National Standard Specification for Audiometers, S3.6–1969.

(3) Pulsed-tone and self-recording audiometers, if used, shall meet the requirements specified in Appendix C: *Audiometric Measuring Instruments.*

(4) Audiometric examinations shall be administered in a room meeting the requirements listed in Appendix D: *Audiometric Test Rooms.*

(5) Audiometer calibration.

(i) The functional operation of the audiometer shall be checked before each day's use by testing a person with known, stable hearing thresholds, and by listening to the audiometer's output to make sure that the output is free from distorted or unwanted sounds. Deviations of 10 decibels or greater require an acoustic calibration.

(ii) Audiometer calibration shall be checked acoustically at least annually in accordance with Appendix E: *Acoustic Calibration of Audiometers.* Test frequencies below 500 Hz and above 6000 Hz may be omitted from this check. Deviations of 15 decibels or greater require an exhaustive calibration.

(iii) An exhaustive calibration shall be performed at least every two years in accordance with sections 4.1.2; 4.1.3.; 4.1.4.3; 4.2; 4.4.1; 4.4.2; 4.4.3; and 4.5 of the American National Standard Specification for Audiometers, S3.6–1969. Test frequencies below 500 Hz and above 6000 Hz may be omitted from this calibration.

(i) Hearing protectors.

(1) Employers shall make hearing protectors available to all employees exposed to an 8-hour time-weighted average of 85 decibels or greater at no cost to the employees. Hearing protectors shall be replaced as necessary.

(2) Employers shall ensure that hearing protectors are worn:

(i) By an employee who is required by paragraph (b)(1) of this section to wear personal protective equipment; and

(ii) By any employee who is exposed to an 8-hour time-weighted average of 85 decibels or greater, and who:

(a) Has not yet had a baseline audiogram established pursuant to paragraph (g)(5)(ii); or

(b) Has experienced a standard threshold shift.

(3) Employees shall be given the opportunity to select their hearing protectors from a variety of suitable hearing protectors provided by the employer.

(4) The employer shall provide training in the use and care of all hearing protectors provided to employees.

(5) The employer shall ensure proper initial fitting and supervise the correct use of all hearing protectors.

(j) Hearing protector attenuation.

(1) The employer shall evaluate hearing protector attenuation for the specific noise environments in which the protector will be used. The employer shall use one of the evaluation methods described in Appendix B: *Methods for Estimating the Adequacy of Hearing Protection Attenuation.*

(2) Hearing protectors must attenuate employee exposure at least to an 8-hour time-weighted average of 90 decibels as required by paragraph (b) of this section.

(3) For employees who have experienced a standard threshold shift, hearing protectors must attenuate employee exposure to an 8-hour time-weighted average of 85 decibels or below.

(4) The adequacy of hearing protector attenuation shall be re-evaluated whenever employee noise exposures increase to the extent that the hearing protectors provided may no longer provide adequate attenuation. The employee shall provide more effective hearing protectors where necessary.

(k) Training program.

(1) The employer shall institute a training program for all employees who are exposed to noise at or above an 8-hour time-weighted average of 85 decibels, and shall ensure employee participation in such program.

(2) The training program shall be repeated annually for each employee included in the hearing conservation program. Information provided in the training program shall be updated to

STANDARDS AND INTERPRETATIONS

be consistent with changes in protective equipment and work processes.

(3) The employer shall ensure that each employee is informed of the following:

(i) The effects of noise on hearing;

(ii) The purpose of hearing protectors, the advantages, disadvantages, and attenuation of various types, and instructions on selection, fitting, use, and care; and

(iii) The purpose of audiometric testing, and an explanation of the test procedures.

(l) Access to information and training materials.

(1) The employer shall make available to affected employees or their representatives copies of this standard and shall also post a copy in the workplace.

(2) The employer shall provide to affected employees any informational materials pertaining to the standard that are supplied to the employer by the Assistant Secretary.

(3) The employer shall provide, upon request, all materials related to the employer's training and education program pertaining to this standard to the Assistant Secretary and the Director.

(m) Recordkeeping.

(1) **Exposure measurements.** The employer shall maintain an accurate record of all employee exposure measurements required by paragraph (d) of this section.

(2) **Audiometric tests.**

(i) The employer shall retain all employee audiometric test records obtained pursuant to paragraph (g) of this section:

(ii) This record shall include:

(a) Name and job classification of the employee;

(b) Date of the audiogram;

(c) The examiner's name;

(d) Date of the last acoustic or exhaustive calibration of the audiometer; and

(e) Employee's most recent noise exposure assessment.

(f) The employer shall maintain accurate records of the measurements of the background sound pressure levels in audiometric test rooms.

(3) **Record retention.** The employer shall retain records required in this paragraph (m) for at least the following periods.

(i) Noise exposure measurement records shall be retained for two years.

(ii) Audiometric test records shall be retained for the duration of the affected employee's employment.

(4) **Access to records.** All records required by this section shall be provided upon request to employees, former employees, representatives designated by the individual employee, and the Assistant Secretary. The provisions of 29 CFR 1910.20(a)–(e) and (g)–(i) apply to access to records under this section.

(5) **Transfer of records.** If the employer ceases to do business, the employer shall transfer to the successor employer all records required to be maintained by this section, and the successor employer shall retain them for the remainder of the period prescribed in paragraph (m)(3) of this section.

(n) Appendices.

(1) Appendices A, B, C, D, and E to this section are incorporated as part of this section and the contents of these Appendices are mandatory.

(2) Appendices F and G to this section are informational and are not intended to create any additional obligations not otherwise imposed or to detract from any existing obligations.

(o) Exemptions. Paragraphs (c) through (n) of this section shall not apply to employers engaged in oil and gas well drilling and servicing operations.

(p) Startup date. Baseline audiograms required by paragraph (g) of this section shall be completed by March 1, 1984.

(Approved by the Office of Management and Budget under Control Number 1218-0048)

[54 F.R. 24333, June 7, 1989]

APPENDIX A: NOISE EXPOSURE COMPUTATION

This Appendix is Mandatory

I. Computation of Employee Noise Exposure

(1) Noise dose is computed using Table G-16a as follows:

(I) When the sound level, L, is constant over the entire work shift, the noise dose, D, in percent, is given by: $D = 100 \ C/T$ where C is the total length of the work day, in hours, and T is the reference duration corresponding to the measured sound level, L, as given in Table G-16a or by the formula shown as a footnote to that table.

(II) When the workshift noise exposure is composed of two or more periods of noise at different levels, the total noise dose over the work day is given by:
$D = 100 \ (C_1/T_1 + C_2/T_2 + \ldots + C_n/T_n)$,
where C_n indicates the total time of exposure at a specific noise level, and T_n indicates the reference duration for that level as given by Table G-16a.

(2) The eight-hour time-weighted average sound level (TWA), in decibels, may be computed from the dose, in percent, by means of the formula: $TWA = 16.61 \log_{10} (D/100) + 90$. For an eight-hour workshift with the noise level constant over the entire shift, the TWA is equal to the measured sound level.

(3) A table relating dose and TWA is given in Section II.

TABLE G-16A

A-weighted sound level, L (decibel)	Reference duration, T (hour)
80	32
81	27.9
82	24.3
83	21.1
84	16.4
85	16
86	13.9
87	12.1
88	10.6
89	9.2
90	8
91	7.0
92	6.1
93	5.3
94	4.6

TABLE G-16A—Continued

A-weighted sound level, L (decibel)	Reference duration, T (hour)
95	4
96	3.5
97	3.0
98	2.6
99	2.3
100	2
101	1.7
102	1.5
103	1.3
104	1.1
105	1
106	0.87
107	0.76
108	0.66
109	0.57
110	0.5
111	0.44
112	0.38
113	0.33
114	0.29
115	0.25
116	0.22
117	0.19
118	0.16
119	0.14
120	0.125
121	0.11
122	0.095
123	0.082
124	0.072
125	0.063
126	0.054
127	0.047
128	0.041
129	0.036
130	0.031

In the above table the reference duration, T, is computed by

$$T = \frac{8}{2^{(L-90)/5}}$$

where L is the measured A-weighted sound level.

II. Conversion Between "Dose" and "8-Hour Time-Weighted Average" Sound Level

Compliance with paragraphs (c)-(r) of this regulation is determined by the amount of exposure to noise in the workplace. The amount of such exposure is usually measured with an audiodosimeter which gives a readout in terms of "dose." In order to better understand the requirements of the amendment, dosimeter readings can be converted to an "8-hour time-weighted average sound level." (TWA).

In order to convert the reading of a dosimeter into TWA, see Table A-1, below. This table applies to dosimeters that are set by the manufacturer to calculate dose or percent expo-

STANDARDS AND INTERPRETATIONS

sure according to the relationships in Table G-16a. So, for example, a dose of 91 percent over an eight hour day results in a TWA of 89.3 dB, and, a dose of 50 percent corresponds to a TWA of 85 dB.

If the dose as read on the dosimeter is less than or greater than the values found in Table A-1, the TWA may be calculated by using the formula: $TWA = 16.61 \log_{10}(D/100) + 90$ where TWA=8-hour time-weighted average sound level and D=accumulated dose in percent exposure.

Table A-1.—Conversion From "Percent Noise Exposure" or "Dose" to "8-Hour Time-Weighted Average Sound Level" (TWA)

Dose or percent noise exposure	TWA
10	73.4
15	76.3
20	78.4
25	80.0
30	81.3
35	82.4
40	83.4
45	84.2
50	85.0
55	85.7
60	86.3
65	86.9
70	87.4
75	87.9
80	88.4
81	88.5
82	88.6
83	88.7
84	88.7
85	88.8
86	88.9
87	89.0
88	89.1
89	89.2
90	89.2
91	89.3
92	89.4
93	89.5
94	89.6
95	89.6
96	89.7
97	89.8
98	89.9
99	89.9
100	90.0
101	90.1
102	90.1
103	90.2
104	90.3
105	90.4
106	90.4
107	90.5
108	90.6
109	90.6
110	90.7
111	90.8
112	90.8
113	90.9
114	90.9

Table A-1.—Conversion From "Percent Noise Exposure" or "Dose" to "8-Hour Time-Weighted Average Sound Level" (TWA)—Continued

Dose or percent noise exposure	TWA
115	91.1
116	91.1
117	91.1
118	91.2
119	91.3
120	91.3
125	91.6
130	91.9
135	92.2
140	92.4
145	92.7
150	92.9
155	93.2
160	93.4
165	93.6
170	93.8
175	94.0
180	94.2
185	94.4
190	94.6
195	94.8
200	95.0
210	95.4
220	95.7
230	96.0
240	96.3
250	96.6
260	96.9
270	97.2
280	97.4
290	97.7
300	97.9
310	98.2
320	98.4
330	98.6
340	98.8
350	99.0
360	99.2
370	99.4
380	99.6
390	99.8
400	100.0
410	100.2
420	100.4
430	100.5
440	100.7
450	100.8
460	101.0
470	101.2
480	101.3
490	101.5
500	101.6
510	101.8
520	101.9
530	102.0
540	102.2
550	102.3
560	102.4
570	102.6
580	102.7

STANDARDS AND INTERPRETATIONS

Table A-1.—Conversions From "Percent Noise Exposure" or "Dose" to "8-Hour Time-Weighted Average Sound level" (TWA)—Continued

Dose or percent noise exposure	TWA
590	102.8
600	102.9
610	103.0
620	103.2
630	103.3
640	103.4
650	103.5
660	103.6
670	103.7
680	103.8
690	103.9
700	104.0
710	104.1
720	104.2
730	104.3
740	104.4
750	104.5
760	104.6
770	104.7
780	104.8
790	104.9
800	105.0
810	105.1
820	105.2
830	105.3
840	105.4
850	105.4
860	105.5
870	105.6
880	105.7
890	105.8
900	105.8
910	105.9
920	106.0
930	106.1
940	106.2
950	106.2
960	106.3
970	106.4
980	106.5
990	106.5
999	106.6

APPENDIX B: METHODS FOR ESTIMATING THE ADEQUACY OF HEARING PROTECTOR ATTENUATION

This Appendix is Mandatory

For employees who have experienced a significant threshold shift, hearing protector attenuation must be sufficient to reduce employee exposure to a TWA of 85 dB. Employers must select one of the following methods by which to estimate the adequacy of hearing protector attenuation.

The most convenient method is the Noise Reduction Rating (NRR) developed by the Environmental Protection Agency (EPA). According to EPA regulation, the NRR must be shown on the hearing protector package. The NRR is then related to an individual worker's noise environment in order to assess the adequacy of the attenuation of a given hearing protector. This Appendix describes four methods of using the NRR to determine whether a particular hearing protector provides adequate protection within a given exposure environment. Selection among the four procedures is dependent upon the employer's noise measuring instruments.

Instead of using the NRR, employers may evaluate the adequacy of hearing protector attenuation by using one of the three methods developed by the National Institute for Occupational Safety and Health (NIOSH), which are described in the "List of Personal Hearing Protectors and Attenuation Data," HEW Publication No. 76-120, 1975, pages 21-37. These methods are known as NIOSH methods #1, #2 and #3. The NRR described below is a simplification of NIOSH method #2. The most complex method is NIOSH method #1, which is probably the most accurate method since it uses the largest amount of spectral information from the individual employee's noise environment. As in the case of the NRR method described below, if one of the NIOSH methods is used, the selected method must be applied to an individual's noise environment to assess the adequacy of the attenuation. Employers should be careful to take a sufficient number of measurements in order to achieve a representative sample for each time segment.

Note.—The employer must remember that calculated attenuation values reflect realistic values only to the extent that the protectors are properly fitted and worn.

When using the NRR to assess hearing protector adequacy, one of the following methods must be used:

(I) When using a dosimeter that is capable of C-weighted measurements:

(A) Obtain the employee's C-weighted dose for the entire workshift, and convert to TWA (see Appendix A, II).

(B) Subtract the NRR from the C-weighted TWA to obtain the estimated A-weighted TWA under the ear protector.

(II) When using a dosimeter that is not capable of C-weighted measurements, the following method may be used:

(A) Convert the A-weighted dose to TWA (see Appendix A).

(B) Subtract 7 dB from the NRR.

(C) Subtract the remainder from the A-weighted TWA to obtain the estimated A-weighted TWA under the ear protector.

(III) When using a sound level meter set to the A-weighting network:

(A) Obtain the employee's A-weighted TWA.

(B) Subtract 7 dB from the NRR, and subtract the remainder from the A-weighted TWA to obtain the estimated A-weighted TWA under the ear protector.

(Iv) When using a sound level meter set on the C-weighting network:

STANDARDS AND INTERPRETATIONS

(A) Obtain a representative sample of the C-weighted sound levels in the employee's environment.

(B) Subtract the NRR from the C-weighted average sound level to obtain the estimated A-weighted TWA under the ear protector.

(v) When using area monitoring procedures and a sound level meter set to the A-weighing network.

(A) Obtain a representative sound level for the area in question.

(B) Subtract 7 dB from the NRR and subtract the remainder from the A-weighted sound level for that area.

(vi) When using area monitoring procedures and a sound level meter set to the C-weighting network:

(A) Obtain a representative sound level for the area in question.

(B) Subtract the NRR from the C-weighted sound level for that area.

APPENDIX C: AUDIOMETRIC MEASURING INSTRUMENTS

This Appendix is Mandatory

1. In the event that pulsed-tone audiometers are used, they shall have a tone on-time of at least 200 milliseconds.

2. Self-recording audiometers shall comply with the following requirements:

(A) The chart upon which the audiogram is traced shall have lines at positions corresponding to all multiples of 10 dB hearing level within the intensity range spanned by the audiometer. The lines shall be equally spaced and shall be separated by at least ¼ inch. Additional increments are optional. The audiogram pen tracings shall not exceed 2 dB in width.

(B) It shall be possible to set the stylus manually at the 10-dB increment lines for calibration purposes.

(C) The slewing rate for the audiometer attenuator shall not be more than 6 dB/sec except that an initial slewing rate greater than 6 dB/sec is permitted at the beginning of each new test frequency, but only until the second subject response.

(D) The audiometer shall remain at each required test frequency for 30 seconds (± 3 seconds). The audiogram shall be clearly marked at each change of frequency and the actual frequency change of the audiometer shall not deviate from the frequency boundaries marked on the audiogram by more than ± 3 seconds.

(E) It must be possible at each test frequency to place a horizontal line segment parallel to the time axis on the audiogram, such that the audiometric tracing crosses the line seg-

ment at least six times at that test frequency. At each test frequency the threshold shall be the average of the midpoints of the tracing excursions.

APPENDIX D: AUDIOMETRIC TEST ROOMS

This Appendix is Mandatory

Rooms used for audiometric testing shall not have background sound pressure levels exceeding those in Table D-1 when measured by equipment conforming at least to the Type 2 requirements of American National Standard Specification for Sound Level Meters, S1.4-1971 (R1976), and to the Class II requirements of American National Standard Specification for Octave, Half-Octave, and Third-Octave Band Filter Sets, S1.11-1971 (R1976).

Table D-1.—Maximum Allowable Octave-Band Sound Pressure Levels for Audiometric Test Rooms

Octave-band center frequency (Hz)	500	1000	2000	4000	8000
Sound pressure level (dB)	40	40	47	57	62

APPENDIX E: ACOUSTIC CALIBRATION OF AUDIOMETERS

This Appendix is Mandatory

Audiometer calibration shall be checked acoustically, at least annually, according to the procedures described in this Appendix. The equipment necessary to perform these measurements is a sound level meter, octave-band filter set, and a National Bureau of Standards 9A coupler. In making these measurements, the accuracy of the calibrating equipment shall be sufficient to determine that the audiometer is within the tolerances permitted by American Standard Specification for Audiometers, S3.6-1969.

(1) Sound Pressure Output Check

A. Place the earphone coupler over the microphone of the sound level meter and place the earphone on the coupler.

B. Set the audiometer's hearing threshold level (HTL) dial to 70 dB.

C. Measure the sound pressure level of the tones that each test frequency from 500 Hz through 6000 Hz for each earphone.

D. At each frequency the readout on the sound level meter should correspond to the levels in Table E-1 or Table E-2, as appropriate, for the type of earphone, in the column entitled "sound level meter reading."

(2) Linearity Check

A. With the earphone in place, set the frequency to 1000 Hz and the HTL dial on the audiometer to 70 dB.

B. Measure the sound levels in the coupler at each 10-dB decrement from 70 dB to 10 dB, noting the sound level meter reading at each setting.

C. For each 10-dB decrement on the audiometer the sound level meter should indicate a corresponding 10 dB decrease.

D. This measurement may be made electrically with a voltmeter connected to the earphone terminals.

(3) Tolerances

When any of the measured sound levels deviate from the levels in Table E-1 or Table E-2 by ±3 dB at any test frequency between 500 and 3000 Hz, 4 dB at 4000 Hz, or 5 dB at 6000 Hz, an exhaustive calibration is advised. An exhaustive calibration is required if the deviations are greater than 10 dB at any test frequency.

Table E-1.—Reference Threshold Levels for Telephonics—TDH-39 Earphones

Frequency, Hz	Reference threshold level for TDH-39 earphones, dB	Sound level meter reading, dB
500.	11.5	81.5
1000.	7	77
2000.	9	79
3000.	10	80
4000.	9.5	79.5
6000.	15.5	85.5

Table E-2.—Reference Threshold Levels for Telephonics—TDH-49 Earphones

Frequency, Hz	Reference threshold level for TDH-49 earphones, dB	Sound level meter reading, dB
500.	13.5	83.5
1000.	7.5	77.5
2000.	11	81.0
3000.	9.5	79.5
4000.	10.5	80.5
6000.	13.5	83.5

APPENDIX F: CALCULATIONS AND APPLICATION OF AGE CORRECTIONS TO AUDIOGRAMS

This Appendix is Non-Mandatory

In determining whether a standard threshold shift has occurred, allowance may be made for the contribution of aging to the change in hearing level by adjusting the most recent audiogram. If the employer chooses to adjust the audiogram, the employer shall follow the procedure described below. This procedure and the age correction tables were developed by the National Institute for Occupational Safety and Health in the criteria document entitled "Criteria for a Recommended Standard . . . Occpational Exposure to Noise," ((HSM)-11001).

For each audiometric test frequency:
(I) Determine from Tables F-1 or F-2 the age correction values for the employee by:

(A) Finding the age at which the most recent audiogram was taken and recording the corresponding values of age corrections at 1000 Hz through 6000 Hz;

(B) Finding the age at which the baseline audiogram was taken and recording the corresponding values of age corrections at 1000 Hz through 6000 Hz

(II) Subtract the values found in step (i)(A) from the value found in step (i)(B).

(III) The differences calculated in step (ii) represented that portion of the change in hearing that may be due to aging.

Example: Employee is a 32-year-old male. The audiometric history for his right ear is shown in decibels below.

Employee's age	Audiometric test frequency (Hz)				
	1000	2000	3000	4000	6000
26.	10	5	5	10	5
*27.	0	0	0	5	5
28.	0	0	0	10	5
29.	5	0	5	15	5
30.	0	5	10	20	10
31.	5	10	20	15	15
*32.	5	10	10	25	20

The audiogram at age 27 is considered the baseline since it shows the best hearing threshold levels. Asterisks have been used to identify the baseline and most recent audiogram. A threshold shift of 20 dB exists at 4000 Hz between the audiograms taken at ages 27 and 32.

(The threshold shift is computed by subtracting the hearing threshold at age 27, which was 5, from the hearing threshold at age 32, which is 25). A retest audiogram has confirmed this shift. The contribution of aging to this change in hearing may be estimated in the following manner:

STANDARDS AND INTERPRETATIONS

Go to Table F-1 and find the age correction values (in dB) for 4000 Hz at age 27 and age 32.

	Frequency (Hz)				
	1000	2000	3000	4000	6000
Age 32	6	5	7	10	14
Age 27	5	4	6	7	11
Difference	1	1	1	3	3

The difference represents the amount of hearing loss that may be attributed to aging in the time period between the baseline audiogram and the most recent audiogram. In this example, the difference at 4000 Hz is 3 dB. This value is subtracted from the hearing level at 4000 Hz, which in the most recent audiogram is 25, yielding 22 after adjustment. Then the hearing threshold in the baseline audiogram at 4000 Hz (5) is subtracted from the adjusted annual audiogram hearing threshold at 4000 Hz (22). Thus the age-corrected threshold shift would be 17 dB (as opposed to a threshold shift of 20 dB without age correction).

Table F-1.—Age Correction Values in Decibels for Males

Years	Audiometric Test Frequencies (Hz)				
	1000	2000	3000	4000	6000
20 or younger	5	3	4	5	8
21	5	3	4	5	8
22	5	3	4	5	8
23	5	3	4	6	9
24	5	3	5	6	9
25	5	3	5	7	10
26	5	4	5	7	10
27	5	4	6	7	11
28	6	4	6	8	11
29	6	4	6	8	12
30	6	4	6	9	12
31	6	4	7	9	13
32	6	5	7	10	14
33	6	5	7	10	14
34	6	5	8	11	15
35	7	5	8	11	15
36	7	5	9	12	16
37	7	6	9	12	17
38	7	6	9	13	17
39	7	6	10	14	18
40	7	6	10	14	19
41	7	6	10	14	20
42	8	7	11	16	20
43	8	7	12	16	21
44	8	7	12	17	22
45	8	7	13	18	23
46	8	8	13	19	24
47	8	8	14	19	24
48	9	8	14	20	25
49	9	9	15	21	26
50	9	9	16	22	27
51	9	9	16	23	28
52	9	10	17	24	29
53	9	10	18	25	30
54	10	10	18	26	31

Table F-1.—Age Correction Values in Decibels For Males—Continued

Years	Audiometric Test Frequencies (Hz)				
	1000	2000	3000	4000	6000
55	10	11	19	27	32
56	10	11	20	28	34
57	10	11	21	29	35
58	10	12	22	31	36
59	11	12	22	32	37
60 or older	11	13	23	33	38

Table F-2.—Age Correction Values in Decibels for Females

Years	Audiometric Test Frequencies (Hz)				
	1000	2000	3000	4000	6000
20 or younger	7	4	3	3	6
21	7	4	4	3	6
22	7	4	4	4	6
23	7	5	4	4	7
24	7	5	4	4	7
25	8	5	4	4	7
26	8	5	5	4	8
27	8	5	5	5	8
28	8	5	5	5	8
29	8	5	5	5	9
30	8	6	5	5	9
31	8	6	6	5	9
32	9	6	6	6	10
33	9	6	6	6	10
34	9	6	6	6	10
35	9	6	7	7	11
36	9	7	7	7	11
37	9	7	7	7	12
38	10	7	7	7	12
39	10	7	8	8	12
40	10	7	8	8	13
41	10	8	8	8	13
42	10	8	9	9	13
43	11	8	9	9	14
44	11	8	9	9	14
45	11	8	10	10	15
46	11	9	10	10	15
47	11	9	10	11	16
48	12	9	11	11	16
49	12	9	11	11	16
50	12	10	11	12	17
51	12	10	12	12	17
52	12	10	12	13	18
53	13	10	13	13	18
54	13	11	13	14	19
55	13	11	14	14	19
56	13	11	14	15	20
57	13	11	15	15	20
58	14	12	15	16	21
59	14	12	16	16	21
60 or older	14	12	16	17	22

State	Office and address	Contact
Alabama	Alabama Consultation Program, P.O. Box 6005, University, Alabama 35486	(205) 348-7136, Mr. William Weems, Director.
Alaska	State of Alaska, Department of Labor, Occupational Safety & Health, 3301 Eagle St., Pouch 7-022, Anchorage, Alaska 99510.	(907) 276-5013, Mr. Stan Godsoe, Project Manager (Air Mail).
American Samoa	Service not yet available.	
Arizona	Consultation and Training, Arizona Division of Occupational Safety and Health, P.O. Box 19070, 1624 W. Adams, Phoenix, Ariz. 85005.	(602) 255-5795, Mr. Thomas Ramaley, Manager.
Arkansas	OSHA Consultation, Arkansas Department of Labor, 1022 High St., Little Rock, Ark. 72202	(501) 371-2992, Mr. George Smith, Project Director.
California	CAL/OSHA Consultation Service, 2nd Floor, 525 Golden Gate Avenue, San Francisco, Calif. 94102.	(415) 557-2870, Mr. Emmett Jones, Chief.
Colorado	Occupational Safety & Health Section, Colorado State University, Institute of Rural Environmental Health, 110 Veterinary Science Building, Fort Collins, Colo. 80523.	(303) 491-6151, Dr. Roy M. Buchan, Project Director.
Connecticut	Division of Occupational Safety & Health, Connecticut Department of Labor, 200 Folly Brook Boulevard, Wethersfield, Conn. 06109.	(203) 566-4550, Mr. Leo Alix, Director.
Delaware	Delaware Department of Labor, Division of Industrial Affairs, 820 North French Street, 6th Floor, Wilmington, Del. 19801.	(302) 571-3908, Mr. Bruno Salvadori, Director.
District of Columbia	Occupational Safety & Health Division, District of Columbia, Department Employment Services, Office of Labor Standards, 2900 Newton Street NE., Washington, D.C. 20018.	(202) 832-1230, Mr. Lorenzo M. White, Acting Associate Director.
Florida	Department of Labor & Employment Security, Bureau of Industrial Safety and Health, LaFayette Building, Room 204, 2551 Executive Center Circle West, Tallahassee, Fla. 32301.	(904) 488-3044, Mr. John C. Glenn, Administrator.
Georgia	Economic Development Division, Technology and Development Laboratory, Engineering Experiment Station, Georgia Institute of Technology, Atlanta, Ga. 30332.	(404) 894-3806, Mr. William C. Howard, Assistant to Director. Mr. James Burson, Project Manager.
Guam	Department of Labor, Government of Guam, 23548 Guam Main Facility, Agana, Guam 96921	(671) 772-6291, Joe R. San Agustin, Director.
Hawaii	Education and Information Branch, Division of Occupational Safety and Health, Suite 910, 677 Ala Moana, Honolulu, Hawaii 96813.	(808) 548-2511, Mr. Don Alper, Manager (Air Mail).
Idaho	OSHA Onsite Consultation Program, Boise State University, Community and Environmental Health, 1910 University Drive, Boise, Idaho 83725.	(208) 385-3929, Dr. Eldon Edmundson, Director.
Illinois	Division of Industrial Services, Dept. of Commerce and Community Affairs, 310 S. Michigan Avenue, 10 Floor, Chicago, Ill. 60601.	(800) 972-4140/4216 (Toll-free in State), (312) 793-3270, Mr. Stan Czwinski, Assistant Director.
Iowa	Bureau of Labor, 307 E. Seventh Street, Des Moines, Iowa 50319	(515) 281-3606, Mr. Allen J. Meier, Commissioner.
Indiana	Bureau of Safety, Education and Training, Indiana Division of Labor, 1013 State Office Building, Indianapolis, Indiana 46204.	(317) 633-5845, Mr. Harold Mills, Director.
Kansas	Kansas Dept. of Human Resources, 401 Topeka Ave., Topeka, Kans. 66603	(913) 296-4086, Mr. Jerry Abbott, Secretary.
Kentucky	Education and Training, Occupational Safety and Health, Kentucky Department of Labor, 127 Building, 127 South, Frankfort, Ky. 40601.	(502) 564-6895, Mr. Larry Potter, Director.
Louisiana	No services available as yet (Pending FY 83).	
Maine	Division of Industrial Safety, Maine Dept. of Labor, Labor Station 45, State Office Building, Augusta, Maine 04333.	(207) 289-3331, Mr. Lester Wood, Director.
Maryland	Consultation Services, Division of Labor & Industry, 501 St. Paul Place, Baltimore, Maryland 21202	(301) 659-4210, Ms. Ileana O'Brien, Project Manager, 7(c)(1) Agreement.
Massachusetts	Division of Industrial Safety, Massachusetts Department of Labor and Industries, 100 Cambridge Street, Boston, Massachusetts 02202.	(617) 727-3567, Mr. Edward Noseworthy, Project Director.
Michigan (Health)	Special Programs Section, Division of Occupational Health, Michigan Dept. of Public Health, 3500 N. Logan, Lansing, Mich. 48909.	(517) 373-1410, Mr. Irving Davis, Chief.
Michigan (Safety)	Safety Education & Training Division Bureau of Safety and Regulation, Michigan Department of Labor, 7150 Harris Drive, Box 30015, Lansing, Michigan 48909.	(517) 322-1809, Mr. Alan Harvie, Chief.
Minnesota	Training and Education Unit, Department of Labor and Industry, 5th Floor, 444 Lafayette Road, St. Paul, Minn. 55101.	(612) 296-2973, Mr. Timothy Tierney, Project Manager.
Mississippi	Division of Occupational Safety and Health, Mississippi State Board of Health, P.O. Box 1700 Jackson, Mississippi 39205.	(601) 982-6315, Mr. Henry L. Laird, Director.
Missouri	Missouri Department of Labor and Industrial Relations, 722 Jefferson Street, Jefferson City, Missouri 65101.	1-(800) 392-0208, (314) 751-3403, Ms. Paula Smith, Mr. Jim Brake.
Montana	Montana Bureau of Safety & Health, Division of Workers Compensation, 815 Front Street, Helena, Montana 59601.	(406) 449-3402, Mr. Ed Gatzemeier, Chief.
Nebraska	Nebraska Department of Labor, State House Station, State Capitol, P.O. Box 94600, Lincoln, Nebraska 68509.	475-8451 Ext. 258, Mr. Joseph Carroll, Commissioner.
Nevada	Department of Occupational Safety and Health, Nevada Industrial Commission, 515 E. Musser Street, Carson City, Nev. 89714.	(702) 885-5240, Mr. Allen Traenkner, Director.
New Hampshire	For information contact	Office of Consultation Programs, Room N3472 200 Constitution Avenue, N.W. Washington, D.C. 20210, Phone (202) 523-8965.
New Jersey	New Jersey Department of Labor and Industry Division of Work Place Standards, CN-054, Trenton, New Jersey 08625.	(609) 292-2313, FTS-8-477-2313, Mr. William Clark, Assistant Commissioner.
New Mexico	OSHA Consultation, Health and Environment Department, Environmental Improvement Division, Occupational Health & Safety Section, 4215 Montgomery Boulevard, NE., Albuquerque, New Mexico 87109.	(505) 842-3387, Mr. Albert M. Stevens, Project Manager.
New York	Division of Safety and Health, New York State Department of Labor, 2 World Trade Center, Room 6995, New York, New York 10047.	(212) 488-7746/7, Mr. Joseph Alleva, Project Manager, DOSH.
North Carolina	Consultation Services, North Carolina Department of Labor, 4 West Edenton Street, Raleigh, N.C. 27601	(919) 733-4885, Mr. David Pierce, Director.
North Dakota	Division of Environmental Research, Department of Health, Missouri Office Building, 1200 Missouri Avenue, Bismarck, N. Dak. 58505.	(701) 224-2348, Mr. Jay Crawford, Director.
Ohio	Department of Industrial Relations, Division of Onsite Consultation, P.O. Box 825, 2323 5th Avenue, Columbus, Ohio 43216.	(800) 282-1425 (Toll-free in State), (614) 466-7485, Mr. Andrew Doehrel, Project Manager.
Oklahoma	OSHA Division, Oklahoma Department of Labor, State Capitol, Suite 118, Oklahoma City. Okla. 73105.	(405) 521-2461, Mr. Charles W. McGion, Director.
Oregon	Consultative Section, Department of Workers' Compensation, Accident Prevention Division, Room 102, Building 1, 2110 Front Street NE., Salem, Oregon 97310.	(503) 378-2890, Mr. Jack Buckland, Supervisor.
Pennsylvania	For information contact	Office of Consultation Programs, Room N3472, 200 Constitution Avenue NW., Washington, D.C. 20210, Phone (202) 523-8965.
Puerto Rico	Occupational Safety & Health, Puerto Rico Department of Labor and Human Resources, 505 Munoz Rivera Ave., 21st Floor, Hato Rey, Puerto Rico 00919.	(809) 754-2134, Mr. John Cinque, Assistant Secretary, (Air Mail).
Rhode Island	Division of Occupational Health, Rhode Island Department of Health, The Cannon Building, 206 Health Department Building, Providence, R.I. 02903.	(401) 277-2438, Mr. James E. Hickey, Chief.
South Carolina	Consultation and Monitoring, South Carolina Department of Labor, P.O. Box 11329, Columbia, S.C. 29211.	(803) 758-8921, Mr. Robert Peck, Director, 7(c)(1), Project.
South Dakota	South Dakota Consultation Program, South Dakota State University, S.T.A.T.E.-Engineering Extension, 201 Pugsley Center-SDSO, Brookings, S. Dak. 57007.	(605) 688-4101, Mr. James Ceglian, Director.
Tennessee	OSHA Consultative Services, Tennessee Department of Labor, 2nd Floor, 501 Union Building, Nashville, Tennessee 37219.	(615) 741-2793, Mr. L. H. Craig Director.
Texas	Division of Occupational Safety and State Safety Engineer, Texas Department of Health and Resources, 1100 West 49th Street, Austin, Texas 78756.	(512) 458-7287, Mr. Walter G. Martin, P.E. Director
Trust Territories	Service not yet available.	
Utah	Utah Job Safety and Health Consultation Service, Suite 4004, Crane Building, 307 West 200 South, Salt Lake City, Utah 84101.	(801) 533-7927/8/9, Mr. H. M. Bergeson, Project Director.
Vermont	Division of Occupational Safety and Health, Vermont Department of Labor and Industry, 118 State Street, Montpelier, Vt. 05602.	(802) 828-2765, Mr. Robert Mcleod, Project Director.
Virginia	Department of Labor and Industry, P.O. Box 12064, 205 N. 4th Street, Richmond, Va. 23241	(804) 786-5875, Mr. Robert Beard, Commissioner.
Virgin Islands	Division of Occupational Safety and Health, Virgin Islands Department of Labor, Lagoon Street, Room 207, Frederiksted, Virgin Islands 00840.	(809) 772-1315, Mr. Louis Llanos, Deputy Director-DOSH.
Washington	Department of Labor and Industry, P.O. Box 207, Olympia, Wash. 98504	(206) 753-6500, Mr. James Sullivan, Assistant Director.
West Virginia	West Virginia Department of Labor, Room 451B, State Capitol, 1900 Washington Street, Charleston, W. Va. 25305.	FTS 8-885-7890, Mr. Lawrence Barker, Commissioner.
Wisconsin (Health)	Section of Occupational Health, Department of Health and Social Services, P.O. Box 309, Madison, Wisconsin 53701.	(608) 266-0417, Ms. Patricia Natzke, Acting Chief.
Wisconsin (Safety)	Division of Safety and Buildings, Department of Industry, Labor and Human Relations, 1570 E. Moreland Blvd., Waukesha, Wis. 53186.	(414) 544-8686, Mr. Richard Michalski, Supervisor.
Wyoming	Wyoming Occupational Health and Safety Department, 200 East 8th Avenue, Cheyenne, Wyo. 82002.	(307) 777-7786, Mr. Donald Owsley, Health and Safety Administrator.

APPENDIX H: AVAILABILITY OF REFERENCED DOCUMENTS

Paragraphs (c) through (o) of 29 CFR 1910.95 and the accompanying appendices contain provisions which incorporate publications by reference. Generally, the publications provide criteria for instruments to be used in monitoring and audiometric testing. These criteria are intended to be mandatory when so indicated in the applicable paragraphs of Section 1910.95 and appendices.

It should be noted that OSHA does not require that employers purchase a copy of the referenced publications. Employers, however, may desire to obtain a copy of the referenced publications for their own information.

The designation of the paragraph of the standard in which the referenced publications appear, the titles of the publications, and the availability of the publication are as follows:

Paragraph designation	Referenced publication	Available from—
Appendix B	"List of Personal Hearing Protectors and Attenuation Data," HEW Pub. No. 76-120, 1975, NTIS-PB267461.	National Technical Information Service, Port Royal Road, Springfield, VA 22161.
Appendix D	"Specification for Sound Level Meters," S1.4-1971 (R1976).	American National Standards Institute, Inc., 1430 Broadway, New York, NY 10018.
§1910.95(k)(2), appendix E	"Specifications for Audiometers," S3.6-1969.	American National Standards Institute, Inc., 1430 Broadway, New York, NY 10018.
Appendix D	"Specification for Octave, Half-Octave and Third-Octave Band Filter Sets," S1.11-1971 (R1976).	Back Numbers Department, Dept. STD, American Institute of Physics, 333 E. 45 St., New York, NY 10017; American National Standards Institute, Inc., 1430 Broadway, New York, NY 10018.

The referenced publications (or a microfiche of the publications) are available for review at many universities and public libraries throughout the country. These publications may also be examined at the OSHA Technical Data Center, Room N2439, United States Department of Labor, 200 Constitution Avenue, NW., Washington, D.C. 20210, (202) 523-9700 or at any OSHA Regional Office (see telephone directories under United States Government—Labor Department).

APPENDIX I: DEFINITIONS

These definitions apply to the following terms as used in paragraphs (c) through (n) of 29 CFR 1910.95.

Action level—An 8-hour time-weighted average of 85 decibels

STANDARDS AND INTERPRETATIONS

measured on the A-scale, slow response, or equivalently, a dose of fifty percent.

Audiogram—A chart, graph, or table resulting from an audiometric test showing an individual's hearing threshold levels as a function of frequency.

Audiologist—A professional, specializing in the study and rehabilitation of hearing, who is certified by the American Speech-Language-Hearing Association or licensed by a state board of examiners.

Baseline audiogram—The audiogram against which future audiograms are compared.

Criterion sound level—A sound level of 90 decibels.

Decibel (dB)—Unit of measurement of sound level.

Hertz (Hz)—Unit of measurement of frequency, numerically equal to cycles per second.

Medical pathology—A disorder or disease. For purposes of this regulation, a condition or disease affecting the ear, which should be treated by a physician specialist.

Noise dose—The ratio, expressed as a percentage, of (1) the time integral, over a stated time or event, of the 0.6 power of the measured SLOW exponential time-averaged, squared A-weighted sound pressure and (2) the product of the criterion duration (8 hours) and the 0.6 power of the squared sound pressure corresponding to the criterion sound level (90 dB).

Noise dosimeter—An instrument that integrates a function of sound pressure over a period of time in such a manner that it directly indicates a noise dose.

Otolaryngologist—A physician specializing in diagnosis and treatment of disorders of the ear, nose and throat.

Representative exposure—Measurements of an employee's noise dose or 8-hour time-weighted average sound level that the employers deem to be representative of the exposures of other employees in the workplace.

Sound level—Ten times the common logarithm of the ratio of the square of the measured A-weighted sound pressure to the square of the standard reference pressure of 20 micropascals. Unit: decibels (dB). For use with this regulation, SLOW time response, in accordance with ANSI S1.4-1971 (R1976), is required.

South level meter—An instrument for the measurement of sound level.

Time-weighted average sound level—That sound level, which if constant over an 8-hour exposure, would result in the same noise dose as is measured.

Appendix C

Noise and Vibration Standards
(Partial Listings)

Part 1: Standards Organizations

Acoustical Society of America (ASA)
Back Number Dept., Dept Std.
American Institute of Physics
333 East 45th Street
New York, NY 10017

Air-Conditioning and Refrigeration Institute (ARI)
1815 North Fort Myer Drive
Arlington, VA 22209

Air Diffusion Council (ADC)
435 North Michigan
Chicago, IL 60611

Air Moving and Conditioning Association (AMCA)
30 West University Drive
Arlington Heights, IL 60004

American Gear Manufacturers Association (AGMA)
1330 Massachusetts Avenue, N.W.
Washington, DC 20005

American National Standards Institute (ANSI)
1430 Broadway
New York, NY 10018

American Society of Heating, Refrigeration, and Air Conditioning Engineers (ASHRAE)
1791 Tullie Circle N.E.
Atlanta, GA 30300

American Society for Testing and Materials (ASTM)
1916 Race Street
Philadelphia, PA 19103

American Textile Machinery Association (ATMA)
1730 M Street N.W.
Washington, DC 20036

Anti-Friction Bearing Manufacturers Association (AFBMA)
60 East 42nd Street
New York, NY 10017

Compressed Air and Gas Institute (CAGI)
2130 Keith Building
Cleveland, OH 44115

Diesel Engine Manufacturers Association (DEMA)
2130 Keith Building
Cleveland, OH 44115

Factory Mutual Systems
184 High Street
Boston, MA 02110

Federal Construction Guide Specification (FCGS)
General Services Administration
Public Building Service
Office of Construction Management

Criteria and Research Branch
19th and F Street, N.W.
Washington, DC 20405

Federal Specifications
Specification Sales (3 FRDS)
Building 197, Washington Navy Yard
General Services Administration
Washington, DC 20407

Hearing Aid Industry Conference, Inc.
75 East Wacker Drive
Chicago, IL 60001

Institute of Electrical and Electronic Engineers
    (IEEE)
445 Hoes Lane
Piscataway, NJ 08854

Instrument Society of America (ISA)
400 Stanwix Street
Pittsburgh, PA 15222

International Conference of Building Officials
    (ICBO)
5360 South Warkman Mill Road
Whittier, CA 90601

International Organization for Standardization
    (ISO)
150 Central Secretariat
Cast Postale CH 1211
Geneva 20 Switzerland

Military Specifications
Commanding Officer
Naval Publications and Forms Center
5801 Tabor Avenue
Philadelphia, PA 19120

National Electrical Manufacturers Association
    (NEMA)
816 15th Street, Ste. 438
Washington, DC 20005

National Fluid Power Association (NEPA)
3333 N. Mayfair Road
Milwaukee, WI 53222

National Machine Tool Builders Association
    (NMTBA)
7901 West Park Drive
McLean, VA 22101

Power Saw Manufacturers Association (PSMA)
Box 7256
Belle View Station
Alexandria, VA 22307

Public Building Services (PBS)
General Services Administration
Public Building Service
Office of Construction Management
Criteria and Research Branch
GSA Building
19th and F Street, N.W.
Washington, DC 20405

Radio Manufacturers Association
1317 F Street N.W.
Washington DC 20004

Society of Automotive Engineers (SAE)
400 Commonwealth Drive
Warrendale, PA 15096

Steel Door Institute (SDI)
2130 Keith Building
Cleveland, OH 19103

Woodworking Machinery Manufacturers Association (WMMA)
1900 Arch Street
Philadelphia, PA 19103

Some Federal Agencies Having Noise Regulations or Standards

Department of Labor, OSHA, Washington, DC
Department of Housing and Urban Development, HUD, Washington, DC
Federal Aviation Administration, FAA, Washington, DC
Department of Transportation and Federal Highway Administration

Part 2: Standards

International Standards Organization (ISO)

ISO—Vibration (Partial List)

1. Terminology
    ISO 1925: 1990 Mechanical vibration—Balancing—Vocabulary—Bilingual edition
    ISO 2041: 1990 Vibration and shock—Vocabulary—Bilingual edition

2. General Standards

ISO 2017: 1982 Vibration and shock—Isolators—Procedure for specifying characteristics

ISO 4866: 1990 Mechanical vibration and shock—Vibration of buildings—Guidelines for the measurement of vibrations and evaluation of their effects on buildings

ISO 7626-1: 1986 Vibration and shock—Experimental determination of mechanical mobility—Part 1: Basic definitions and transducers

ISO 7626-2: 1990 Vibration and shock—Experimental determination of mechanical mobility—Part 2: Measurements using single-point translation excitation with an attachment vibration exciter

ISO 7919-1: 1986 Mechanical vibration of non-reciprocating machines—Measurements on rotating shafts and evaluation—Part 2: General guidelines

3. Test Equipment

ISO 2954: 1975 Mechanical vibration of rotating and reciprocating machinery—Requirements for instruments for measuring vibration severity

ISO 5348: 1987 Mechanical vibration and shock—Mechanical mounting of accelerometers

ISO 8568: 1989 Mechanical shock—Testing machines—Characteristics and performance

4. Human Exposure to Vibration and Shock

ISO 2631-1: 1985 Evaluation of human exposure to whole-body vibration—Part 1: General requirements

ISO 2631-2: 1989 Evaluation of human exposure to whole-body vibration—Part 2: Continuous and shock-induced vibrations in buildings (1 to 80 Hz)

ISO 2631-3: 1985 Evaluation of human exposure to whole-body vibration—Part 3: Evaluation of exposure to whole-body z-axis vertical vibration in the frequency range 0.1 to 0.63 Hz

ISO 8041: 1990 Human response to vibration—Measuring instrumentation

ISO—Acoustics (Partial List)

1. Basic Standards for Acoustics

ISO 31-7: 1978 Quantities and units of acoustics

ISO 131: 1979 Acoustics—Expression of physical and subjective magnitudes of sound or noise in air

ISO 266: 1975 Acoustics—preferred frequencies for measurements

ISO 354: 1985 Acoustics—Measurement of sound absorption in a reverberation room

ISO 532: 1975 Acoustics—Method for calculating loudness level

2. Methods of Noise Measurement

ISO 1996-1: 1982 Acoustics—Description and measurement of environmental noise—Part 1: Basic quantities and procedures

ISO 1996-2: 1987 Acoustics—Description and measurement of environmental noise—Part 2: Acquisition of data pertinent to land use

ISO 1996-3: 1987 Acoustics—Description and measurement of environmental noise—Part 3: Application to noise limits

ISO 3740: 1980 Acoustics—Determination of sound power levels of noise sources—Guidelines for the use of basic standards and for the preparation of noise test codes

ISO 3741: 1988 Acoustics—Determination of sound power levels of noise sources—Precision methods for broadband sources in reverberation rooms

ISO 3742: 1988 Acoustics—Determination of sound power levels of noise sources—Precision methods for discrete-frequency and narrow-band sources in re-verberation rooms

ISO 3743: 1988 Acoustics—Determination of sound power levels of noise sources—Engineering methods for special reverberation test rooms

ISO 3744: 1981 Acoustics—Determination of sound power levels of noise sources—Engineering methods for free-field conditions over a reflecting plane

ISO 3745: 1977 Acoustics—Determination of sound power levels of noise sources—Precision methods for anechoic and semianechoic rooms

ISO 3746: 1979 Acoustics—Determination of sound power levels of noise sources—Survey method

ISO 3747: 1987 Determination of sound power levels of noise sources—Survey method using a reference sound source

ISO 6081: 1986 Acoustics—Noise emitted by machinery and equipment—Guidelines for the preparation of test codes of engineering grade requiring noise measurements at the operator's or bystander's position

3. Human Exposure to Noise

ISO 1999: 1990 Acoustics—Determination of occupational noise exposure and estimation of noise-induced hearing impairment

ISO 4869: 1981 Acoustics—Measurement of sound attenuation of hearing protectors—Subjective method

ANSI S1.4 1971 American National Specification for Sound-Level Meters

ANSI S1.7 1970 Sound Absorption of Acoustical Materials in Reverberation Rooms, Method of Test for

ANSI S1.21 1972 American National Standard Methods for the Determination of Sound-Power Levels of Small Sources in Reverberation Rooms

ANSI S1.17 1975 Method for Rating the Sound-Power Spectra of Small Stationary Noise Sources

ANSI S6.4 1973 Computing the Effective Perceived Noise Level for Flyover Aircraft Noise, Definitions and Procedures for

ANSI S3.15 Method for Measurement of Community Noise

ANSI S2.47 American National Standard Vibration of Buildings—Guidelines for the Measurement of Vibrations and Evaluation of Their Effects on Buildings

S12.4-1986 American National Standard Method for Assignment of High-Energy Implosive Sounds with Respect to Residential Communities

S12.36-1990 American National Standard Survey Methods for the Determination of Sound Power Levels of Noise Sources

S12.40-1990 American National Standard Sound Level Descriptors for Determination of Compatible Land Use

S12.13-1991 *Draft* American National Standard for Evaluating the Effectiveness of Hearing Conservation Programs

S1.25-1991 Specification for Personnel Noise Dosimeter

American National Standards Institute (ANSI)

ANSI S1.1 1960 American National Standard Acoustical Terminology (R1971)

ANSI S1.2 1962 American National Standard Method of the Physical Measurement of Sound (partially revised by S1.13 1971 and by S1.2 1972)

American Society for Testing and Materials (ASTM)

ASTM C432-66 Standard Method of Test for Sound Absorption of Acoustical Materials in Reverberation Rooms (ANSI S1.7 1970)

ASTM E90-75 Standard Recommended Practice for Laboratory Measurement of

Airborne Sound Transmission Loss of Building Partitions

ASTM E336-71 Standard Recommended Practice for Measurement of Airborne Sound Insulation in Buildings

ASTM E966-84 Standard Recommended Practice for Measurement of Airborne Noise Insulation of Building Facades

ASTM E413-73 Standard Classification for Determination of Sound Transmission Class

Appendix D

A Generic Hearing Conservation Program

The following is an example of a Hearing Conservation Program. It should be reviewed by the company's safety personnel and modified as required.

A Hearing Conservation Program for the XYZ Company

Policy

It will be the policy of the XYZ Company to protect all employees exposed to noise levels of 85 dBA or greater. This will be done by developing, installing, and maintaining a hearing conservation program (HCP) that has at least the requirements as set out in DOL OSHA 29CFR1910.95, Occupational Noise Exposure. The HCP will be in force until feasible engineering controls can be developed and installed to reduce noise levels to 85 dBA or less. If controls are not feasible, in some or all areas, the HCP will continue to protect the employees' hearing.

Introduction

The purpose of an effective HCP is to prevent permanent noise-induced hearing loss resulting from occupational noise exposure. To ensure compliance with the noise exposure regulations the HCP will be composed of the following elements:

I. Noise surveys
II. Education and training
III. Engineering and/or administrative control
IV. Hearing protection devices
V. Audiometric testing and evaluation

Listed in Table D.1 are the noise exposure levels
218

(NELs) which shall be used to determine employee noise exposures. The NELs are based on the best available information from industrial experience. The NELs are sound pressure levels and exposure durations that represent conditions under which it is believed nearly all workers may be repeatedly exposed without adverse effect on their ability to hear and understand normal speech. The values should be used as guidelines in the control of noise exposure. Due to individual susceptibility, they should not be regarded as fine lines between safe and dangerous levels. These values apply to total duration of exposure per working day regardless of whether this is one continuous exposure or a number of short-term exposures.

All employees exposed to noise at or above 50 percent of the NEL shall be included in the HCP. For an eight-hour workday, a 50 percent noise exposure is equivalent to 85 dBA. This level is referred to as the time-weighted average (TWA) noise exposure. In those cases where work shifts differ from eight hours, the TWA noise exposure for inclusion in the HCP shall be adjusted accordingly (e.g., 82 dBA is equivalent to 50 percent of the NEL for 12-hour workdays, etc.). It is recommended that all personnel with borderline noise exposures, or those with infrequent exposures at or above 50 percent of the NEL, be included in the HCP (for example, maintenance personnel).

I. Noise Surveys
  A. Preliminary survey
    1. Purpose
      To determine which areas or job activities need a more detailed survey.

2. Survey procedure
   Conduct a walk-through of the plant with a sound level meter. Typically, if an area has any sound levels above 85 dBA (82 dBA where employees work 12-hour shifts), then it should be scheduled for an exposure survey. In most instances, the preliminary survey can be bypassed and the exposure survey initiated.
3. Documentation
   Locations not represented in an exposure survey must be inspected and documented by sound level meter survey as being a low-noise hazard work site (see 1.A.2. above). Major equipment changes or other work site modifications may necessitate a new sound level meter survey. Otherwise, periodic spot tests to validate the previous survey(s) will be sufficient.

B. Exposure survey
1. Purpose
   To identify all employees to be included in the HCP and to determine actual or representative noise exposures for all employees whenever possible. These data will assist in the selection of hearing protection.
2. Survey methods
   The noise survey can be completed with (a) personal monitoring, (b) area sampling, or (c) a combination of (a) and (b).
   a. Personal monitoring—sound level and stopwatch
      Measure sound levels 6–12 in. from the employee's ear. The stopwatch is used to measure the time an employee being monitored actually spends at one location or is exposed to a specific sound level. After this information has been collected for the employee's workday, it is then used to calculate the noise exposure.
   b. Personal monitoring—noise dosimeters
      This is an electronic device worn by the employee. The dosimeter automatically averages varying sound levels during a given time period (eight-hour workday, etc.). This method is the most accurate available for assessing noise exposure. Follow manufacturer's instruction for calibration and use.

Use representative monitoring to efficiently determine employees' noise exposures. This enables the surveyor to monitor a number of individuals from a group of employees. This works well if the group engages in similar work with exposure to similar noise sources. The result is used to determine the noise exposure for the work group. To ensure that the data are truly representative of exposure, conduct the measurements at least twice before assigning noise exposures.

3. Area sampling
   This is a method of sampling with a sound level meter to determine the overall sound level at various locations throughout a plant. If personal monitoring is impractical, the employees to be included in the HCP can be identified through area sampling. While area monitoring is the least accurate from an exposure standpoint, it is the most conservative as far as those to be included in the HCP. For purposes of identifying employees to be included in the HCP, a "worst case" method is necessary. In this situation, employees who work in areas with sound levels above 85 dBA (82 dBA for 12-hour shifts) would automatically be included in the program. The maximum area sound level is used to evaluate which hearing protectors are effective enough to attenuate the noise to 85 dBA or less. Methods for evaluating hearing protectors are described in section IV.E.
4. Documentation
   Federal regulations require that all monitoring and area sampling results are to be maintained for 30 years. The medical group or contractor service conducting the audiometric exams should be informed of the employee's most recent noise exposure assessment. All employees should be informed of the noise exposure determined for their job along with an explanation or interpretation of the results.

II. Education and training
A. Purpose
   The central purpose is motivation of employees to actively participate in the HCP and to cooperate by wearing their hearing

protectors. The reasons for having a hearing conservation program must be clearly explained so the need to protect hearing is understood and appreciated.

B. Implementation

Training must occur at least annually. Training may be conducted in one session or as many separate sessions as necessary to cover all the topics. It must include at least the following items:

1. The effects of noise on hearing
2. The purpose, advantages, disadvantages, and attenuation of various types of hearing protectors
3. The selection, fitting, and proper use and care of their protectors
4. The purpose and procedures of the audiometric tests

C. Documentation

Require each attendee to acknowledge, by signature, that he or she received training. This record should be maintained. The type of training materials used should be included, and a brief outline of the topics covered should also be kept on file.

III. Engineering and/or administrative controls

A. Engineering controls

1. Purpose

To reduce employee noise exposure by the installation of engineering controls. These controls involve the initial design or retrofit of existing equipment or the use of quiet spaces for the employees.

2. Feasibility factors

Management should take into consideration the existing technology as well as constraints such as production, maintenance, physical, and economic. Also, the noise control solutions must be compatible with the specific manufacturing process.

3. Purchase specifications

Include noise specifications when purchasing new equipment. Vendors supplying equipment should be advised that specified low noise levels will be considered in the selection process. Require suppliers to provide information on the noise levels of currently available equipment.

4. Engineering noise survey

Conduct a detailed noise study to determine feasibility of engineering controls for noise reduction. Noise control strategies require objective analysis from a practical and economic basis.

B. Administrative controls

1. Purpose

To determine if any administrative decision or method in place of engineering controls is feasible to reduce the noise exposure of employees.

2. Examples of administrative controls

a. Arrange work schedules so employees will work a major portion of their shift at or near the noise exposure levels and not be exposed to higher levels.
b. Use worker rotation and job scheduling, where possible, to keep individual noise exposure within permissible time limits.
c. If a noisy machine is not needed for full-time production, operate it a portion of each day rather than all day for part of the week.
d. Operate occasional high noise level equipment at times when a minimum number of employees will be exposed.

IV. Hearing protection devices

A. Purpose

Hearing protection devices are to be used when engineering and/or administrative control measures are neither feasible nor adequate to reduce noise.

B. Availability of hearing protectors

Employers must make hearing protectors available to all employees exposed at or above 50 percent of the NEL (action level) at no cost to the employees. Hearing protectors must be replaced free of charge when necessary (and within reason) at no direct cost to the employee. Employees must be presented with two or more different types or models of hearing protectors when making their selection. The selection of the proper fit or size should be administered by an individual such as an industrial hygienist, nurse, hearing conservationist, physician, or safety engineer who is properly trained in fitting of protectors. Employees must be shown how to use and care for their protectors. Employees must be supervised on the job to ensure that they continue to wear protectors correctly.

C. Mandatory use

The use of personal hearing protection devices is required for all affected employees whenever any of the following exist:

1. All workers who are exposed to or above 50 percent of the NEL (85 dBA for an eight-hour workday)
2. All areas designated by the employer as "hearing protection required," regardless of the amount of time spent in that area
3. A job activity that requires exposure to suspected elevated noise levels that have not been measured, and the job is not a recurring one (e.g., temporary construction or maintenance projects using air hammers, explosive guns, generator tests)

D. Warning signs

Ensure that all areas requiring the use of hearing protection devices are properly designated. When signs are used to designate an area, the following wording is recommended: "High Noise Area, Hearing Protection Required."

E. Selection of hearing protection devices

An important factor in establishing an effective HCP is the requirement by management that hearing protection devices (HPDs) be worn and be worn properly by employees. Another critical element of a successful program is the proper selection of HPDs suitable for the specific noise environments in which they will be utilized.

The selection of HPDs should be such that the hearing protector attenuation will reduce an employee's noise exposure or TWA to 85 dBA or less. The most convenient method is to use the noise reduction rating (NRR) developed by the Environmental Protection Agency. When using the NRR the Hearing Conservation Amendment to the Noise Standard (29 CFR 1910.95(c)) mandates the use of one of the following methods for assessing HPD adequacy.

1. When using a dosimeter that is capable of C-weighted measurements:
   a. Obtain the employee's C-weighted dose for the entire work shift, and convert to a TWA.
   b. Subtract the NRR from the C-weighted TWA to obtain the estimated A-weighted TWA under the HPD.

2. When using a dosimeter that is capable of A-weighted measurements:
   a. Convert the A-weighted dose to a TWA.
   b. Subtract 7 dB from the NRR.
   c. Subtract the remainder from the A-weighted TWA to obtain the estimated A-weighted TWA under the HPD.

3. When using a sound level meter set to the A-weighting network:
   a. Obtain the employee's A-weighted TWA.
   b. Subtract 7 dB from the NRR.
   c. Subtract the remainder from the A-weighted TWA to obtain the estimated TWA under the HPD.

4. It is important to remember that the calculated attenuation values used to determine the NRR reflect realistic values only to the extent that protectors are properly fitted and worn.

5. Instead of using the NRR, the regulation allows employers to evaluate the adequacy of HPD attenuation by using one of the three methods developed by the National Institute for Occupational Safety and Health (NIOSH). These methods are described in the "List of Personal Hearing Protectors and Attenuation Data," HEW Publication No. 76-120, 1975, pages 21–37.

V. Audiometric testing

A. Purpose

Preventing occupational hearing loss through analysis of the audiometric test results and determining the effectiveness of the HCP. Steps to check the HCP effectiveness are:

1. Detecting significant threshold shifts in an employee's hearing ability during the course of his or her employment
2. Providing a record of an employee's hearing acuity
3. Evaluating the effectiveness of engineering noise control measures by measuring the hearing thresholds of employees working near the treated equipment

4. Identifying plant areas requiring an engineering noise control study
5. Helping to provide justification for noise control expenditures
6. Identifying weaknesses in the hearing protection program such as inadequate hearing protectors, lack of proper use, and/or ineffective education and training of employees

B. Periodic testing
1. All employees included in the HCP shall have a baseline audiometric test. The baseline audiogram is a reference audiogram against which future hearing tests are compared for hearing conservation purposes.
2. All employees included in the HCP shall have a periodic test at least annually after obtaining a baseline audiogram.
3. It is recommended that all employees be away from workplace noise for at least 14 hours prior to their audiometric test. If the audiometric tests are to be conducted during the employees' work shift, then each employee to be tested should be instructed at the beginning of the work shift to wear hearing protection while on the job in "high-noise areas" at least up to the time of the hearing test.
4. The medical professional responsible for the supervision of the audiometric testing program shall determine the follow-up procedures necessary whenever an individual employee exhibits a standard threshold shift. Changes in hearing acuity that exceed an average of 10 dB or more at 2000, 3000, and 4000 Hz in either ear, relative to the baseline audiogram, are considered to be a standard threshold shift (STS).
5. When deemed practical, it is recommended that all employees included in the HCP have an audiometric examination prior to leaving or retiring from the company, if their last audiometric test preceded the retirement date by more than six months.

C. Audiometers
1. Specifications for audiometers
Audiometric examination shall be performed using an audiometer that conforms to the requirements for wide-range pure-tone discrete-frequency audiometers prescribed by the *American National Standard Specifications for Audiometers*, ANSI S3.6-1969 (R-1973 or latest revision). If a pulse tone audiometer is used, the on-time of the tone shall be at least 200 msec. The instrument used shall be either a manual audiometer or any other audiometer testing system of equal or greater accuracy and effectiveness.

2. Audiometer calibration
a. Acoustically
Audiometer calibration shall be checked acoustically at least annually to determine that the audiometer is within the tolerance permitted by ANSI S3.6-1969 (R-1973 or latest revision). This procedure should only be attempted by a properly trained and equipped individual. Usually the audiometer manufacturer can complete this calibration.

b. Biologically
A biological calibration shall be made prior to each day's use of the audiometer. This procedure shall consist of:
i. Testing at least one person having a known stable audiometric curve that does not exceed 10-dB hearing threshold level at any frequency and comparing the test results with the known curve.
ii. Registering the subject's response to distortions and/or unwanted sounds from the audiometer.
iii. Whenever the results of the "daily use" biological calibration indicate hearing level differences *greater than* (+ or −)5 dB at any frequency, the signal is distorted, or if there are attenuator or tone switch transients (e.g., clicks, noises, hums) the audiometer shall be removed from service and subjected to an acoustical calibration prior to any further testing. Only after the problem with the audiometer is corrected to within permitted tolerances can it be put back into service.

D. Background noise in audiometric test areas
The area designated for the audiometric testing shall be as free from noise and vibration as possible. The sound pressure level in any octave band when measured in the audiometric booth, or room in its absence, where subjects are actually tested shall not exceed the values given below.

Maximum Allowable Background Noise Levels[a]

Octave Band Center Frequency (Hz)	Sound Pressure Level (dBre 0.0002 N/M^2)
500	40
1000	40
2000	47
4000	57
8000	62

[a]OSHA CFR 1910.95 Occupational Noise Standard Appendix D: Audiometric Test Rooms

E. Documentation
Audiometer calibration records and background noise levels in test booths or rooms should be maintained with the audiometric test results. When contract services are used, these records must be provided by them and be kept on file.

Examples of Noise Exposure Calculation

The calculation of noise exposure is fairly straightforward once the noise measurements and duration time have been recorded. The first step is to calculate the noise dose (*D*) in percent of the NEL using the expression:

$$D = 100 \left[\frac{C_1}{T_1} + \frac{C_2}{T_2} + \cdots \frac{C_n}{T_n} \right]$$

where the terms C_1 through C_n indicate the total time a worker is exposed to a specific noise level. Now the T_1 through T_n terms are the reference duration times for each noise level as given in Table D.1.

Once the noise dose has been calculated, the next step is to look up the eight-hour time-weighted average exposure in dBA from Table D.2. Use of the noise dose equation is shown in the following examples.

Table D.1   Noise Exposure Levels (NELs) to Continuous Noise

Sound level (dBA)	Noise exposure duration per workday (hr; theoretical)	(min)
80	32.0	1920
81	27.9	1674
82	24.3	1458
83	21.1	1266
84	18.4	1104
85	16.0	960
86	13.9	834
87	12.1	726
88	10.6	636
89	9.2	552
90	8.0	480
91	7.0	420
92	6.2	372
93	5.3	318
94	4.6	276
95	4.0	240
96	3.5	210
97	3.0	180
98	2.6	156
99	2.3	138
100	2.0	120
101	1.7	102
102	1.5	90
103	1.4	84
104	1.3	78
105	1.0	60
106	0.87	52.2
107	0.76	45.6
108	0.66	39.6
109	0.57	34.2
110	0.50	30
111	0.44	26.4
112	0.38	22.4
113	0.33	19.8
114	0.29	17.4
115	0.25	15
116	0.22	13.2
117	0.19	11.4
118	0.16	9.6
119	0.14	8.4
120	0.125	7.5
121	0.110	6.6
122	0.095	5.7
123	0.082	4.9
124	0.072	4.3
125	0.063	3.8
126	0.054	3.2
127	0.047	2.8
128	0.041	2.4
129	0.036	2.1
130	0.031	1.8

Table D.2   Conversion from "Percent Noise Exposure" or "Dose" to Eight-Hour Time-Weighted Average (TWA) Sound Level

Percent Noise Exposure or Dose (%)	TWA (dBA)	Percent Noise Exposure or Dose (%)	TWA (dBA)
10	73.4	106	90.4
15	76.3	107	90.5
20	78.4	108	90.6
25	80.0	109	90.6
30	81.3	110	90.7
35	82.4	111	90.8
40	83.4	112	90.8
45	84.2	113	90.9
50	85.0	114	90.9
55	86.3	115	91.1
60	86.3	116	91.1
65	86.9	117	91.1
70	87.4	118	91.3
75	87.9	119	91.3
80	88.4	120	91.3
81	88.5	125	91.6
82	88.6	130	91.6
83	88.7	135	92.2
84	88.7	140	92.4
85	88.8	145	92.7
86	88.9	150	92.9
87	89.0	155	93.2
88	89.1	160	93.4
89	89.2	165	93.6
90	89.2	170	93.8
91	89.3	175	94.0
92	89.4	180	94.2
93	89.5	185	94.4
94	89.6	190	94.6
95	89.6	195	94.8
96	89.7	200	95.0
97	89.8	210	95.4
98	89.9	220	95.7
99	89.9	230	96.0
100	90.0	240	96.3
101	90.1	250	96.6
102	90.1	260	96.9
103	90.2	270	97.2
104	90.3	280	97.4
105	90.4	290	97.7
300	97.9	710	104.1
310	98.2	720	104.2
320	98.4	730	104.3
330	98.6	740	104.4
340	98.8	750	104.5
350	99.0	760	104.6
360	99.2	770	104.7
370	99.4	780	104.8
380	99.6	790	104.9
390	99.8	800	105.0

Table D.2   (Continued)

Percent Noise Exposure or Dose (%)	TWA (dBA)	Percent Noise Exposure or Dose (%)	TWA (dBA)
400	100.0	810	105.1
410	100.2	820	105.2
420	100.4	830	105.3
430	100.5	840	105.4
440	100.7	850	105.4
450	100.8	860	105.5
460	101.0	870	105.6
470	101.2	880	105.7
480	101.3	890	105.8
490	101.5	900	105.8
500	101.6	910	105.9
510	101.8	920	106.0
520	101.9	930	106.1
530	102.0	940	106.2
540	102.2	950	106.2
550	102.3	960	106.3
560	102.4	970	106.4
570	102.6	980	106.5
580	102.7	990	106.5
590	102.8	999	106.6
600	102.9		
610	103.0		
620	103.2		
630	103.3		
640	103.4		
650	103.5		
660	103.6		
670	103.7		
680	103.8		
690	103.9		
700	104.0		

Example 1: A group of industrial employees are exposed to continuous noise according to the following schedule.

Exposure level (dBA)	Time or duration of exposure (hr)
85	3
90	2
92	1
95	2

Problem: Is this group of employees overexposed according to noise regulations? What is their level of exposure in dBA?

Solution: Using the equation:

$$D = C_1/T_1 + C_2/T_2 + \cdots + C_n/T_n$$

we can determine their daily noise dose. In this problem, the workers are exposed to 85 dBA for 3 hours ($C_1 = 3$). From Table D.2, we note that the permissible exposure time for 85 dBA is 16 hours ($T_1 = 16$), 90 dBA is 8 hours ($T_2 = 8$), 92 dBA is 6.2 hours ($T_3 = 6.2$), and 95 dBA is 4 hours ($T_4 = 4$). To calculate their noise dose, we simply substitute into the previous equation and add up the fractions:

$$D = (3/16 + 2/8 + 1/6.2 + 2/4)$$

$$D = 1.0988$$

Since D exceeds unity, or one (1), this value is in excess of the permissible levels, and thus, these workers are overexposed.

To determine the level of exposure, convert 1.0988 into a percentage:

$$(1.0988) \times (100\%)$$

$$= 109.88\%, \text{ approximately } 110\%$$

Next, using Table D.2, look up the value for approximately 110 percent: at 110 percent the sound level is 90.7 dBA. (Note: This value is often referred to as an eight-hour time-weighted average (TWA) noise exposure in dBA.)

TWA noise exposure = 16.61 log (110%/100%)
$$+ 90 \text{ dBA}$$
$$= 16.61 \log (1.1) + 90 \text{ dBA}$$
$$= 16.61(0.0414) + 90 \text{ dBA}$$
$$= 0.7 + 90 \text{ dBA}$$
$$= 90.70 \text{ dBA}$$

Example 2: An operator one day at a plant spends 6 hr and 20 min of his 8-hr shift in a soundproof control room where the sound level is 70 dBA. Because of a problem with a compressor, it is necessary for him to spend 1 hr and 40 min working on the compressor. While working in the plant, he is exposed to a continuous sound level of 107 dBA.

Problem: For this operator's 8-hr workday, is he overexposed according to the workplace noise standard?

Solution: Inspecting Table D.1, one will notice that no time exposure is indicated for any values less than 80 dBA. It can be assumed that any sound levels less than 80 dBA will contribute a negligible percentage to the total noise dose; therefore, ignore adding exposure times for any sound levels less than 80 dBA into the equation. Next, we notice that this operator is exposed to 107 dBA for 1 hr and 40 min (1 hr and 40 min = 1 + 40/60 = 1.67 hr). Thus, $C_1 = 1.67$ and from Table D.1 we get $T_1 = 0.76$ hr for a sound level of 107 dBA. Using the equation for noise dose we have:

$$C_1/T_1 = 1.67/0.76 = 2.193 \quad \text{or} \quad 219.3\%$$

Obviously, this operator is overexposed despite a relatively short time spent in the plant when compared to his total workday.

From Table D.2, we find that 219% = 95.7 dBA, or to be more precise:

$$\text{TWA noise exposure} = 16.61 \log_{10} (219/100) + 90$$
$$= 95.65 \text{ dBA}$$

Appendix E

Typical Design Specifications

The following is an *example* specification only, and is not intended for construction since job conditions will be *different*.

**Noise Control Specification
For
Emergency Generator
Sets Located in 9th Floor
Equipment Room**

Prepared by

XYZ CONSULTING FIRM
Pleasant Drive
Suite A
Any Town, USA

Contents

Section 01010
Summary of Work

Part 1: General

1.01 Description

A. The work to be done under this contract consists of the furnishing of all labor, supervision, materials (unless specified as owner furnished), equipment, tools, appliances, and services necessary for the work as shown on the drawings, and as specified in these technical provisions.

B. Scope of work

1. Furnish and install acoustical enclosures, ventilation silencers, engine exhaust pipe isolation, cooling water pipe isolation to radiator, engine vibration isolation (see Section 15200, Vibration Isolation), and other materials on the 9th floor equipment room.

2. Furnish and install two (2) acoustical enclosures around emergency generator sets.

3. Furnish and install intake and exhaust air ventilation for each enclosure.

4. Furnish and install a set of vibration isolators for each engine.

5. Furnish and install an exhaust silencer for each engine.

6. Furnish and install a floating floor system that will support the engine and acoustical enclosures.

C. Contract drawings: The work shall conform to the contract drawings which form a part of these specifications.

1.02 Quality assurance

A. Codes and standards

1. OSHA
   Occupational Safety and Health Act CFR 1910.95

2. ANSI S1.13 (current year)
   Methods for the Physical Measurement of Sound Pressure Levels

3. ANSI S1.4 (current year)
   Specifications for Sound Level Meters

4. ANSI S-5.1 CAGI Test Code Standard (current year) CAGI—Pneurop Test Code for measurement of sound from pneumatic equipment

B. Contractor responsibility

1. The contractor shall have had experience in installing noise control materials and shall have personnel, skill, and organization to provide efficient and effective completion of the work.

2. The contractor shall acquaint himself with all matters and conditions concerning the site and the work to be done and shall pursue this work so that all phases of the work to be done can be coordinated without delays or damage.

1.03 Product delivery, storage, and handling

A. Protection of materials: Contractor shall take such precautions as necessary to protect all materials from damage.

B. Failure to comply shall be sufficient cause for rejection of materials in question.

C. The contractor shall be responsible for protection of any owner-furnished equipment provided for his installation.

D. All damage resulting from operating or storage and handling of material and equipment shall be repaired by contractor to full satisfaction of owner at contractor's expense.

1.04 Job conditions

A. Work schedule: It shall be the responsibility of the contractor to schedule and coordinate all work with minimum of delay and with minimum interference with owner's production activities.

B. Examination of site

1. Attention is directed to the fact that the drawings furnished by the owner may not indicate (and the specifications may not dictate) each and every item of work to be accompanied.

2. Bidders on work, before submitting proposals, shall visit and examine the site of the work to satisfy themselves as to nature and scope of the work to be done and all details involved.

3. Submission of a proposal will be taken as evidence that such examination has been made and various features noted.

4. Later claims for extra compensation on account of additional labor, materials, or equipment required or, on account of difficulties encountered, which should have been seen, will not be recognized, and all such items shall be properly disposed of by contractor at his expense.

C. Structural and space conditions

1. Specifications and drawings are intended to encompass systems that will not interfere with structural, electrical, and architectural design of building and which will fit into available spaces.

2. It is not within the scope of the drawings to indicate all of the contractor's responsibilities. It is the contractor's responsibility to install the work to conform to the structure, avoid obstructions and interferences with other trades, preserve headroom, and keep openings and passageways clear.

D. Housekeeping

1. During the work the contractor shall

   a. Accomplish the work in a neat and orderly fashion.

   b. Collect debris periodically, as required, to maintain working area in an uncluttered state.

   c. Take precautions to prevent contaminants (dust, dirt, sawdust, etc.) from spreading to areas outside the immediate working area or from setting on equipment/fixtures in the work area.

2. Completing the work for all areas that will be occupied by owner's personnel between contractor work periods the contractor shall

   a. Remove all debris generated by the work.

   b. Clean all floors, fixtures, equipment, etc., of any (all) contaminants generated by the work.

c. In general, render the area ready for owner's occupancy.

Section 13010
Special Construction
Noise Control

Part 1: General

1.01 Quality assurance: Materials or equipment specified by reference to published standards of manufacturers shall comply with the requirements of current specifications or standards listed. In cases of conflict between reference specification or standards, most stringent requirements will govern. Not all standards may be listed. All acoustical materials to be used within plant environment shall be suitably protected to keep contaminants in the air out of the material that might degrade the overall performance.

1.02 Applicable publications
   A. OSHA—29 CFR 1910.95, Occupational Noise Exposure
   B. ASTM C423 (current year)—sound absorption of acoustical material in reverberation rooms.
   C. ASTM E90 (current year)—sound transmission loss testing of interiors

1.03 General requirements: The contractor shall verify all measurements and shall take all field measurements necessary prior to fabrication. Contractor shall be responsible to ensure that all material and equipment fit and are completely operational to owner's satisfaction. Materials and parts necessary to complete each item, even though such work is not definitely shown specified, shall be included. All work shall be done in a workmanlike manner and shall be in good condition upon completion.

1.04 Submittals: Shop drawings shall be submitted showing all equipment locations and installation details including material thicknesses and paint finish as required.

Part 2: Products

2.01 General
   A. Acceptable manufacturers of acoustical enclosures
      1. Alturdyne, San Diego, CA
         (619) 565-7131
         6-in. enclosure panels

      2. Industrial Acoustics Corporation, Bronx, NY
         (212) 931-8000
         Model Noise Lock II Panels
      3. United McGill Corporation, Griffin, CA
         (404) 228-9864
      4. VAW Systems Limited, Winnipeg Manitoba, Canada
         Model T-G
   B. Acceptable manufacturers of ventilation silencers
      1. Industrial Acoustics Corp., Bronx, NY
         (212) 931-8000
         Model S
      2. United McGill Corporation, Griffin, CA
         (404) 228-9864
         Model U3
      3. VAW Systems Limited, Winnipeg Manitoba, Canada
         (204) 783-4311
         Model XXX
   C. Acceptable manufacturers of engine exhaust silencers
      1. Beaird Industries, Shreveport, LA
         (318) 865-6351
         Maxim Silencers Model M51
      2. Universal Silencer Company, Stoughton, WI
         (608) 873-4272
         Model ENA
      3. Burgess-Manning, Buffalo, NY
         Model BEO
   D. Acceptable material or manufacturers of acoustical lagging
      1. Childers Products Co., Houston, TX
         (713) 691-3661
         MUFLAG
      2. Standard gypsum board 5/8 in. thick

2.02 Product description
   A. Acoustical enclosure
      1. The acoustical enclosures shall be 4 or 6 in. thick, self-supporting, and mounted on the floating floor slab provided by others.
      2. Access hatches and/or opening shall be provided as directed by the owner and shall not reduce the noise reduction capabilities. All hinges, latches, seals, and gaskets shall be of heavy-duty design and capacity attached with heavy-duty fasteners.

3. All penetration by pipes and ducts shall be flexible and provide an airtight seal to preserve the acoustical integrity.
4. The acoustical enclosure, as a system, including all penetrations, openings, ventilation silencers, and duct work, shall reduce the noise level of one engine to the levels listed below when measured at 3 ft from the enclosure, on all sides, in accordance with the procedure found in Attachment A.

quired in addition to or in lieu of the thermal protection of 1100°F specified by the owner.

C. Ventilation silencers

1. Provide ventilation silencers (splitter panel type) to direct cooling air into and out of the enclosure. The silencers shall be of manufacturer's standard construction. The silencers shall be sized in accordance with cooling air requirements of the engine. A fan may be re-

Frequency (Hz)	63	125	250	500	1000	2000	4000	8000	dBA
L_p (dB re 20: micro Pa)	83	82	74	66	59	48	37	37	70

5. The enclosure(s) shall have interior perforated liner and acoustical absorbing material suitably protected to keep oil, dirt, and other contaminates out. The acoustical material shall be of blanket form (not loose fill) properly supported. Internal braces, septums, and noise barrier masses shall be provided as required to meet the noise level requirements in paragraph 4 above.

B. Engine exhaust silencer(s)

1. The silencers shall have a critical-grade noise attenuation capability, manufacturer's standard design, with a minimum of three chambers, steel shell, and standard paint system.
2. The silencer(s) shall be sized in accordance with the engine manufacturer's recommendations for pressure drop to include the inlet connection and length of exhaust pipe.
3. The silencer shall be located as close to the engine as possible or at a distance that is recommended by the engine manufacturer.
4. A metal flexible connection shall be used between the engine and silencer inlet connection.
5. The silencer and exhaust pipe shall be suspended or otherwise supported using spring isolators as found in section 15200, Vibration Control.
6. If required to meet the noise level, requirements in paragraph 2.02.A.4, acoustical lagging, shall be installed to reduce radiated noise. This may be re-

quired to provide force ventilation. If required, this shall be installed as directed by the owner.
2. The silencers shall be attached to the top or side panels located to provide cross ventilation or as directed by the owner or space constraints.
3. The silencer shall be a part of the total noise reduction system and shall be selected to meet the L_p in paragraph 2.02.A.4. This also includes radiated noise from duct work. Acoustical lagging shall be used, if required.
4. An elbow with turning vanes shall be installed between the silencer and duct work to direct the air at minimum pressure drop. The elbow shall be constructed of 16-ga H.R.S.

D. Acoustical lagging

1. Materials shall be applied to the engine exhaust pipe and ventilation duct work as required to meet system noise level requirements.
2. A fiberglass or mineral fiber (4 lb/ft^3 maximum density) shall be applied and suitably secured to the surface(s).
3. The outer cover shall be either an aluminum-lead laminate or 1/2-in. sheet rock.
4. All outer material shall be secured in place and all joints sealed in accordance with the manufacturer's recommendations.
5. All materials, sealants, and adhesives must be submitted to the owner for compliance with their safety standards.

Part 3: Execution

3.01 Inspection

    A. The contractor shall examine the area and/or drawings where the material shall be installed for conflicts that might exist. The contractor will be responsible for ensuring that all equipment fits and is functional upon completion.

    B. Contractor shall advise the owner's representative of any conflicts that will prevent the installation of the materials on the plans and specifications.

3.02 Installation

    A. Contractor shall install the materials in a neat, workmanlike manner and will not disrupt work in the area. This shall be coordinated with the owner's representative.

    B. All materials shall be new and be stored and protected to keep from being damaged while waiting for installation. Damaged materials shall be rejected.

    C. The contractor shall leave the area each day in the condition found by cleaning up all the material and debris.

    D. The enclosure system shall be installed as a sealed system with all joints, panels, connectors, and openings properly treated to meet the resulting noise level requirements.

    E. The owner's representative shall at his discretion conduct such tests as required to verify the performance of the materials for the end result of the installed system(s).

Section 13080
Sound-Isolating Floor System
Part 1: General

1.01 Summary

    A. Scope of work

        1. Provide isolation products for a jack-up type floating floor system in the penthouse mechanical equipment room.

        2. Provide supervisory personnel from the isolation manufacturer to perform inspections and other field services at designated times.

        3. Construct the floating floors in accordance with an installation procedure submitted by the isolator manufacturer for this project.

    B. Related sections

        1. Refer to 03100 for concrete framework.

        2. Refer to 03200 for concrete reinforcement.

        3. Refer to 03300 for cast-in-place concrete.

        4. Refer to 15 (***) for isolating floor drains.

    C. Substitution of equipment and materials

        1. Refer to General Conditions and Supplementary Conditions

        2. Substitutions of equipment and materials shall meet or exceed the "quality" of the products which are listed in these specifications. If required, submit samples for testing by an independent testing laboratory to determine their suitability for the job and pay for the services of the testing laboratory.

1.02 Description

    A. Design parameters

        1. The floating floor system shall consist of a 6-in.-thick concrete slab 110 lb/ft^3 lightweight which is isolated from and supported 4 in. above the structural slab by jack-up type resilient pad isolators. (*Note the following is not to be included in spec*: For this sample specification the actual material and spacing for this job were used. A typical floating floor may have 4-in.-thick 150-lb/ft^3 standard concrete with a 2-in. airspace. The 2-in. airspace has a natural frequency of about 18 Hz while the 4-in. airspace will have a natural frequency of about 12 Hz. The spacing will be selected based upon the type of building construction and its natural frequency, and the speed of the equipment being isolated. The natural frequencies of the airspace must not coincide with the natural frequency of the building and equipment mounted on the floating floor. If they do a resonant frequency could occur that might transmit excessive vibrations. For large areas, air vents may be required in the isolated floor.)

        2. The floating floor slab shall be isolated from adjoining walls, columns, and curbs by means of perimeter isolation boards.

        3. All perimeters of the floating floor shall be caulked to prevent sound flanking around the edges of the raised floor slab.

B. Performance requirements
   1. The natural frequency of the floating floor shall not coincide with the natural frequency of the structural floor.
   2. The submitted floating floor system shall have an STC-70 rating or greater as verified at an independent laboratory.
   3. Background noise in the occupied space below the floating floor installation, with HVAC equipment not operating, shall not exceed the NC-35 noise criterion curve due to airborne sound transmitted through the floor.

1.03 Submittals
Refer to Section (***) for format, timing, and quantities of submittals.
(*Not to be included in specification*: Since this will usually be a portion of a larger specification there will be other portions of the specification that will tell the contractor the format, timing, and quantities of the submittal information.)
A. Manufacturer's catalog sheets describing the jack-up isolators, perimeter isolation board materials, and acoustical caulking
B. Load and deflection curves of isolators
C. Test reports: from an independent laboratory
   1. Showing that the isolator neoprene compound material meets or exceeds the AASHO specifications listed under Products
   2. Showing that the submitted system has been tested in accordance with ASTM E90-70 and meets the STC performance requirement.
D. Calculations: prepared by the isolator manufacturer
   1. Showing the relationship between uniform dead loads, uniform live loads, and the concentrated loads of this project located on floating floors
   2. Showing that the performance requirement of natural frequency is met by the submitted system
E. Floor system construction procedure: prepared by the isolator manufacturer
F. Shop drawings: prepared by the isolator manufacturer
   1. Installation of temporary waterproofing and bond break material
   2. Installation of perimeter isolation board
   3. Placement of isolators

4. Size, type, and spacing of concrete reinforcement
5. Relationship of isolation system components to adjacent parts of the building structure

1.04 Quality assurance
A. Floating floor system components shall be designed and fabricated by a manufacturer with at least five years' experience in the design and fabrication of jack-up floating floor systems.
B. The floating floor installer shall have successfully completed five (5) jack-up type floating floor installations of similar size and scope and shall be certified by the isolator manufacturer to install the submitted system.

1.05 Product delivery, storage, and handling
A. Protect materials from exposure to weather.
B. Deliver in manufacturer's unopened containers or bundles, adequately identified by name, brand, type, etc.
C. Store inside a dry, ventilated space.

1.06 Site conditions
A. Provide temporary coverings and whatever else may be necessary to protect other work from moisture, deterioration, and soiling caused from the floating floor construction.
B. If unsatisfactory job conditions are discovered which hinder or raise questions about the installation of the floating floor, do not proceed with the work until conditions have been corrected in a manner acceptable to the field representative of the isolation manufacturer.
C. Ventilation requirements: Provide natural or mechanical means of ventilation to properly dry interior spaces.

1.07 Sequencing and scheduling
A. Coordinate work with other trades and coordinate scheduling with the construction supervisor to minimize delays.

Part 2: Products

2.01 Isolators
A. Provide jack-up type neoprene isolators as required to allow no more than 0.3 in. of deflection and a dynamic frequency of no more than 10 Hz throughout the load range.
B. The isolators shall consist of cast iron housings with integral steel jack screws of

2-in.-thick Bridge Bearing Grade DuPont neoprene isolation pads.

C. The neoprene isolation pads shall have the following AASHO Table B Bridge Bearing properties:

Neoprene physical properties	Grade (Durometer A)		
	40	50	60
Original:			
a. Hardness (ASTM D-676)	40.5	50.5	60.5
b. Minimum tensile strength (ASTM D-412)	2000 psi	2500 psi	2500 psi
c. Minimum elongation at break	450%	400%	350%
Oven aging tests, 70 hr/212°F, maximum change (ASTM D-573):			
a. Hardness	+15	+15	+15
b. Tensile strength	15	15	15
c. Elongation at break	−40	−40	−40
Ozone aging test, 1 ppm in air by volume; 20% strain; 100 2F (ASTM D-1149):			
a. 100 hr		No cracks	
Compression set:			
a. 22 hr at 158°F (ASTM D-395-method B)	30% max	25% max	25% max

D. Approved products: Mason Industries FSN series Jack-up Mountings and Kinetics Noise Control Floor Liftslab System.

2.02 Temporary waterproofing and bond break material: Provide two (2) plies of polyethylene sheeting.

2.03 Perimeter isolation board:

A. Provide 1-in.-thick 10#/ft^3 fiberglass, 1/2-in.-thick neoprene sponge, or a minimum of 1/2-in.-thick resilient polyethylene. At columns the perimeter isolation shall be a minimum of 3-in.

B. Approved products: Mason Industries AFG-10 Perimeter Isolation Board

2.04 Perimeter caulking compound

A. Provide a caulking compound that is non-hardening, nondrying, and nonbleeding

B. Approved products
1. Mason Industries Type CC-75 Caulking Compound
2. USG Acoustical Sealant

Part 3: Execution

3.01 Preparation: Coordinate related work being performed under Related Sections in preparation for placement of isolation materials.

3.02 Installation: Install the floating floor systems according to the installation procedure and drawings submitted by the isolator manufacturer.

3.03 Field quality control

A. Notify the architect and obtain field services from the isolator manufacturer as follows.

B. Inspect and approve all surfaces prior to placement of any isolation materials.

C. Inspect and approve the installation of perimeter isolation boards.

D. Inspect and approve the installation of temporary waterproofing and bond break material.

E. Assist in the initial placement of isolators to ensure that the submitted installation procedure is being strictly adhered to.

F. Inspect and approve the final placement of all isolation materials prior to the pouring of the concrete topping slab.

G. Assist in the initial stages of raising the floor to ensure that the submitted installation procedure is being followed.

H. Perform a final inspection of the completed floating floor systems after the installation of sealants. Verify that no portion of the floating floor is in rigid contact with the building structure.

3.04 Cleaning

A. Exercise every effort to keep the premises clean and neat from debris on a daily basis.

B. Remove temporary protective coverings, etc., from door frames, windows, and other surfaces. Repair surfaces that have been stained, marred, or otherwise damaged during the work. When work is completed, remove unused materials, containers, and equipment.

3.05 Acceptance

    A. Upon completion of the work, the owner may elect to verify the floating floor performance as part of the acceptance procedure.

    B. Should the installed floating floor not provide required noise isolation and the cause is determined to be defective workmanship or faulty materials, the contractor shall remedy defective workmanship and materials to the satisfaction of the owner and without additional cost to the owner.

    C. Final acceptance of the installation will be based on the performance of the floating floor with all equipment operating.

Section 15200
Vibration Control

Part 1: General

1.1 Related requirements

    A. The conditions of the contract, including the General Conditions and Division 1, General Requirements, apply to work covered by this section.

    B. Comply with Division 15 sections, as applicable. Refer to other divisions for coordination of work. Refer to Division 13 for isolated "floating" floors, if required. Refer to Division 16 for electrical transformers, unit substations, and switch gear to be isolated.

1.2 Description: Provide labor, materials, equipment, tools, and services and perform operations required for, and reasonably incidental to, the providing and adjustment of vibration control devices and systems.

1.3 Submittals

    A. Provide complete submittal data for each type of device or each system of vibration control. Submittal data shall include, but not be limited to, the following

       1. For each vibration isolator indicate the size, type, and deflection, the total supported weight, disturbing frequency and efficiency, and other information required to verify compliance with these specifications.

       2. For steel bases and reinforced concrete inertia blocks, include details that show completely the reinforcing steel required to provide rigid bases for the isolated equipment.

       3. For each type of device and each system, include procedures for installation and adjustment.

    B. Substitutions of "internally isolated" mechanical equipment in lieu of equipment isolated as specified in this section must be approved for each individual unit of equipment, a minimum of ten (10) days prior to the project bid date.

Part 2: Products

2.1 General design features

    A. Types of structural steel rails and bases, inertia blocks or housekeeping pads, curb-mounted bases, types of isolators, and required isolator deflections shall be as specified under Schedule of Isolated Equipment.

    B. Provide vibration isolators and bases that have been designed and treated for resistance to corrosion.

       1. Steel components shall be PVC coated or phosphate and painted with industrial-grade enamel.

       2. Nuts, bolts, and washers shall be zinc electroplated or cadmium plated.

       3. Thoroughly clean structural bases of welding slag and prime with zinc-chromate or metal etching primer. Apply a finish coat of industrial enamel over the primer.

    C. Where exposed to weather, provide isolators coated with neoprene or Bitumastic paint, having PVC-coated, hot-dip galvanized, or zinc-electroplated steel parts. Etch enamel for outdoor installation. Etch and paint aluminum components with industrial-grade enamel for outdoor installation.

    D. Provide spring isolators capable of 50 percent overtravel beyond the schedule deflection before becoming solid.

    E. Provide height-saving brackets designed for an operating clearance of 1 in. under the supported inertia blocks and designed so that the isolators can be installed and removed when the operating clearance is 1 in. or less. For use with spring isolators having $2\frac{1}{2}$-in.

deflection or more, provide height-saving brackets of the precompression type to limit bolt length between the top of the isolator and the underside of the bracket.

F. Each isolator shall limit the length of the exposed adjustment bolt between the top of the isolator and the underside of a bracket or isolation base to a maximum range of one to two (1–2) inches.

G. Select isolators supporting a given piece of equipment for approximately equal spring deflection.

H. Isolators for equipment installed outdoors shall be designed to provide adequate restraint due to normal wind conditions and to withstand wind loads of 30 lb/ft^2 applied to any exposed surface of the equipment, without failure.

2.2 Isolation materials

A. Fiberglass pads and shapes: Provide glass fiber of not more than 0.18 mil in diameter, produced by multiple-flame attenuation process, molded with manufacturer's standard fillers and binders through 10 compression cycles at three (3) times rated load-bearing capacity, to achieve natural frequency of not more than 12 Hz in thickness and shapes required for use in vibration isolation units.

B. Neoprene pads: Provide oil-resistant neoprene sheets of manufacturer's standard hardness and cross-ribbed pattern designed for neoprene-in-shear-type vibration isolation and in thickness required.

C. Cork/neoprene pads: Provide close-grained composition cork sheet, laminated between two (2) sheets of ribbed oil-resistant neoprene in thickness required.

D. Vibration isolation springs: Provide wound-steel compression springs of high-strength, heat-treated, spring alloy steel with outside diameter not less than 0.8 times operating height, with lateral stiffness not less than vertical stiffness, and designed to reach solid height before exceeding rated fatigue point of steel.

2.3 Isolator types

A. Type 1: A pad-type mounting consisting of two (2) layers of $\frac{3}{8}$-in.-thick ribbed or waffled neoprene pads bonded to a 16-ga galvanized steel separator plate. Bolting is not required. Pads shall be sized for approximately 20 to 40 psi load or a deflection of 0.10 in. or 0.16 in. Shall be Kinetics Type NGOD, Mason Type WSW, or approved equal.

B. Type 2: A base-mount-type isolator with an elastomeric mounting having a steel base plate with mounting holes and a threaded insert at the top of the mounting for attaching equipment. Metal parts shall be completely embedded in the elastomeric materials. The elastometer may be neoprene or high-quality synthetic rubber with antiozone and antioxidant additives. Mountings shall be designed for approximately $\frac{1}{4}$-in. deflection and loaded so that deflection does not exceed 15 percent of the free height of the mounting. Shall be Kinetics Type RD, Mason Type ND, or approved equal.

C. Type 2a: A rubber-suspension-type isolator with an elastomeric hanger, consisting of a rectangular steel box and an elastomeric isolation element that shall be of neoprene or high-quality synthetic rubber with antiozone and antioxidant additives. The elements shall be designed for approximately $\frac{1}{4}$-in. deflection and loaded so that deflection does not exceed 15 percent of the free height of the element. The design shall prevent metal-to-metal contact between the hanger rod and the steel box. Shall be Kinetics Type RH, Mason Type HD, or approved equal.

D. Type 3: Provide a base-mount-type isolator with an adjustable freestanding, open-spring mounting with combination leveling bolt and equipment fastening bolt. The spring (or springs) shall be rigidly attached to the mounting base plate and to the spring compression plate. To assure stability, the outside diameter shall be a minimum of 0.8 times the vertical operating height. The isolator shall be designed for a minimum Kx/Ky (horizontal to vertical spring rate) of 1.0. A neoprene pad having a minimum thickness of $\frac{1}{4}$ in. shall be bonded to the bottom of the base plate. Base plates shall be sized to limit pad loading to 100 psi. Shall be Kinetics Type FSD, Mason Type SF, or approved equal.

E. Type 3a: A suspension-type isolator with a spring hanger consisting of a rectangular steel box, coil springs, spring cups, neoprene-impregnated fabric washer, steel washer, and neoprene insert designed to prevent metal-to-metal contact between the hanger rod and bottom of the hanger box. The hanger box shall be capable of supporting a load of 400 percent of rated load without noticeable deformation or failure. Shall be Kinetics Type SH, Mason Type HS, or approved equal.

F. Type 3b: A suspension-type isolator with a spring hanger as described in Type 3a with the addition of an elastomeric element at the top of the box for acoustic isolation. The design shall prevent metal-to-metal contact between the hanger rod and the top of the hanger box. The elastomeric element shall meet the design requirements for Type 2 mountings. Shall be Kinetics Type SRH, Mason Type DNHS, or approved equal.

G. Type 4: Restrained spring isolator, a base-mount-type isolator with an adjustable open-spring isolator having one or more coil springs rigidly attached to a top compression plate and a base plate. A ribbed or waffled neoprene pad having a minimum thickness of $\frac{1}{4}$ in. shall be bonded to the bottom of the base plate. This isolator shall fit within a welded steel enclosure consisting of a top plate and rigid lower housing serving as a blocking device during installation. Restraining bolts shall connect the top plate and lower housing to prevent the isolated equipment from rising when drained of water. Neoprene grommets shall be provided to prevent metal-to-metal contact between the restraining bolts and isolator housing. Base plates shall be sized to limit pad loading to 100 psi and springs shall be designed for a minimum Kx/Ky (horizontal to vertical spring rate) of 1.0. Shall be Kinetics Type FLS, Mason Type SLR, or approved equal.

H. Type 5: Thrust restraint, a spring isolator hanger similar to Type 3a installed in pairs to resist the thrust caused by air pressure.

I. Type 6: Sheet junction isolation material shall be $\frac{1}{2}$-in.-thick closed cell neoprene, ASTM Grade SEC40 in sheets cut to fit penetrations as required.

J. Pipe riser expansion/isolation hanger: A spring-type hanger installed at riser suspension points to control load shifts as the riser expands and contracts as well as vibration. Shall be Mason Type HES or approved equal.

K. Pipe riser anchor: An all-directional anchor welded to pipe clamps to provide noise and vibration break. Shall be Mason Type ADA or approved equal.

2.4 Flexible pipe connectors

A. For nonferrous piping, provide bronze hose covered with bronze wire braid with copper tube ends or bronze flanged ends, braze-welded to hose.

B. For ferrous piping, provide stainless steel hose covered with stainless steel wire braid with NPT steel nipples or 150 psi, ANSI flanges, welded to hose.

C. Provide neoprene flexible pipe connectors with galvanized steel flanges. Select connectors with temperature and pressure ratings to suit intended service. Shall be Mason Type MFNC, MFNEC, or approved equal.

2.5 Base types

A. In addition to the following schedule equipment bases, install floor-mounted mechanical equipment on concrete housekeeping pads as specified in Section 15050.

B. Type A: No base required, isolators directly attached to equipment.

C. Type B: Structural steel rails or base. Where rails or beams are indicated for use with isolator units to support equipment, provide structural steel support members with isolator support brackets and anchor bolt holes designed by the vibration isolation materials manufacturer. Structural steel bases shall comply with ANSI/ASTM A36 and shall have a minimum depth equal to 10 percent of the longest span between isolators, but not less than four (4) inches or as indicated on the drawings. Sizes and shapes shall be as required for equipment to be supported. Isolator support brackets shall be welded to the structural beam base as required to provide the lowest possible mounting height of supported equipment. Anchor bolt holes shall be prelocated and drilled into equipment bases for bolt-down equipment. Steel beams shall provide a rigid, distortion-free mounting base for supported equipment without excessive differential motion between driving and driven equipment components.

D. Type C: Inertia base. Provide reinforced concrete inertia blocks, including perimeter steel pouring form, reinforcing bars welded in place, bolting templates, and height-saving brackets for mounting of the isolators. Each inertia block shall have a thickness of at least six (6) inches or greater as required to provide a rigid mounting for equipment. The weight of each inertia block shall be no less than 150 percent of the weight of the equipment supported. Inertia blocks shall be sized to extend no less than four (4) inches beyond the base of the supported equipment in each direction and shall be T-shaped where necessary to conserve space. Inertia blocks for pumps shall support the suction elbows on end suction pumps and both the suction and discharge elbows on horizontal split-case pumps. Perimeter steel members shall be structural channels having a minimum depth of 10 percent of the longest span, but no less than six (6) inches. Shall be Kinetics Type C18-H, Mason Type KSO or BMK, or approved equal.

E. Type D: Roof curb isolators. Provide fabricated frame units sized to match roof curbs as shown, consisting of prefabricated extruded aluminum rail system incorporating 1-in. deflection type 3, freestanding stable isolation springs that are shaped and positioned to prevent metal-to-metal contact. Provide continuous airtight and waterproof seal between extrusions. Isolation rails shall be designed and supplied by the vibration isolation manufacturer. Aluminum rail sections shall include an integral slot anchoring springs to the bottom section but allowing horizontal adjustment. Spring elements shall be as specified in this section for type 3 springs and shall be selected and located to maintain a level rail assembly and uniform spring deflection when equipment is installed. Include provisions for anchorage of frame unit to roof curb and for anchorage of equipment to unit.

F. Type E: Fabricated equipment bases. Where supplementary bases are indicated for use with isolator units to support equipment (base not integral with equipment) provide a welded unit, fabricated of structural steel shapes, plates, and bars complying with ANSI/ASTM A36 as shown. Provide welded support brackets at points indicated and anchor base to spring isolator units. Except as otherwise indicated, arrange brackets to result in the lowest possible mounting height for equipment. Provide bolt holes in base matching anchor bolt holes in equipment. Where indicated, provide auxiliary steel base for support of motor, mounted on equipment base with slotted anchor bolt holes for adjustment of motor position. Where sizes of base framing members are not indicated, fabricate base with depth of structure not less than 10 percent times longest span of base rigidly braced to support equipment without deflection or distortions that would be detrimental to equipment or equipment performance.

Part 3: Execution

3.1 General

A. This work in general shall include but not necessarily be limited to the following.

1. Isolate all mechanical and electrical equipment from the building structure by means of appropriately selected noise and vibration isolators.

2. Piping over 1 in. o.d. located in mechanical equipment rooms connected to Vibration-isolated equipment shall be isolated from the building structure by means of noise and vibration isolation hangers. Provide the first three (3) hangers or support points in each direction from each piece of isolated equipment, with vibration isolation hangers or supports having the same static deflection as the equipment isolators.

3. Duct work that is rigidly attached to isolated air-moving equipment in mechanical equipment rooms shall be isolated from the building structure by means of noise and vibration isolation hangers or mounts for a minimum of 50 ft.

4. Isolate piping and duct vertical risers from the building structure by means of noise and vibration isolation guides and supports.

B. Piping and duct work isolated in accordance with these specifications shall freely pass through walls and floors without rigid connections.

C. Provide acoustic seals at walls, ceilings, or floor openings for all resiliently supported piping or other elements connected to isolated moving equipment as detailed or scheduled. In addition to use for vibration isolation, all other conduit, piping, and duct penetrations into adjacent spaces or chases shall be sealed to minimize transmission of *airborne* sound. As a minimum, the seals shall consist of $\frac{3}{4}$-in.-thick neoprene with a tight fitted opening for the conduit, pipe, or duct sealed with USG acoustical sealant.

3.2 Performance of isolators

  A. General: Comply with minimum static deflections recommended by the American Society of Heating, Refrigerating and Air-Conditioning Engineers, including definitions of critical and noncritical locations, for selection and application of vibration isolation materials and units as indicated.

  B. Manufacturer's recommendations: Except as otherwise indicated, comply with manufacturer's recommendations for selection and application of vibration isolation materials and units.

3.3 Schedule of isolated equipment: Below is a schedule of isolated equipment for this project. Any equipment, system, construction, or condition that may be altered, added, or changed, or that is not specifically considered herein or on the drawings, shall be treated in the same manner as specified for similar equipment, systems, or construction and shall comply with the noise and vibration isolation requirements of these specifications.

Isolator Schedule

Equipment	Type	Deflection (in.)
1. Engines	4	2
2. Water piping	3a-3	2
3. Ventilation ducts	3a	2
4. Gas turbine inlet duct	3	1
5. Gas turbine exhaust duct	3a	3

Note: Engine isolators attached directly to engine skid.

3.4 Installation

  A. General: Except as otherwise indicated, comply with manufacturer's instruction for installation and load application to vibration isolation materials and units. Adjust to ensure that units do not exceed rated operating deflection or bottom out under loading and are not short-circuited by other contacts or bearing points. Remove space blocks and similar devices, if any, intended for temporary protection against overloading during installation.

  B. Anchor and attach units to substrate and equipment as required for secure operation and to prevent displacement by normal forces and as indicated.

  C. Adjust leveling devices as required to distribute loading uniformly onto isolators. Shim units as required where leveling devices cannot be used to distribute loading properly.

  D. Install inertia base frames on isolator units as indicated so that a minimum 1-in. clearance below base will result when frame is filled with concrete and supported equipment has been installed and loaded for operation.

  E. Provide a concrete housekeeping pad below vibration isolators. Refer to Section 15050.

  F. Piping and conduit in equipment rooms shall be suspended from rigid structural beams or intermediate structural members attached to rigid structure beams. Do not suspend pipes, conduits, or equipment from the slab.

  G. Install noise and vibration isolation hangers as close to the supporting structure as possible without the hanger box or frame contacting the supporting structure. Do not bolt the hanger box directly to the structure.

  H. Weld riser isolator units in place as required to prevent displacement from loading and operations.

  I. Install flexible pipe connectors on equipment side of shutoff valves horizontally and parallel to equipment shafts wherever possible.

  J. Electrical connections to resiliently mounted mechanical and electrical equipment shall include a long loop of flexible conduit at least 20 diameters in length, between rigid electrical wiring conduit and any resiliently mounted equipment.

3.5 Testing: Upon completion of the installation and after the system is put into operation, inspect the systems of vibration control and correct any discrepancies or maladjustments. If necessary, instrumentation tests and measurements shall be made to determine the source, cause, and path of any objectionable vibration; after such tests are completed, take proper steps to correct the objectionable condition.

3.6 Certification: Provide a written report from the vibration control equipment manufacturer, or his designated representative, certifying the correctness of the installation and compliance with the requirements of this section. This report shall include actual measured equipment isolator deflections for each major item of equipment.

Attachment A
Typical Equipment
Sound Measurement Procedure

Emergency Generator Sets
9th Floor M.E.R.

Prepared by

XYZ CONSULTING FIRM
Pleasant Drive
Suite A
Any Town, USA

Contents

I. Introduction
  A. During the equipment development program, a high-priority item has been sound. Studies have been completed to characterize the sound levels and identify the sound sources when operating in a normal manner.
  B. The criteria established for the equipment is shown in paragraph 22A.4 of Section 13010 on the specification at three feet from the equipment surface when operating in a normal manner.

  C. This procedure will outline a methodology for measuring the sound levels in the owner's or purchaser's facility to verify that the specified sound levels have been achieved.
II. Purpose
  A. The sound level to be measured will be taken from the equipment as installed.
  B. The measured data will be used to determine compliance with the noise specification.
  C. This sound measurement procedure will be used by the owner as part of the acceptance tests and the purchaser as part of the acceptance tests. Test results will be supplied to the owner.
III. Equipment to be tested: Acoustical enclosure system housing the 1250-kilowatt generator set(s) located in the ninth-floor equipment room
IV. Reference standards
  A. All measurements will be made in accordance with the following standards:
    1. ANSI S1.4-1983, Specification for Sound Level Meters
    2. ANSI S1.13-1971, Methods for the Measurement of Sound Pressure Levels
  B. This measurement procedure has been developed in accordance with ANSI standard S12.1-1983, Guidelines for the Preparation of Standard Procedure to Determine Noise Emissions from Sources.
V. Definitions
  Acoustic—An adjective used in conjunction with a basic property of sound, such as "acoustic energy."
  Acoustical treatment—The use of acoustical absorbent, acoustical isolation, or any changes or additions to the structure to correct acoustical faults or improve the acoustical environment.
  Ambient sound—The existing background sounds in a space usually interpreted as excluding the sound from the equipment under consideration. Ambient sound may be expressed as a sound pressure level in octave frequency bands or as A-weighted sound levels.
  ANSI—American National Standards Institute.
  A-weighted—The value of a sound pressure signal after it has been passed through the A-weighted filter response described in Ameri-

can National Standards Institute S1.4-1983, Specification for Sound Level Meters. A-weighting is often used because its value correlates well with subjective interpretations of loudness, annoyance, etc. A-weighted sound levels are expressed in decibels and are designated by the symbol dBA.

Level, sound noise—A measure of sound pressure level as determined by electrical equipment meeting ANSI requirements. Unless specifically stated otherwise, levels refer to root mean square of sinusoidally varying level.

Level meter, sound—An electrical instrument for determining sound pressure level.

Level meter, sound, Type 2—An electrical instrument for measuring sound pressure level in accordance with ANSI S1.4-1983 specification for sound level meters and within tolerance outlined in the standard (± 2 dB slow response).

Logarithmically averaged sound pressure levels (L_p)

$$L_p = 10 \log 1/N \sum_{i=1}^{N} (10)^{L_i/10}$$

where

N = the total number of measurements
L_i = level at each observation

Normal-manner operation—The equipment in full operation as found in the installed condition.

Sound pressure level (SPL)

$$L_p = 20 \log_{10} (p/p_{ref})$$

where p is the rms pressure in a sound wave, expressed in Pascal (Pa), and p_{ref} is the reference pressure of 2×10^{-5} Pa (Pa = N/m^2); L_p is expressed in decibel (dB) units. Single microphones and most other sound-measuring equipment are capable of measuring sound pressure level direction (i.e., with a single measurement).

Windscreen—A shield placed around a microphone to prevent turbulent eddies (from wind with enough velocity to cause turbulent flow over the microphone) from impinging on the diaphragm, which would cause pressure variation similar to that produced by a high noise level.

VI. Sound measurements

A. All measurements made for this procedure will be in decibels (dB), slow response as required by the specification, and rounded to the nearest whole integer. All measurements will be made at a point 36 in. from any surface of the machine envelope and 54 in. above the floor at five locations:
   1. Center of each side (2).
   2. Center line of enclosure (2).
   3. 45° axis of center line at end toward existing office space (2). If space does not permit all locations to be measured, the owner shall establish the final locations.

B. Ambient sound measurements shall be made at each location with all equipment off. If the difference between the equipment measured level and the ambient measured level is less than or equal to 10 dBA, corrections shall be made according to the following table. The corrected dBA value is the true equipment level to be recorded.

Corrections for Ambient L_p (Source ANSI 1.13-1971 par. 11.2.2)

Difference (in decibels) between sound pressure level measured with the sound source operating and ambient sound pressure level alone	Correction (in decibels) to be subtracted from L_p measured with source operating to obtain L_p due to sound source alone
4	2.2
5	1.7
6	1.3
7	1.0
8	0.8
9	0.6
10	0.4
11	0.3
12	0.3
13	0.2

C. Measurements will be made using a Type I sound level meter and octave band filter set, properly calibrated before and after each measurement session. The measurements shall be accurate within ± 2 dBA. The sound level meter or its microphone with windscreen installed shall be placed on a tripod with the microphone at approximately 70° up from a horizontal plane.

D. Any support equipment known to be included in the equipment room shall not be operating during the verification performance test.

E. Measurements shall be made only after it is determined that the equipment is operating in a normal manner and all other test conditions have been satisfied. This shall be determined by the test engineer.

F. The "normal-manner" operation shall be used for all sound measurements.

G. The measurements shall be taken at each location. If any location exceeds the specified level with tolerances included, the measurement shall be repeated for verification.

H. All data shall be recorded on appropriate forms along with a sketch of the area.

I. Octave band noise plots shall be made for visual comparison to the criteria.

VII. Test report: A standard data sheet shall be used providing space for all pertinent information about the equipment as well as sound levels at all locations. The data to be recorded shall include but not be limited to the following

A. Serial number of the unit being tested.

B. Speed and operating mode.

C. Sound levels at all locations.

D. Name of test engineer conducting measurements.

E. Date and time of test.

F. Temperature and humidity. (Significant deviations in sound level can occur if the instrument is not acclimated to room conditions.)

G. Sketch of the area with microphone positions shown in plan and elevation with appropriate distances.

H. Sound level meter model, type, serial number.

I. Calibrator, model, type, serial number, and latest calibration date.

J. List of support equipment.

VIII. Acoustical data sheet (modify to suit)

Item numbers _____ Specification no. ___

Description of equipment _____

Location (circle applicable items)

Outdoors   Walled 1 2 3 4 sides
Indoors

Exposure time (if other than 8 hours per day):

_____

Maximum dBA   Design _____
noise emission
at: Start-up _____

NOTE: Seller shall specify the operating condition which will produce the maximum noise emission.

INSTRUCTION TO SUPPLIER

Please enter the lowest guaranteed noise level for each octave band as follows:

In column A for standard equipment
In column B for factory-installed noise control features
In column C for equipment with field-installed sound attenuating appurtenances and modifications

Supplier to complete

Frequency (Hz)	A—Standard	B—Factory	C—Field
63			
125			
250			
500			
1000			
2000			
4000			
8000			
dBA			
Lin			

Special noise characteristics:

ALL NOISE LEVELS ABOVE ARE EXPRESSED AS L_p dB (re 20 micro Pa)

Appendix F

Buyer's Guide to Products for Noise and Vibration Control

Sound Absorptive Materials

1. Felts
2. Foams
3. Glass fiber
4. Mineral fiber
5. Perforated sheet metal
6. Spray-on coatings
7. Wall treatments

Acoustic Standards, Inc., 146 Emerson St., Orange, CA 92665, (714) 282-5600 (2, 7)

Acoustic Systems, 415 E. St. Elmo Rd., Austin, TX 78745, (800) 531-5412 (7)

Acoustical Composites, Inc., 325 East Grand River, Brighton, MI 48116, (313) 227-3322 (2–5, 7)

AL International LP, 777 West St., P. O. Box 9114, Mansfield, MA 02048, (508) 339-7300 (1)

Amber/Booth Co. Inc., 1403 N. Post Oak, Houston, TX 77055, (713) 688-1228 (2, 6)

American Acoustical Products, 9 Cochituate St., Natick, MA 01760, (508) 655-0870 (1–4, 7)

AZ-USA, Inc., 1401 W. 76th St., Richfield, MN 55423, (800) 842-9790 (2, 7)

Barley-Earhart Co., 233 Divine Hwy., Portland, MI 48875, (517) 647-4117 (1–5)

Bisco Products, Inc., 2300 E. Devon Ave., Elk Grove Village, IL 60007, (708) 640-0800 (2)

Blachford Inc., 1855 Stephenson Hwy., Troy, MI 48007, (313) 689-7800 (2, 3, 7)

BRD Noise and Vibration Control Inc., 112 Fairview Ave., Wind Gap, PA 18091, (215) 863-6300 (2, 3, 5, 7)

Chestnut Ridge Foam, Inc., P.O. Box 781, Latrobe, PA 15235, (800) 234-2734 (2)

CHR Division of Furon Co., 407 East St., New Haven, CT 06509, (203) 777-3631 (2)

Commercial Acoustics Div., Metal Form Mfg., 2639 E. Adams, Phoenix, AZ 85034, (602) 273-1361 (5)

Comprador, Inc., 3 DeWalt Rd., Newark, DE 19711, (302) 427-2550 (2, 3)

The E. J. Davis Company, 10 Dodge Ave., P.O. Box 326, North Haven, CT 06473, (203) 239-5391 (1–4)

E-A-R Specialty Composites, 7911 Zionsville Rd., Indianapolis, IN 46268, (317) 872-1111 (2, 3)

Eckel Industries, Inc., 155 Fawcett St., Cambridge, MA 02138, (617) 491-3221 (2–4, 7)

Empire Acoustical Systems, 317 Park Ave. W., Mansfield, OH 44906, (419) 522-8050 (5, 7)

ESS Pollution Control Products, Inc., 9145 N. Dixie Dr., Dayton, OH 45414, (513) 454-5540 (2, 6, 7)

Ferguson Perforating & Wire Co., P.O. Box 2038, Providence, RI 02905, (401) 941-8876 (5)

Forbo-Vicracoustics, Inc., 135 Lions Dr., Valmont Industrial Park, Hazelton, PA 18201, (717) 459-3490 (7)

General Acoustics Corporation, 12248 Santa Monica Blvd., Los Angeles, CA 90025, (213) 820-1531 (2, 5)

Carroll George, Inc., 15th St. S., P.O. Box 144, Northwood, IA 50459, (515) 324-2231 (2)

(Reprinted by permission from *Sound and Vibration*, July 1991.)

Harrington & King Perforating Co., 5655 W. Fillmore St., Chicago, IL 60644, (312) 626-1800 (5)

Illbruck, Inc., 3800 Washington Ave. N., Minneapolis, MN 55412, (612) 521-3555 (2)

Industrial Acoustics Company, 1160 Commerce Ave., Bronx, NY 10462, (212) 931-8000 (5, 7)

Industrial Noise Control, Inc., 1411 Jeffrey Dr., Addison, IL 60101, (708) 620-1998 (2, 5, 7)

Interior Acoustics, Inc., 176 Rte. 206 S., Somerville, NJ 08876, (908) 874-4155 (3, 7)

Isolatek International, 41 Furnace St., Stanhope, NJ 07874, (201) 347-1200 (4, 6, 7)

Kinetics Noise Control, 6300 Irelan Pl., P.O. Box 655, Dublin, OH 43017, (614) 889-0480 (2, 3, 5, 7)

King Noise Control, Inc., 11820-D Kempersprings Dr., Cincinnati, OH 45240, (513) 851-1500 (2, 5, 7)

Manville Sales Corporation, 17187 N. Laurel Park Dr., #208, Livonia, MI 48152, (313) 462-1170 (3)

Mason Industries Inc., 350 Rabro Dr., Hauppauge, NY 11788, (516) 348-0282 (6)

M.B.I. Products Co., 5309 Hamilton Ave., Cleveland, OH 44114, (216) 431-6400 (3, 4, 7)

Midwest Acoust-A-Fiber, Inc., 759 Pittsburgh Dr., Delaware, OH 43015, (614) 369-3624 (1–4, 7)

3M Company, Industrial Specialties Div., Bldg. 220-7E-01, 3M Center, St. Paul, MN 55105, (612) 733-4076 (2)

MPC, Inc., 835 Canterbury Rd., Westlake, OH 44145, (216) 835-1405 (3, 4, 7)

Noise Control Products, Inc., 55-B Albany Ave., Amityville, NY 11701, (516) 789-8192 (2, 7)

Noise Reduction Corporation, Rte. 2, Box 152, Redwood Falls, MN 56283, (507) 644-3067 (2)

Noisetec, S.A., Pol. Ind. La Ferreria, calle N, Nave 17,08110 Montcada I Reixac, Spain 564.43.51 (1–3, 5–7)

Patterson Associates, 935 Summer St., Lynnfield, MA 01940, (617) 246-0888 (2–4, 7)

Perstorp Components Inc., 4307 Arden St., Ft. Wayne, IN 46804, (219) 432-4752 (1–3)

Gordon J. Pollock & Associates, Inc., 19120 Detroit Rd., Rocky River, OH 44116, (216) 333-8710 (5, 7)

Polymer Technologies Inc., 7006 Pencader Dr., Newark, DE 19702, (302) 738-9001 (2, 5)

The Proudfoot Company, Inc., P.O. Box 338, Botsford, CT 06404, (203) 459-0031 (3, 7)

Pyrok Inc., 136 Prospect Park W., Brooklyn, NY 11215, (718) 788-1225 (6, 7)

RPG Diffusor Systems, Inc., 12003 Wimbleton St., Largo, MD 20772, (301) 249-5647 (1)

Singer Safety Company, 2300 W. Logan Blvd., Chicago, IL 60647, (800) 621-0089 (2, 3)

Sound Reduction Corp., 16601 St. Clair Ave., Cleveland, OH 44110, (216) 481-1900 (3, 4, 7)

The Soundcoat Company, Inc., One Burt Dr., Deer Park, NY 11729, (516) 242-2200 (2, 3, 7)

Soundown Corporation, 45 Congress St., Salem, MA 01970, (800) 359-1036 (2, 3, 5)

Tech Products Corporation, 5030 Linden Ave., Dayton, OH 45432, (513) 252-3661 (2, 3)

Technetics Corp., 1600 Industrial Dr., Deland, FL 32724, (904) 736-7373 (1, 5)

Technicon Industries, Inc., 4412 Republic Dr., Concord, NC 28027, (704) 788-1131 (2, 3, 6)

Tectum Inc., 105 S. Sixth St., Newark, OH 43055, (614) 345-9691 (3, 7)

Tex Tech Industries, P.O. Box 8, Main St., North Monmouth, ME 04265, (207) 933-4404 (1)

United McGill Corporation, 2400 Fairwood Ave., P.O. Box 820, Columbus, OH 43216, (614) 443-5520 (2–7)

United Process Inc., Sound Seal Div., 279 Silver St., Agawam, MA 01001, (413) 789-1770 (2, 3, 6, 7)

Vibrasciences Inc., 234 Front Ave., West Haven, CT 06516, (203) 934-6113 (2, 7)

Sound Absorptive Systems

1. Ceiling systems
2. Masking noise generators
3. Panels
4. Unit absorbers
5. Wall treatments

Acon, Inc., 4600 Webster St., Dayton, OH 45414, (513) 276-2111 (3)

Acoustic Standards, Inc., 146 Emerson St., Orange, CA 92665, (714) 282-5600 (3–5)

Acoustic Systems, 415 E. St. Elmo Rd., Austin, TX 78745, (800) 531-5412 (3, 5)

Acoustical Composites, Inc., 325 East Grand River, Brighton, MI 48116, (313) 227-3322 (1, 3, 5)

AL International LP, 777 West St., P.O. Box 9114, Mansfield, MA 02048, (508) 339-7300 (3)

American Acoustical Products, 9 Cochituate St., Natick, MA 01760, (508) 655-0870 (3–5)

AZ-USA, Inc., 1401 W. 76th St., Richfield, MN 55423, (800) 842-9790 (5)

Barley-Earhart Co., 233 Divine Hwy., Portland, MI 48875, (517) 647-4117 (3)

Boet American Company, 8895 North Military Trail,

#305C, Palm Beach Gardens, FL 33410, (407) 622-7113 (5)

BRD Noise and Vibration Control Inc., 112 Fairview Ave., Wind Gap, PA 18091, (215) 863-6300 (1, 3-5)

Commercial Acoustics Div., Metal Form Mfg., 2639 E. Adams, Phoenix, AZ 85034, (602) 273-1361 (3)

The E. J. Davis Company, 10 Dodge Ave., P.O. Box 326, North Haven, CT 06473, (203) 239-5391 (4)

Dynasound, Inc., 6439 Atlantic Blvd., Norcross, GA 30071, (404) 242-8176 (2)

E-A-R Specialty Composites, 7911 Zionsville Rd., Indianapolis, IN 46268, (317) 872-1111 (3-5)

Eckel Industries, Inc., 155 Fawcett St., Cambridge, MA 02138, (617) 491-3221 (1, 3-5)

Empire Acoustical Systems, 317 Park Ave. W., Mansfield, OH 44906, (419) 522-8050 (3, 5)

ESS Pollution Control Products, Inc., 9145 N. Dixie Dr., Dayton, OH 45414, (513) 454-5540 (1, 3-5)

Forbo-Vicracoustics, Inc., 135 Lions Dr., Valmont Industrial Park, Hazelton, PA 18201, (717) 459-3490 (3, 5)

General Acoustics Corporation, 12248 Santa Monica Blvd., Los Angeles, CA 90025, (213) 820-1531 (3-5)

Illbruck, Inc., 3800 Washington Ave. N., Minneapolis, MN 55412, (612) 521-3555 (1, 4, 5)

Industrial Acoustics Company, 1160 Commerce Ave., Bronx, NY 10462, (212) 931-8000 (1, 3-5)

Industrial Noise Control, Inc., 1411 Jeffrey Drive, Addison, IL 60101, (708) 620-1998 (1, 3, 5)

Interior Acoustics, Inc., 176 Rte. 206 S., Somerville, NJ 08876, (908) 874-4155 (1-5)

Isolatek International, 41 Furnace St., Stanhope, NJ 07874, (201) 347-1200 (1, 5)

Kinetics Noise Control, 6300 Irelan Place, P.O. Box 655, Dublin, OH 43017, (614) 889-0480 (3-5)

King Noise Control, Inc., 11820-D Kempersprings Dr., Cincinnati, OH 45240, (513) 851-1500 (3, 5)

Mason Industries Inc., 350 Rabro Dr., Hauppauge, NY 11788, (516) 348-0282 (1)

M.B.I. Products Co., 5309 Hamilton Ave., Cleveland, OH 44114, (216) 431-6400 (1, 3-5)

Midwest Acoust-A-Fiber, Inc., 759 Pittsburgh Dr., Delaware, OH 43015, (614) 369-3624 (1, 3-5)

MPC, Inc., 835 Canterbury Rd., Westlake, OH 44145, (216) 835-1405 (1, 3, 5)

Neiss Corporation, P.O. Box 478, Rockville, CT 06066, (203) 872-8528 (2, 3)

Noise Control Products, Inc., 55-B Albany Ave., Amityville, NY 11701, (516) 789-8192 (1, 3, 5)

Noise Reduction Corporation, Rte. 2, Box 152, Redwood Falls, MN 56283, (507) 644-3067 (1)

Noisetec, S.A., Pol. Ind. La Ferreria, calle N, Nave 17,08110 Montcada I Reixac, Spain, 564.43.51 (1, 3-5)

Patterson Associates, 935 Summer St., Lynnfield, MA 01940, (617) 246-0888 (3-5)

Perstorp Components Inc., 4307 Arden St., Ft. Wayne, IN 46804, (219) 432-4752 (5)

Gordon J. Pollock & Associates, Inc., 19120 Detroit Rd., Rocky River, OH 44116, (216) 333-8710 (1, 3-5)

Polymer Technologies Inc., 7006 Pencader Drive, Newark, DE 19702, (302) 738-9001 (1, 3, 5)

The Proudfoot Company, Inc., P.O. Box 338, Botsford, CT 06404, (203) 459-0031 (2-5)

Rathbone Products, Inc., P.O. Box 10356, 1180 Corrugated Way, Columbus, OH 43201, (614) 297-7782 (5)

Rink, 2620 N. Flowingwells Road, Tucson, AZ 85703, (602) 622-7601 (3, 4)

RPG Diffusor Systems, Inc., 12003 Wimbleton St., Largo, MD 20772, (301) 249-5647 (1, 3-5)

Singer Safety Company, 2300 West Logan Blvd., Chicago, IL 60647, (800) 621-0089 (1, 3, 4)

Sound Fighter Systems, Inc., P.O. Box 6075, Shreveport, LA 71136, (318) 861-6640 (3, 5)

Sound Reduction Corp., 16601 St. Clair Ave., Cleveland, OH 44110, (216) 481-1900 (1, 3, 5)

Soundown Corporation, 45 Congress St., Salem, MA 01970, (800) 359-1036 (3)

Target Enterprises Inc., Roxbury, VT 05669, (802) 728-3081 (4)

Tech Products Corporation, 5030 Linden Ave., Dayton, OH 45432, (513) 252-3661 (3-5)

Technicon Industries, Inc., 4412 Republic Dr., Concord, NC 28027, (704) 788-1131 (4, 5)

Tectum Inc., 105 S. Sixth St., Newark, OH 43055, (614) 345-9691 (1)

TLT-Babcock, Inc., 3480 W. Market St., Akron, OH 44333, (216) 867-8540 (3)

United McGill Corporation, 2400 Fairwood Ave., P.O. Box 820, Columbus, OH 43216, (614) 443-5520 (2-5)

United Process Inc., Sound Seal Div., 279 Silver St., Agawam, MA 01001, (413) 789-1770 (3-5)

Vibrasciences Inc., 234 Front Ave., West Haven, CT 06516, (203) 934-6113 (1, 3)

VSM Corporation, 7515 Northfield Rd., Cleveland, OH 44146, (216) 439-5400 (3-5)

Sound Barrier Materials

1. Pipe lagging
2. Plain and mass-loaded plastics
3. Sealants and sealing tapes
4. Sheet glass, metal and plastic

Acoustic Standards, Inc., 146 Emerson St., Orange, CA 92665, (714) 282-5600 (2)

Acoustical Composites, Inc., 325 E. Grand River, Brighton, MI 48116, (313) 227-3322 (2, 3)

Amber/Booth Co. Inc., 1403 N. Post Oak, Houston, TX 77055, (713) 688-1228 (1)

American Acoustical Products, 9 Cochituate St., Natick, MA 01760, (508) 655-0870 (1–4)

Barley-Earhart Co., 233 Divine Hwy., Portland, MI 48875, (517) 647-4117 (1, 2, 4)

Blachford Inc., 1855 Stephenson Hwy., Troy, MI 48007, (313) 689-7800 (2)

BRD Noise and Vibration Control Inc., 112 Fairview Ave., Wind Gap, PA 18091, (215) 863-6300 (1, 2)

Comprador, Inc., 3 DeWalt Rd., Newark, DE 19711, (302) 427-2550 (2)

The E. J. Davis Company, 10 Dodge Ave., P.O. Box 326, North Haven, CT 06473, (203) 239-5391 (1)

Duracote Corporation, 350 N. Diamond St., P.O. Box 512, Ravenna, OH 44266, (216) 296-9600 (2)

E-A-R Specialty Composites, 7911 Zionsville Rd., Indianapolis, IN 46268, (317) 872-1111 (1, 2)

Eckel Industries, Inc., 155 Fawcett St., Cambridge, MA 02138, (617) 491-3221 (1, 2)

Empire Acoustical Systems, 317 Park Ave. W., Mansfield, OH 44906, (419) 522-8050 (4)

ESS Pollution Control Products, Inc., 9145 N. Dixie Dr., Dayton, OH 45414, (513) 454-5540 (1, 2, 4)

Carroll George, Inc., 15th St. S., P.O. Box 144, Northwood, IA 50459, (515) 324-2231 (2)

Illbruck, Inc., 3800 Washington Ave. N., Minneapolis, MN 55412, (612) 521-3555 (3)

Industrial Noise Control, Inc., 1411 Jeffrey Dr., Addison, IL 60101, (708) 620-1998 (1)

Kinetics Noise Control, 6300 Irelan Pl., P.O. Box 655, Dublin, OH 43017, (614) 889-0480 (1, 2)

King Noise Control, Inc., 11820-D Kempersprings Dr., Cincinnati, OH 45240, (513) 851-1500 (1, 2)

Midwest Acoust-A-Fiber, Inc., 759 Pittsburgh Dr., Delaware, OH 43015, (614) 369-3624 (4)

Noise Reduction Corporation, Rte. 2, Box 152, Redwood Falls, MN 56283, (507) 644-3067 (1–3)

Noisetec, S.A., Pol. Ind. La Ferreria, calle N, Nave 17,08110 Montcada I Reixac, Spain 564.43.51 (1, 2)

Patterson Associates, 935 Summer Street, Lynnfield, MA 01940, (617) 246-0888 (2, 3)

Perstorp Components Inc., 4307 Arden St., Ft. Wayne, IN 46804, (219) 432-4752 (2)

Polymer Technologies Inc., 7006 Pencader Dr., Newark, DE 19702, (302) 738-9001 (2)

The Proudfoot Company, Inc., P.O. Box 338, Botsford, CT 06404, (203) 459-0031 (1)

Rink, 2620 N. Flowingwells Rd., Tucson, AZ 85703, (602) 622-7601 (4)

Singer Safety Company, 2300 West Logan Blvd., Chicago, IL 60647, (800) 621-0089 (2)

The Soundcoat Company, Inc., One Burt Dr., Deer Park, NY 11729, (516) 242-2200 (1–3)

Soundown Corporation, 45 Congress St., Salem, MA 01970, (800) 359-1036 (1, 2, 4)

Taracorp Industries, 1200 16th St., Granite City, IL 62040, (800) 851-3300 (4)

Tech Products Corporation, 5030 Linden Ave., Dayton, OH 45432, (513) 252-3661 (1, 2)

Technicon Industries, Inc., 4412 Republic Dr., Concord, NC 28027, (704) 788-1131 (1–3)

United McGill Corporation, 2400 Fairwood Ave., P.O. Box 820, Columbus, OH 43216, (614) 443-5520 (2, 3)

United Process Inc., Sound Seal Div., 279 Silver St., Agawam, MA 01001, (413) 789-1770 (1, 2, 4)

Vibrasciences Inc., 234 Front Ave., West Haven, CT 06516, (203) 934-6113 (1)

Sound Barrier Systems

1. Curtains
2. Doors
3. Operable partitions
4. Panels
5. Seals
6. Transportation noise barriers
7. Walls
8. Windows

Acoustic Standards, Inc., 146 Emerson St., Orange, CA 92665, (714) 282-5600 (1, 2, 4–8)

Acoustic Systems, 415 E. St. Elmo Rd., Austin, TX 78745, (800) 531-5412 (2, 4, 7, 8)

Acoustical Composites, Inc., 325 E. Grand River, Brighton, MI 48116, (313) 227-3322 (1, 4, 6, 7)

Amber/Booth Co. Inc., 1403 N. Post Oak, Houston, TX 77055, (713) 688-1228 (1, 4)

American Acoustical Products, 9 Cochituate St., Natick, MA 01760, (508) 655-0870 (1, 4, 5)

Amweld Building Products, Inc., 1500 Amweld Dr., Niles, OH 44446, (216) 527-4385 (2)

Barley-Earhart Co., 233 Divine Hwy., Portland, MI 48875, (517) 647-4117 (1, 4, 6)

Bisco Products, Inc., 2300 E. Devon Ave., Elk Grove Village, IL 60007, (708) 640-0800 (1, 4, 6)

Boet American Company, 8895 N. Military Trail, #305C, Palm Beach Gardens, FL 33410, (407) 622-7113 (2, 8)

BRD Noise and Vibration Control Inc., 112 Fairview Ave., Wind Gap, PA 18091, (215) 863-6300 (1, 2, 4, 7)

Commercial Acoustics Div., Metal Form Mfg., 2639 E. Adams, Phoenix, AZ 85034, (602) 273-1361 (2, 4, 7)

The E. J. Davis Company, 10 Dodge Ave., P.O. Box 326, North Haven, CT 06473, (203) 239-5391 (1)

DeVAC, Inc., 4365 Willow Dr., Hamel, MN 55340, (612) 478-2300 (8)

E-A-R Specialty Composites, 7911 Zionsville Rd., Indianapolis, IN 46268, (317) 872-1111 (1, 4)

Eckel Industries, Inc., 155 Fawcett St., Cambridge, MA 02138, (617) 491-3221 (1-4, 6-8)

Empire Acoustical Systems, 317 Park Ave. W., Mansfield, OH 44906, (419) 522-8050 (4, 6, 7)

ESS Pollution Control Products, Inc., 9145 N. Dixie Dr., Dayton, OH 45414, (513) 454-5540 (1, 2, 4-8)

Forbo-Vicracoustics, Inc., 135 Lions Dr., Valmont Industrial Park, Hazelton, PA 18201, (717) 459-3490 (4, 7)

General Acoustics Corporation, 12248 Santa Monica Blvd., Los Angeles, CA 90025, (213) 820-1531 (1, 3, 4, 7)

Industrial Acoustics Company, 1160 Commerce Ave., Bronx, NY 10462, (212) 931-8000 (2-4, 6-8)

Industrial Noise Control, Inc., 1411 Jeffrey Dr., Addison, IL 60101, (708) 620-1998 (1-4, 7, 8)

Interior Acoustics, Inc., 176 Rte. 206 S., Somerville, NJ 08876, (908) 874-4155 (1, 4)

Isolatek International, 41 Furnace St., Stanhope, NJ 07874, (201) 347-1200 (7)

Jamison Door Company, P.O. Box 70, Hagerstown, MD 21741, (301) 733-3100 (2, 8)

Kinetics Noise Control, 6300 Irelan Pl., P.O. Box 655, Dublin, OH 43017, (614) 889-0480 (1)

King Noise Control, Inc., 11820-D Kempersprings Dr., Cincinnati, OH 45240, (513) 851-1500 (1, 2, 4, 8)

Krieger Steel Products, 4896 Gregg Rd., P.O. Box 308, Pico Rivera, CA 90660, (213) 695-0645 (2)

Mason Industries Inc., 350 Rabro Dr., Hauppauge, NY 11788, (516) 348-0282 (7)

M.B.I. Products Co., 5309 Hamilton Ave., Cleveland, OH 44114, (216) 431-6400 (1, 4)

Midwest Acoust-A-Fiber, Inc., 759 Pittsburgh Dr., Delaware, OH 43015, (614) 369-3624 (4, 7)

Minor Rubber Co., Inc., 49 Ackerman St., Bloomfield, NJ 07003, (800) 433-6886 (5)

MPC, Inc., 835 Canterbury Rd., Westlake, OH 44145, (216) 835-1405 (4, 7)

Neiss Corporation, P.O. Box 478, Rockville, CT 06066, (203) 872-8528 (1-4, 7, 8)

Noise Control Products, Inc., 55-B Albany Ave., Amityville, NY 11701, (516) 789-8192 (2, 4, 7, 8)

Noise Reduction Corporation, Rte. 2, Box 152, Redwood Falls, MN 56283, (507) 644-3067 (1)

Noisetec, S.A., Pol. Ind. La Ferreria, calle N, Nave 17,08110 Montcada I Reixac, Spain 564.43.51 (1-4, 7, 8)

Overly Manufacturing Company, P.O. Box 70, Greensburg, PA 15601, (412) 834-7300 (2, 4, 5, 8)

Patterson Associates, 935 Summer St., Lynnfield, MA 01940, (617) 246-0888 (1-4, 6, 7)

Pioneer Industries, 401 Washington Ave., Carlstadt, NJ 07072, (201) 933-1900 (2)

Gordon J. Pollock & Associates, Inc., 19120 Detroit Rd., Rocky River, OH 44116, (216) 333-8710 (1-4, 6, 7)

Polymer Technologies Inc., 7006 Pencader Dr., Newark, DE 19702, (302) 738-9001 (1, 4)

Protective Door Industries, 2250 Western Ave., Park Forest, IL 60466, (708) 481-3400 (2, 8)

The Proudfoot Company, Inc., P.O. Box 338, Botsford, CT 06404, (203) 459-0031 (1-4, 6, 7)

Pyrok Inc., 136 Prospect Park W., Brooklyn, NY 11215, (718) 788-1225 (4, 6)

Reinforced Earth Company, 2010 Corporate Ridge, Suite 1000, McLean, VA 22102, (703) 821-1175 (6)

Rink, 2620 N. Flowingwells Road, Tucson, AZ 85703, (602) 622-7601 (4, 7)

Singer Safety Company, 2300 W. Logan Blvd., Chicago, IL 60647, (800) 621-0089 (1)

Sound Construction & Engineering, 522 Cottage Grove Rd., Bloomfield, CT 06002, (203) 243-1428 (2, 4)

Sound Fighter Systems, Inc., P.O. Box 6075, Shreveport, LA 71136, (318) 861-6640 (2, 4, 6, 7)

Sound Technologies, Inc., P.O. Box 600, Michigan City, IN 46360, (219) 879-2600 (2, 4, 8)

The Soundcoat Company, Inc., One Burt Dr., Deer Park, NY 11729, (516) 242-2200 (1, 4–6)

Soundown Corporation, 45 Congress St., Salem, MA 01970, (800) 359-1036 (1, 4, 5)

Specialty Doors, Inc., 269 W. 154th St., South Holland, IL 60473, (708) 339-4331 (2, 4, 8)

Tech Products Corporation, 5030 Linden Ave., Dayton, OH 45432, (513) 252-3661 (1, 3, 4, 7)

Technicon Industries, Inc., 4412 Republic Dr., Concord, NC 28027, (704) 788-1131 (1, 4, 5)

United McGill Corporation, 2400 Fairwood Ave., P.O. Box 820, Columbus, OH 43216, (614) 443-5520 (1–4, 7, 8)

United Process Inc., Sound Seal Div., 279 Silver St., Agawam, MA 01001, (413) 789-1770 (1, 4, 6)

Vibrasciences Inc., 234 Front Ave., West Haven, CT 06516, (203) 934-6113 (1)

Vibration Isolation Products, 11275 San Fernando Rd., San Fernando, CA 91340, (818) 896-1191 (5)

Vibron Limited, 1720 Meyerside Dr., Mississauga, ON, Canada L5T 1A3, (416) 670-4922 (4, 7)

VSM Corporation, 7515 Northfield Rd., Cleveland, OH 44146, (216) 439-5400 (4, 7)

Composite Materials

1. Barrier/fiber composites
2. Barrier/foam composites
3. Masonry units

Acoustic Standards, Inc., 146 Emerson St., Orange, CA 92665, (714) 282-5600 (2)

Acoustical Composites, Inc., 325 E. Grand River, Brighton, MI 48116, (313) 227-3322 (1, 2)

American Acoustical Products, 9 Cochituate St., Natick, MA 01760, (508) 655-0870 (1, 2)

AZ-USA, Inc., 1401 W. 76th St., Richfield, MN 55423, (800) 842-9790 (2)

Barley-Earhart Co., 233 Divine Hwy., Portland, MI 48875, (517) 647-4117 (1, 2)

Bisco Products, Inc., 2300 E. Devon Ave., Elk Grove Village, IL 60007, (708) 640-0800 (2)

Blachford Inc., 1855 Stephenson Hwy., Troy, MI 48007, (313) 689-7800 (1, 2)

Bowman Metal Deck, P.O. Box 260, Pittsburgh, PA 15230, (412) 429-7502 (6)

BRD Noise and Vibration Control Inc., 112 Fairview Ave., Wind Gap, PA 18091, (215) 863-6300 (1, 2)

Chestnut Ridge Foam, Inc., P.O. Box 781, Latrobe, PA 15235, (800) 234-2734 (2)

Comprador, Inc., 3 DeWalt Rd., Newark, DE 19711, (302) 427-2550 (1, 2)

The E. J. Davis Company, 10 Dodge Ave., P.O. Box 326, North Haven, CT 06473, (203) 239-5391 (1, 2)

Duracote Corporation, 350 N. Diamond St., P.O. Box 512, Ravenna, OH 44266, (216) 296-9600 (1, 2)

E-A-R Specialty Composites, 7911 Zionsville Rd., Indianapolis, IN 46268, (317) 872-1111 (1, 2)

Eckel Industries, Inc., 155 Fawcett St., Cambridge, MA 02138, (617) 491-3221 (2)

Empire Acoustical Systems, 317 Park Ave. W., Mansfield, OH 44906, (419) 522-8050 (1)

ESS Pollution Control Products, Inc., 9145 N. Dixie Dr., Dayton, OH 45414, (513) 454-5540 (1, 2)

General Acoustics Corporation, 12248 Santa Monica Blvd., Los Angeles, CA 90025, (213) 820-1531 (1)

Carroll George, Inc., 15th St. S., P.O. Box 144, Northwood, IA 50459, (515) 324-2231 (2)

Illbruck, Inc., 3800 Washington Ave. N., Minneapolis, MN 55412, (612) 521-3555 (2)

Industrial Noise Control, Inc., 1411 Jeffrey Dr., Addison, IL 60101, (708) 620-1998 (1, 2)

Interior Acoustics, Inc., 176 Rte. 206 S., Somerville, NJ 08876, (908) 874-4155 (1)

Kinetics Noise Control, 6300 Irelan Pl., P.O. Box 655, Dublin, OH 43017, (614) 889-0480 (1, 2)

King Noise Control, Inc., 11820-D Kempersprings Dr., Cincinnati, OH 45240, (513) 851-1500 (1)

Manville Sales Corporation, 17187 N. Laurel Park Dr., #208, Livonia, MI 48152, (313) 462-1170 (1)

M.B.I. Products Co., 5309 Hamilton Ave., Cleveland, OH 44114, (216) 431-6400 (1)

Midwest Acoust-A-Fiber, Inc., 759 Pittsburgh Dr., Delaware, OH 43015, (614) 369-3624 (1, 2)

Noise Reduction Corporation, Rte. 2, Box 152, Redwood Falls, MN 56283, (507) 644-3067 (2)

Noisetec, S.A., Pol. Ind. La Ferreria, calle N, Nave 17,08110 Montcada I Reixac, Spain 564.43.51 (1, 2)

Patterson Associates, 935 Summer St., Lynnfield, MA 01940, (617) 246-0888 (1, 2)

Perstorp Components Inc., 4307 Arden St., Ft. Wayne, IN 46804, (219) 432-4752 (1, 2)

Polymer Technologies Inc., 7006 Pencader Dr., Newark, DE 19702, (302) 738-9001 (2)

The Proudfoot Company, Inc., P.O. Box 338, Botsford, CT 06404, (203) 459-0031 (1, 3)

Singer Safety Company, 2300 West Logan Blvd., Chicago, IL 60647, (800) 621-0089 (1, 2)

The Soundcoat Company, Inc., One Burt Dr., Deer Park, NY 11729, (516) 242-2200 (1, 2)

Soundown Corporation, 45 Congress St., Salem, MA 01970, (800) 359-1036 (1, 2)

Tech Products Corporation, 5030 Linden Ave., Dayton, OH 45432, (513) 252-3661 (1, 2)

Technicon Industries, Inc., 4412 Republic Dr., Concord, NC 28027, (704) 788-1131 (1, 2)

Tex Tech Industries, P.O. Box 8, Main St., North Monmouth, ME 04265, (207) 933-4404 (1)

United McGill Corporation, 2400 Fairwood Ave., P.O. Box 820, Columbus, OH 43216, (614) 443-5520 (1, 2)

United Process Inc., Sound Seal Div., 279 Silver St., Agawam, MA 01001, (413) 789-1770 (1, 2)

Vibrasciences Inc., 234 Front Ave., West Haven, CT 06516, (203) 934-6113 (2)

Composite Systems

1. Curtains
2. Enclosures/quiet rooms
3. Open-plan partitions
4. Panels
5. Quilted composites
6. Roof decks

Acon, Inc., 4600 Webster St., Dayton, OH 45414, (513) 276-2111 (2, 4)

Acoustic Standards, Inc., 146 Emerson St., Orange, CA 92665, (714) 282-5600 (1, 2, 4, 6)

Acoustic Systems, 415 E. St. Elmo Rd., Austin, TX 78745, (800) 531-5412 (2, 4)

Acoustical Composites, Inc., 325 E. Grand River, Brighton, MI 48116, (313) 227-3322 (1–5)

American Acoustical Products, 9 Cochituate St., Natick, MA 01760, (508) 655-0870 (1, 4, 5)

Barley-Earhart Co., 233 Divine Hwy., Portland, MI 48875, (517) 647-4117 (1, 4)

Bisco Products, Inc., 2300 E. Devon Ave., Elk Grove Village, IL 60007, (708) 640-0800 (5)

BRD Noise and Vibration Control Inc., 112 Fairview Ave., Wind Gap, PA 18091, (215) 863-6300 (1–5)

Comprador, Inc., 3 DeWalt Rd., Newark, DE 19711, (302) 427-2550 (2, 4)

E-A-R Specialty Composites, 7911 Zionsville Rd., Indianapolis, IN 46268, (317) 872-1111 (1, 2, 4, 5)

Eckel Industries, Inc., 155 Fawcett St., Cambridge, MA 02138, (617) 491-3221 (1, 2, 4)

Empire Acoustical Systems, 317 Park Ave. W., Mansfield, OH 44906, (419) 522-8050 (2, 4)

ESS Pollution Control Products, Inc., 9145 N. Dixie Dr., Dayton, OH 45414, (513) 454-5540 (1, 2, 4–6)

Forbo-Vicracoustics, Inc., 135 Lions Dr., Valmont Industrial Park, Hazelton, PA 18201, (717) 459-3490 (3, 4)

General Acoustics Corporation, 12248 Santa Monica Blvd., Los Angeles, CA 90025, (213) 820-1531 (1–3, 5)

I.D.E. Processes Corporation, Noise Control Div., 106 81st Ave., Kew Gardens, NY 11415, (718) 544-1177 (2, 4)

Industrial Acoustics Company, 1160 Commerce Ave., Bronx, NY 10462, (212) 931-8000 (2, 4)

Industrial Noise Control, Inc., 1411 Jeffrey Dr., Addison, IL 60101, (708) 620-1998 (1, 2, 4, 5)

Interior Acoustics, Inc., 176 Rte. 206 S., Somerville, NJ 08876, (908) 874-4155 (1, 4, 5)

Kinetics Noise Control, 6300 Irelan Pl., P.O. Box 655, Dublin, OH 43017, (614) 889-0480 (1, 5)

King Noise Control, Inc., 11820-D Kempersprings Dr., Cincinnati, OH 45240, (513) 851-1500 (1, 2, 4, 5)

M.B.I. Products Co., 5309 Hamilton Ave., Cleveland, OH 44114, (216) 431-6400 (4, 6)

Midwest Acoust-A-Fiber, Inc., 759 Pittsburgh Dr., Delaware, OH 43015, (614) 369-3624 (3–6)

MPC, Inc., 835 Canterbury Rd., Westlake, OH 44145, (216) 835-1405 (4)

Neiss Corporation, P.O. Box 478, Rockville, CT 06066, (203) 872-8528 (1, 2, 4)

Noise Control Products, Inc., 55-B Albany Ave., Amityville, NY 11701, (516) 789-8192 (2, 4)

Noise Reduction Corporation, Rte. 2, Box 152, Redwood Falls, MN 56283, (507) 644-3067 (1, 2)

Noisetec, S.A., Pol. Ind. La Ferreria, calle N, Nave 17,08110 Montcada I Reixac, Spain 564.43.51 (1, 2, 4)

Patterson Associates, 935 Summer St., Lynnfield, MA 01940, (617) 246-0888 (1–5)

Gordon J. Pollock & Associates, Inc., 19120 Detroit Rd., Rocky River, OH 44116, (216) 333-8710 (1–5)

Polymer Technologies Inc., 7006 Pencader Dr., Newark, DE 19702, (302) 738-9001 (1, 5)

The Proudfoot Company, Inc., P.O. Box 338, Botsford, CT 06404, (203) 459-0031 (1–5)

Pyrok Inc., 136 Prospect Park W., Brooklyn, NY 11215, (718) 788-1225 (4)

Rathbone Products, Inc., P.O. Box 10356, 1180 Corrugated Way, Columbus, OH 43201, (614) 297-7782 (3)

Rink, 2620 N. Flowingwells Rd., Tucson, AZ 85703, (602) 622-7601 (2, 4)

Singer Safety Company, 2300 W. Logan Blvd., Chicago, IL 60647, (800) 621-0089 (1, 2, 4, 5)

Sound Fighter Systems, Inc., P.O. Box 6075, Shreveport, LA 71136, (318) 861-6640 (2)

Sound Technologies, Inc., P.O. Box 600, Michigan City, IN 46360, (219) 879-2600 (2)

The Soundcoat Company, Inc., One Burt Dr., Deer Park, NY 11729, (516) 242-2200 (1, 5)

Soundown Corporation, 45 Congress St., Salem, MA 01970, (800) 359-1036 (1, 5)

Tech Products Corporation, 5030 Linden Ave., Dayton, OH 45432, (513) 252-3661 (1, 2, 4, 5)

Technicon Industries, Inc., 4412 Republic Dr., Concord, NC 28027, (704) 788-1131 (1, 5)

Tectum Inc., 105 S. Sixth St., Newark, OH 43055, (614) 345-9691 (6)

United McGill Corporation, 2400 Fairwood Ave., P.O. Box 820, Columbus, OH 43216, (614) 443-5520 (1–5)

United Process Inc., Sound Seal Div., 279 Silver St., Agawam, MA 01001, (413) 789-1770 (1, 2, 4, 5)

Vibrasciences Inc., 234 Front Ave., West Haven, CT 06516, (203) 934-6113 (1, 4)

Vibron Limited, 1720 Meyerside Dr., Mississauga, ON, Canada L5T 1A3, (416) 670-4922 (2)

VSM Corporation, 7515 Northfield Rd., Cleveland, OH 44146, (216) 439-5400 (4)

Vibration Damping Materials

1. Adhesives
2. Constrained-layer composites
3. Mastics
4. Sheets
5. Tapes

Acoustic Standards, Inc., 146 Emerson St., Orange, CA 92665, (714) 282-5600 (3)

Acoustical Composites, Inc., 325 E. Grand River, Brighton, MI 48116, (313) 227-3322 (2)

Amber/Booth Co. Inc., 1403 N. Post Oak, Houston, TX 77055, (713) 688-1228 (3)

American Acoustical Products, 9 Cochituate St., Natick, MA 01760, (508) 655-0870 (2–5)

Barry Controls, 40 Guest St., Brighton, MA 02178, (617) 787-1555 (2)

Bisco Products, Inc., 2300 E. Devon Ave., Elk Grove Village, IL 60007, (708) 640-0800 (4, 5)

Blachford Inc., 1855 Stephenson Hwy., Troy, MI 48007, (313) 689-7800 (3, 4)

BRD Noise and Vibration Control Inc., 112 Fairview Ave., Wind Gap, PA 18091, (215) 863-6300 (2–4)

CHR Division of Furon Co., 407 East St., New Haven, CT 06509, (203) 777-3631 (4)

Comprador, Inc., 3 DeWalt Rd., Newark, DE 19711, (302) 427-2550 (1, 2, 4)

Duracote Corporation, 350 N. Diamond St., P.O. Box 512, Ravenna, OH 44266, (216) 296-9600 (2)

E-A-R Specialty Composites, 7911 Zionsville Rd., Indianapolis, IN 46268, (317) 872-1111 (1–5)

Eckel Industries, Inc., 155 Fawcett St., Cambridge, MA 02138, (617) 491-3221 (5)

ESS Pollution Control Products, Inc., 9145 N. Dixie Dr., Dayton, OH 45414, (513) 454-5540 (3, 4)

General Acoustics Corporation, 12248 Santa Monica Blvd., Los Angeles, CA 90025, (213) 820-1531 (2)

Industrial Noise Control, Inc., 1411 Jeffrey Dr., Addison, IL 60101, (708) 620-1998 (1, 2, 4)

Kinetics Noise Control, 6300 Irelan Place, P.O. Box 655, Dublin, OH 43017, (614) 889-0480 (1, 3, 4)

King Noise Control, Inc., 11820-D Kempersprings Dr., Cincinnati, OH 45240, (513) 851-1500 (2, 4)

Lord Corporation, Industrial Products Div., 1952 West Grandview Blvd., Erie, PA 16514, (814) 868-5424 (1)

Manville Sales Corporation, 17187 N. Laurel Park Dr., #208, Livonia, MI 48152, (313) 462-1170 (3)

Mason Industries Inc., 350 Rabro Dr., Hauppauge, NY 11788, (516) 348-0282 (3)

MB Dynamics, Inc., 25865 Richmond Rd., Cleveland, OH 44146, (216) 292-5850 (4)

Midwest Acoust-A-Fiber, Inc., 759 Pittsburgh Dr., Delaware, OH 43015, (614) 369-3624 (3)

3M Company, Industrial Specialties Div., Bldg. 220-7E-01, 3M Center, St. Paul, MN 55105, (612) 733-4076 (2)

Noise Reduction Corporation, Rte. 2, Box 152,

Redwood Falls, MN 56283, (507) 644-3067 (2, 5)

Noisetec, S.A., Pol. Ind. La Ferreria, calle N, Nave 17, 08110 Montcada I Reixac, Spain 564.43.51 (1, 2, 4)

Patterson Associates, 935 Summer St., Lynnfield, MA 01940, (617) 246-0888 (1–4)

Perstorp Components Inc., 4307 Arden St., Ft. Wayne, IN 46804, (219) 432-4752 (1–4)

Polymer Dynamics Inc., P.O. Box 4400, Allentown, PA 18105, (800) USA-4PD1 (4)

Polymer Technologies Inc., 7006 Pencader Dr., Newark, DE 19702, (302) 738-9001 (1, 2, 4)

Rogers Corporation, PORON Div., 1 Technical Dr., Rogers, CT 06263, (203) 774-9605 (4, 5)

The Rubber Group, 100 Montgomery St., Belleville, NJ 07109, (800) 255-4884 (3, 4)

Shock Tech Inc., 55 Whitney Rd., Mahwah, NJ 07430, (201) 848-1000 (2, 4)

Singer Safety Company, 2300 W. Logan Blvd., Chicago, IL 60647, (800) 621-0089 (2, 3)

The Soundcoat Company, Inc., One Burt Dr., Deer Park, NY 11729, (516) 242-2200 (1, 2, 4)

Soundown Corporation, 45 Congress St., Salem, MA 01970, (800) 359-1036 (1, 2)

Stock Drive Products, A DSG Company, 2101 Jericho Tpke., New Hyde Park, NY 11040, (516) 328-3300 (4)

Tech Products Corporation, 5030 Linden Ave., Dayton, OH 45432, (513) 252-3661 (2–4)

Technicon Industries, Inc., 4412 Republic Dr., Concord, NC 28027, (704) 788-1131 (2–4)

United McGill Corporation, 2400 Fairwood Ave., P.O. Box 820, Columbus, OH 43216, (614) 443-5520 (1, 3, 4)

United Process Inc., Sound Seal Div., 279 Silver St., Agawam, MA 01001, (413) 789-1770 (1–4)

Vibrasciences Inc., 234 Front Ave., West Haven, CT 06516, (203) 934-6113 (2–4)

Vibration Isolation Products, 11275 San Fernando Rd., San Fernando, CA 91340, (818) 896-1191 (2, 4)

Vibration Isolation Systems

1. Active isolators
2. Bases
3. Elastomeric
4. Floating floors
5. Machinery mounts
6. Pipe connectors
7. Pneumatic
8. Seismic
9. Steel spring
10. Vibration dampers

Acoustic Standards, Inc., 146 Emerson St., Orange, CA 92665, (714) 282-5600 (2, 5, 8)

Aeroflex International, 35 S. Service Rd., Plainview, NY 11803, (516) 694-6700 (2, 4, 5, 8–10)

Akzo Industrial Systems Co., P.O. Box 7249, Asheville, NC 28802, (704) 665-5050 (3, 4)

Amber/Booth Co. Inc., 1403 N. Post Oak, Houston, TX 77055, (713) 688-1228 (1–10)

American Acoustical Products, 9 Cochituate St., Natick, MA 01760, (508) 655-0870 (3, 10)

ANAMET Inc./Anaconda Metal Hose, 698 S. Main St., Waterbury, CT 06725, (203) 574-8500 (6)

Barley-Earhart Co., 233 Divine Hwy., Portland, MI 48875, (517) 647-4117 (10)

Barry Controls, 40 Guest St., Brighton, MA 02178, (617) 787-1555 (1–10)

BRD Noise and Vibration Control Inc., 112 Fairview Ave., Wind Gap, PA 18091, (215) 863-6300 (1–6, 8–10)

Digisonix, 8401 Murphy Dr., Middleton, WI 53562, (608) 831-3999 (1)

E-A-R Specialty Composites, 7911 Zionsville Rd., Indianapolis, IN 46268, (317) 872-1111 (3, 5, 10)

Eckel Industries, Inc., 155 Fawcett St., Cambridge, MA 02138, (617) 491-3221 (4)

Enidine Inc., 7 Centre Dr., Orchard Park, NY 14127, (716) 667-7000 (5, 7, 10)

ESS Pollution Control Products, Inc., 9145 N. Dixie Dr., Dayton, OH 45414, (513) 454-5540 (3, 5, 10)

Firestone Industrial Products Co., 1700 Firestone Blvd., Noblesville, IN 46060, (317) 773-0650 (7)

Freudenberg-NOK, 47690 East Anchor Ct., Plymouth, MI 48170, (313) 451-0020 (3, 10)

Goodyear Tire & Rubber Co., P.O. Box 185, Greensburg, OH 44232, (216) 896-5000 (7)

Industrial Acoustics Company, 1160 Commerce Ave., Bronx, NY 10462, (212) 931-8000 (4)

Kinetic Systems Inc., 20 Arboretum Rd., P.O. Box K, Boston, MA 02131, (800) 992-2884 (1, 2, 7, 8, 10)

Kinetics Noise Control, 6300 Irelan Pl., P.O. Box 655, Dublin, OH 43017, (614) 889-0480 (2–5, 8–10)

King Noise Control, Inc., 11820-D Kempersprings Dr., Cincinnati, OH 45240, (513) 851-1500 (2, 5, 9, 10)

Lord Corporation, Industrial Products Div., 1952 West Grandview Blvd., Erie, PA 16514, (814) 868-5424 (1, 3, 5, 8, 10)

M/RAD Corporation, 71 Pine St., Woburn, MA 01801, (617) 935-5940 (1–4, 7–10)

Mason Industries Inc., 350 Rabro Dr., Hauppauge, NY 11788, (516) 348-0282 (1–10)

MB Dynamics, Inc., 25865 Richmond Rd., Cleveland, OH 44146, (216) 292-5850 (1, 2)

Minor Rubber Co., Inc., 49 Ackerman St., Bloomfield, NJ 07003, (800) 433-6886 (5, 10)

3M Company, Industrial Specialties Div., Bldg. 220-7E-01, 3M Center, St. Paul, MN 55105, (612) 733-4076 (5, 10)

Noisetec, S.A., Pol. Ind. La Ferreria, calle N, Nave 17,08110 Montcada I Reixac, Spain 564.43.51 (3–6, 9, 10)

Polymer Dynamics Inc., P.O. Box 4400, Allentown, PA 18105, (800) USA-4PD1 (5)

Rogers Corporation, PORON Div., 1 Technical Dr., Rogers, CT 06263, (203) 774-9605 (3)

The Rubber Group, 100 Montgomery St., Belleville, NJ 07109, (800) 255-4884 (3)

Shock Tech Inc., 55 Whitney Rd., Mahwah, NJ 07430, (201) 848-1000 (1–3, 5, 6, 8, 10)

Sound Technologies, Inc., P.O. Box 600, Michigan City, IN 46360, (219) 879-2600 (2, 4)

Soundown Corporation, 45 Congress St., Salem, MA 01970, (800) 359-1036 (3, 4)

Stock Drive Products, A DSG Company, 2101 Jericho Tpke., New Hyde Park, NY 11040, (516) 328-3300 (2, 3, 5, 9, 10)

Sunnex Inc., 87 Crescent Rd., Needham, MA 02194, (800) 445-7869 (5)

Target Enterprises Inc., Roxbury, VT 05669, (802) 728-3081 (3)

Tech Products Corporation, 5030 Linden Ave., Dayton, OH 45432, (513) 252-3661 (1–3, 5–10)

Tex Tech Industries, P.O. Box 8, Main St., North Monmouth, ME 04265, (207) 933-4404 (10)

United Process Inc., Sound Seal Div., 279 Silver St., Agawam, MA 01001, (413) 789-1770 (6)

Vibrasciences Inc., 234 Front Ave., West Haven, CT 06516, (203) 934-6113 (2–10)

Vibration Eliminator Co. Inc., 10-28 47th Ave., Long Island City, NY 11101, (718) 729-2500 (1–10)

Vibration Isolation Products, 11275 San Fernando Rd., San Fernando, CA 91340, (818) 896-1191 (1, 3, 5, 6, 8–10)

Vibration Mountings and Controls, 113 Main St., Bloomingdale, NJ 07403, (201) 838-1780 (2–6, 8, 9)

Vibro-Dynamics Corporation, 2443 Braga Dr., Broadview, IL 60153, (708) 345-2050 (5)

Vibron Limited, 1720 Meyerside Dr., Mississauga, ON, Canada L5T 1A3, (416) 670-4922 (2–5, 8, 9)

Silencers

1. Active attenuators
2. Ducts
3. Duct silencers
4. Electric motor silencers
5. Fan silencers
6. Filter silencers
7. General industrial silencers
8. High-pressure discharge silencers
9. Intake and exhaust silencers
10. Pulsation dampers
11. Splitter/louver silencers

Acoustic Standards, Inc., 146 Emerson St., Orange, CA 92665, (714) 282-5600 (3, 5, 7, 9)

Active Noise and Vibration Technologies, 3811 Wier Ave., Phoenix, AZ 85040, (602) 470-0020 (1, 9)

Aerzen USA Corp., 313 National Rd., Exton, PA 19341, (215) 524-9870 (9, 10)

Allied Witan Company, 13805 Progress Pkwy., Cleveland, OH 44133, (216) 237-9630 (6–10)

Amber/Booth Co. Inc., 1403 N. Post Oak, Houston, TX 77055, (713) 688-1228 (3, 8, 10)

Arrow Pneumatics, Inc., 500 Oakwood Rd., Lake Zurich, IL 60047, (708) 438-9100 (6, 9)

Atlas Minerals & Chemicals, Inc., P.O. Box 38, Farmington Rd., Mertztown, PA 19539, (215) 682-7171 (7)

Barry Controls, 40 Guest St., Brighton, MA 02178, (617) 787-1555 (1, 5, 8, 10)

Beaird Industries, Inc., P.O. Box 31115, Shreveport, LA 71130, (318) 865-6351 (5–9)

Boet American Company, 8895 N. Military Trail, #305C, Palm Beach Gardens, FL 33410, (407) 622-7113 (7–9, 11)

BRD Noise and Vibration Control Inc., 112 Fairview Ave., Wind Gap, PA 18091, (215) 863-6300 (3, 5–7, 9–11)

Burgess-Manning Inc., 227 Thorn Ave., Orchard Park, NY 14127, (716) 662-6540 (2–11)

Commercial Acoustics Div., Metal Form Mfg., 2639 E. Adams, Phoenix, AZ 85034, (602) 273-1361 (3, 5, 7, 9, 11)

DeMarco MAX VAC Corporation, P.O. Box 46129, Chicago, IL 60646, (312) 685-5957 (9)

Digisonix, 8401 Murphy Dr., Middleton, WI 53562, (608) 831-3999 (1, 3, 5–7, 9, 11)

Eckel Industries, Inc., 155 Fawcett St., Cambridge, MA 02138, (617) 491-3221 (5, 11)

ESS Pollution Control Products, Inc., 9145 N. Dixie Dr., Dayton, OH 45414, (513) 454-5540 (2, 3, 5–7, 9, 11)

Fluid Kinetics Corporation, P.O. Box 1888, 2368 Eastman Ave., Ventura, CA 93004, (805) 644-5587 (7–11)

General Acoustics Corporation, 12248 Santa Monica Blvd., Los Angeles, CA 90025, (213) 820-1531 (3, 5, 7–9, 11)

I.D.E. Processes Corporation, Noise Control Div., 106 81st Ave., Kew Gardens, NY 11415, (718) 544-1177 (2, 3, 5, 7, 9, 11)

Industrial Acoustics Company, 1160 Commerce Ave., Bronx, NY 10462, (212) 931-8000 (3, 5–9, 11)

Industrial Noise Control, Inc., 1411 Jeffrey Dr., Addison, IL 60101, (708) 620-1998 (2, 3)

King Noise Control, Inc., 11820-D Kempersprings Dr., Cincinnati, OH 45240, (513) 851-1500 (3, 5, 7, 9)

Lord Corporation, Industrial Products Div., 1952 West Grandview Blvd., Erie, PA 16514, (814) 868-5424 (3–5, 7–10)

Manville Sales Corporation, 17187 N. Laurel Park Dr., #208, Livonia, MI 48152, (313) 462-1170 (3)

M.B.I. Products Co., 5309 Hamilton Ave., Cleveland, OH 44114, (216) 431-6400 (3, 5)

Midwest Acoust-A-Fiber, Inc., 759 Pittsburgh Dr., Delaware, OH 43015, (614) 369-3624 (2, 3, 5, 11)

Minor Rubber Co., Inc., 49 Ackerman St., Bloomfield, NJ 07003, (800) 433-6886 (10)

Noise Cancellation Technologies, Inc., 800 Summer St., Stamford, CT 06901, (203) 961-0500 (1)

Noise Control Products, Inc., 55-B Albany Ave., Amityville, NY 11701, (516) 789-8192 (3–7, 9)

Noisetec, S.A., Pol. Ind. La Ferreria, calle N, Nave 17,08110 Montcada I Reixac, Spain 564.43.51 (3, 5, 7–11)

Patterson Associates, 935 Summer St., Lynnfield, MA 01940, (617) 246-0888 (2, 3, 7, 9)

Polymer Technologies Inc., 7006 Pencader Dr., Newark, DE 19702, (302) 738-9001 (3, 5, 7, 9, 11)

Rink, 2620 N. Flowingwells Rd., Tucson, AZ 85703, (602) 622-7601 (3, 5, 7, 9, 11)

Sound Construction & Engineering, 522 Cottage Grove Rd., Bloomfield, CT 06002, (203) 243-1428 (2, 3, 7, 9)

Sound Fighter Systems, Inc., P.O. Box 6075, Shreveport, LA 71136, (318) 861-6640 (2, 3, 5, 7, 9)

Sound Technologies, Inc., P.O. Box 600, Michigan City, IN 46360, (219) 879-2600 (3, 5, 9, 11)

The Soundcoat Company, Inc., One Burt Dr., Deer Park, NY 11729, (516) 242-2200 (5, 6, 9)

Soundown Corporation, 45 Congress St., Salem, MA 01970, (800) 359-1036 (2, 9)

Sunnex Inc., 87 Crescent Rd., Needham, MA 02194, (800) 445-7869 (8)

Tech Products Corporation, 5030 Linden Ave., Dayton, OH 45432, (513) 252-3661 (1, 5, 9)

Technetics Corp., 1600 Industrial Dr., Deland, FL 32724, (904) 736-7373 (3, 5, 8, 9)

Tex Tech Industries, P.O. Box 8, Main St., North Monmouth, ME 04265, (207) 933-4404 (3)

TLT-Babcock, Inc., 3480 W. Market St., Akron, OH 44333, (216) 867-8540 (2, 3, 5, 7–9, 11)

United McGill Corporation, 2400 Fairwood Ave., P.O. Box 820, Columbus, OH 43216, (614) 443-5520 (2–9, 11)

Universal Silencer, P.O. Box 411, Hwy., 51 W., Stoughton, WI 53589, (608) 873-4272 (2, 3, 5–9, 11)

Vibration & Noise Engineering Corp., 2655 Villa Creek Dr., #185, Dallas, TX 75234, (214) 243-1951 (3, 5–11)

Vibron Limited, 1720 Meyerside Dr., Mississauga, ON, Canada L5T 1A3, (416) 670-4922 (3, 5–9, 11)

VSM Corporation, 7515 Northfield Rd., Cleveland, OH 44146, (216) 439-5400 (3, 5, 6, 8, 9)

Index